The Glasgow Beekeepers

Glasgow and District Beekeepers' Association Centenary Book

Edited by Taylor Hood

Northern Bee Books

The Glasgow Beekeepers
Glasgow and District Beekeepers' Association

© Glasgow and District Beekeepers' Association

All rights reserved. No part of this publication may be reproduced, stored in a retrieval system, transmitted in any form or by any means electronic, mechanical, including photocopying, recording or otherwise without prior consent of the copyright holders.

ISBN 978-1-912271-28-3

Published by Northern Bee Books, 2018
Scout Bottom Farm
Mytholmroyd
Hebden Bridge HX7 5JS (UK)

Design and artwork by DM Design and Print

Printed by Lightning Source, UK

The Glasgow Beekeepers

Glasgow and District Beekeepers' Association

Edited by Taylor Hood

Contents

1. Introduction 1

2. Glasgow and District History – Our History - Charles Irwin 3
One Hundred Years and Still going Strong - Taylor Hood 7

3. Ian Craig – Recollections 28
Observations at Hive front 30
Aids to colony manipulations 35
Is there adequate room in a single National brood chamber for the requirements of a good queen? 37
Double Brood Chamber Management 39
Queen Excluders, Supers and Supering 48
Beekeepers Year 51
Swarm Prevention 83
Swarm Control 87
Queen Rearing 89
A Simple Method of Queen Production Using a Standard Size Nucleus Hive .. 96

4. Taylor Hood - Queen rearing using mating mini nucs 101

5. Taylor Hood - Charlie Irwin on temporary and permanent Observation Hives 110

6. Eric McArthur - The Pleasures of Beekeeping 118
Simple Management 121
Swarm Control Without Tears 129
The Fickle Tilia 132
Preparation for Heather Flow 133
At the Heather 135
After the Heather 136
Winter and Early Spring 138
Early Spring 140
December 143
Winter Activities 145
January 146
February 148
Nature Awakens – March 150
May 153
June – Swarms 155
Taking A Swarm 158
August - Hive Products 160
Preparing for Winter 164
Effective Utilisation of Mature "Heather Bees" After the Heather Flow 166
November 168
Bee Protected 171
Beekeeping the Growing Pains 174

Good Beekeeping or Bad	176
Spring Management – Survival	180
March Spring Management	182
Spring Management – Survival (continued)	184
Colony Expansion	186
June	189
Preparation for The Heather	191
August	194
Bee Thinking	196
Queen Rearing for the Hobbyist	198
October	200
November	203
December	206
Strange Ongoings in Mid-February	208
May 1983	209
June	211
July	212
August	213
September	215
October	217
November	218
December	219
The Monthly Round – March 1984	220
March 1987	222
April	224
May	225
June	226
July	227
August	229
September	230
October	232
November	234
December	236
January	237
February	239
March	240
April	242
June	244
July	246
August	248
September	250
October	251
November	252
December	254

January	256
February	257
March	259
April	260
May	262
June	263
July	265
August	267
September	269
October	271
November	273
January	274
February	276
April	278
May	280
June	284
July	285
August	288
October	291
November/December	293
January	296
February	298
March	300
April	303
May	305
June	308
July	310
August	312
September	314
October	316
November	318
December	320
Bees and Things Post Varroa – December	322
Pollen – Simple Food for the Bees	324
Who would be a Beekeeper?	327
To Drift or not to Drift	329
The Drone Trigger	331
Multipurpose Nucleus Formation	332
More on Oxalic Acid Sublimation	334
Oxalic Acid Sublimator Mark II and II	336
Alternative Anti Varroa Hive Cleansing Revisited	338
A New Slant on Chalk Brood Disease	342
7. Taylor Hood – The Shepherd Method of Swarm Control	347

1. Introduction

This book is a series of Articles compiled mainly from the articles in the *Scottish Beekeeper* magazine.

It is a book about practical beekeeping.

The vast majority of articles have been written by Ian Craig MBE, Eric McArthur and Charles Irwin, who are members of the Glasgow and District Beekeepers' Association and have made a huge contribution to Scottish beekeeping over the years. All three are Expert Beekeepers and if their experience was measured in beekeeping years (1 year for each year a beehive kept) it would amount to thousands. This book only covers the areas they have published, their knowledge is even more encompassing. Ian as Education Convener of the Scottish Beeleepers Association, helped educate at least 2 generations of beekeepers – through workshops on microscopy, honey and wax as well as through his Association talks. Eric and Charlie have mentored numerous people passing on their expertise. All 3 being involved in running the beginner classes on beekeeping in the Glasgow area.

This book, hopefully will not just be a book to mark the centenary of the Glasgow and District Beekeepers Association but also a book to mark the contribution these beekeepers have made as well as being a reference book and source of information regarding beekeeping.

Ian's section is very much about his method of beekeeping. Follow this double brood system and you cannot go far wrong.

Eric's section ranges from the nectar sources of the South West of Scotland to how he manages his colonies so that his colonies flourish and produce honey. It does not cover everything he has written. Eric should be considered as a progressive beekeeper who can on occasions be quite controversial. This was and is still is Eric's way to get you to think about issues and to get you to take action. Eric's section spans from the late 1970's until 2017. He wrote under at least three pseudonyms as well as his own name. This is a selection of the body of his work and does not include any of his translations of foreign papers which have had great significance in making people aware of the varroa problem and how to tackle, this pest.

Charlie's section is on the History of Glasgow and District Beekeepers' Association and Observation hives. Charlie is an extremely skilful beekeeper and it is unfortunate that much of his management skills can only be learned by watching him in action. Charlie is a knowledgeable, thoughtful and gentle beekeeper and these characteristics are very evident in his beekeeping.

I would like to thank Ian, Eric and Charles for allowing us to publish their work and the Trustees of the Scottish Beekeepers Association to reprint the articles previously published in *The Scottish Beekeeper* magazine.

2. Our History
Charles Irwin

The minute books of the Association prior to 1986 are missing so information on the early history is gleaned from reports in the Moir Library.

Glasgow and District Beekeepers Association was founded in 1918 by Mr Peter Bebbington. The meetings at that time were held in the Christian Institute in Waterloo Street, Glasgow, the first President being a Mr Richard Whyte. It is recorded that in 1921 the then President Mr Alexander Steven a Scottish Beekeepers' Association Expert Beekeeper, was Beemaster to Princess Louises' Hospital, Erskine House for maimed soldiers, assisting them to add to their incomes by "successful keeping of bees." Mr Steven was an important lecturer who raised the profile of Glasgow and District Beekeepers Association.

At the Scottish Beekeepers Association A.G.M. on 3rd February 1923 a scheme of insurance for members prepared by Glasgow and District Beekeepers was adopted. During 1924 Mr Robert Howie a Glasgow and District member who was the SBA "Propaganda Committee" Convener broadcast beekeeping information on BBC Radio (wireless as it was referred to at that time).

The Association held regular meetings in the Christian Institute and in the summer season apiary visits were arranged. On 5th July 1924 there was a visit to the apiary of Mr H M Stitch, Kilbarchan. On the 16th August 1924, twenty members visited the apiary of Rev. John Beveridge, Gartmore, on a very wet day. The members were obviously adventurous considering the distance from Glasgow they travelled and the type of transport and the state of the roads at that time. Although it is not recorded, at that time it would have been possible to have made part of the journey by rail.

The Association received lectures and demonstrations on subjects such as – Honey Judging, Bee diseases, Dissection of Bees for diagnosis, Microscopy for identification of both Adult and Larval diseases such as acarine, nosema and foulbrood.

The meeting, 16th January 1925 brought a record attendance for a lecture by Mr John Anderson MA BSc on Biology and Beekeeping.

During 1925 five members gained the Beemaster Certificate. The President at this time was Mr H. Malcolm McCallum, the Secretary Miss Annie Adam and the

Treasurer Mr A. Stevenson. Various apiary visits were arranged and lectures given to Guide and Scout groups. In addition to examining hives during visits, demonstrations of wax rendering and honey extracting were given. It is interesting that in the photographs taken at visits very rarely is a veil seen. The membership of the Association in 1925 stood at 124 and members recorded average honey yields of 51$^1/_2$llbs.

At the end of February 1926 the sad news recorded that Mr Richard Whyte (the first President) had died in a house fire with his son. During 1926 five additional lecturers were given for beginners and the normal lectures were geared towards encouraging good beekeeping, meetings at this time being weekly.

In 1927 three Glasgow Members held the SBA Expert Beemaster Certificate and 10 held the Beemaster Certificate. At the AGM that year the Association elected a lady President, Miss Annie Adam. During the summer of 1927 a further two members gained Beemaster Certificates and one the Honey Judge Certificate.

In 1928 a Honey Show was held in the Singer Hall, Clydebank, and by the following year the Honey Show was being held in conjunction with the Horticultural Show in Kelvin Hall, Glasgow for three days in September. At the end of 1928 the Association had 196 members and had 4 summer outings in addition to the usual meetings which were held in the Christian Institute.

In 1929 it was reported that the heather on the Glasgow and District site on Stockiemuir (about 14 miles North of Glasgow) was no longer yielding and a new site was required.

In 1930 members attended an illustrated lecture on the use and management of "skyscraper" hives. Glasgow and District now had 4 SBA Honey Judges. Mr Howie was still broadcasting on beekeeping topics on BBC radio, the topic for May being "The Honey Market." On the 23rd August, Glasgow and District joined the Cardross Association on a visit to a heather site. One of the Auchincruive Beekeeping Advisors was present to give instruction and advice. At the 1930 Honey Show there was a good number of entries for extracted honey classes, however a number of sections were disqualified for poor quality (not fully sealed) and others poorly presented.

In 1931, a Mr Gracie addressing members, spoke against importing French bees. There was also controversy, between the North and West of Scotland, regarding the number of brood frames required in a hive. The West favouring 10 and the North not less than 20. Miss A Adam of Glasgow and District accepted an invitation to become insurance convener of the SBA. The membership of

Glasgow and District had risen to 216 paid up members and it was reported that members had gained 1 Beemaster, 4 Expert Beemasters and 3 Honey Judges during 1930.

In *The Scottish Beekeeper* magazine of July 1931 there is a photograph of a visit to an apiary in Lesmahagow. Forty Members are among the hives and not a veil in sight! In December 1931 Mr John Smith of Kilmarnock gave a talk on crystallised honey and of his early experiments of seeding honey.

In January 1932, Mr A Limond spoke on the "Importance of the Drone in Queen Rearing."

Records for May 20th 1933 show 45 members attended a visit to Mr Walker in Kilmaurs, Ayrshire, where they were shown the Stewarton Hive in use. On the same day they visited Mr Stevens apiary followed by tea in the Town Hall, after which there were talks on beekeeping. In October 1933 it is reported that Dr and Mrs Tennent both gained Expert Beemaster Certificates. The Glasgow Honey Show of 1933 had moved back to the Singer Hall in Clydebank and there was a good number of entries. The Association annual subscription was 2/6 (12^1/$_2$ pence).

The February 1934 *The Scottish Beekeeper* magazine carried the obituary for Mr Robert Y Howie lecturer in Art at Glasgow Training Centre for Teachers. Mr Howie was SBA President 1928 to 1930. In April 1934 the Glasgow and District Secretary, Mr James Marshall, announced Glasgow and District Beekeepers Association was to donate £1-1-0 (1 guinea) each year for 3 years for research into brood diseases at Rothamstead.

In the first quarter of 1935 a special series of lectures by Mr Struthers of Auchincruive College was held in Giffnock School and at the AGM on 27th March 1935 it was noted that there was a drop in membership due to the formation of Paisley Beekeepers Association and Lanarkshire Beekeepers Association. Mrs Shepherd was now Secretary and Dr Tennent representative to the SBA. In May, Mr Peter Forbes gained the Beemaster Certificate. In the summer there were as usual apiary visits, and in December an article in *The Scottish Beekeeper* magazine by Mrs Shepherd on a two brood box system of management using a Snelgrove Board.

4th July 1936 one hundred and seven attended a visit to Mr and Mrs Shepherd's apiary, and 17 candidates were examined for the Beemaster Certificate by Mr Limond of Ayr and gained their certificates that day. The 1936 Honey Show was back at Kelvin Hall in August and in December Mr Shepherd gave a lecture on the Snelgrove method of management using the Snelgrove Board.

The AGM on 24th March 1937 had an attendance of 40 out of a membership of 88 and the Association funds had a balance on hand of £7 15s 10½ d. Honorary President Dr Anderson, Mr Gadsby, Mr Hodge, Hon. Vice President Rev. Mr Beveridge, President Miss Allen, Vice President Mr Black, Treasurer Mr Shepherd and Secretary Mrs Shepherd. Later in the year Mr John Walker from Kilmaurs, Ayrshire gave a lecture titled, "My Life among the Bees." Amongst the information he imparted was that there were two types of Stewarton Hive namely the Stewarton and the Renfrew Stewarton. Mr and Mrs Shepherd were "leading lights" in the Association and were the inventors of the Shepherd Tube, a swarm control device which led the Virgin Queen and young bees when emerging from the hive for orientation flights from the upper box and deposited them at the entrance to the bottom box from where they joined the original colony which had been dequeened. October 1937 the meeting place now being The Central Hall (this being in Bath Street, Glasgow). Miss Allen introduced the speaker Mr Cunningham and his topic, "The Best Bees for Scotland." In November 1937, Miss Allen, who had lived for some years in Holland, gave a talk "Among the Bees in Holland and the use of Skeps in Bee Sheds."

In 1938 meetings continued in Central Halls and 1938 saw the Empire Exhibition in Bellahouston Park, Glasgow. This was a large trade exhibition at which the SBA had a prominent pavilion. Glasgow and District Members were well represented in manning the displays. At this time Dr Tennent of Glasgow and District Association was President of the SBA. In Glasgow there was a parallel organisation "The Glasgow Bee Club" which worked closely with Glasgow and District Association collaborating with apiary outings.

In 1972 at the Bee Club AGM a vote was taken to amalgamate with Glasgow and District Association.

The war years and the nineteen fifties and sixties will require further research. Somewhere during this period the Kelvin Valley Beekeepers swarmed off to form their own Association then about 1981 Eastwood Association was formed. Both group meeting nearer to the residences of their members.

One Hundred Years and Still going Strong
Taylor Hood

History of Glasgow Beekeepers' Association 1918 -2018.

At the end of World War 1 in 1918 a meeting was held for beekeepers and people interested in beekeeping and the Glasgow and District Beekeepers' Association (GDBKA) was set up. The first president was Mr Richard Whyte who died March 1926 after his house, Drumspillan, Pinwherry, went on fire. He saved one of his sons, but he and his other son died in the flames as he tried to save him. In his obituary it states that Mr Whyte was an inventor of various appliances which were of great service to agriculture. (I am not sure if he was the inventor of the Whyte Queen Mating Box and the Whyte Queen Introducing and Nucleus Cage).

The president in 1921 was Mr Alec Steven who died suddenly in August 1921. Mr Steven was an expert Beemaster and was the Beemaster to the Princess Louise's (Queen Victoria's daughter) Hospital for Limbless Sailors and Soldiers, Erskine House, Erskine. Mr. Alec Steven was a member of the training and employment subcommittee at the hospital. The disabled men were taught beekeeping using a graded and structured system and, when sufficiently advanced were given a colony of bees to look after under guidance. By 1920 nearly 400 men had received instruction, encouraging the great majority of them to make a start in beekeeping. In Mr Steven's obituary it said that he had assisted many ex-service men in adding considerably to their incomes by the successful keeping of bees. On Mr Steven's death it was probably Mr John Scouller who became president of the Association.

In 1924 the president was Mr Robert Y Howie an Expert Beemaster.

During 1924 the Association visited the apiaries of Mr H M Stitch of Kilbarchan (a member of GDBKA) and the Rev. John Beveridge at Gartmore.

Mr H M Stich of Paisley wrote in the *The Scottish Beekeeper*. In 1924 he wrote about the mating of queens of different races and also under the heading of "Cullings from Foreign Journals", he wrote about and translated articles published in French and German, as well as doing talks at local association meetings.

During the 1920s and 30s Mr Howie was involved in BBC broadcasts on beekeeping e.g. on May 2nd, 1933 Mr Howie did a 20 minute broadcast on Beekeeping, Its Dreams and Realities. Mr Howie also did association talks. In January 1927 Mr Howie did a talk for beginners on subduing, handling, uniting, feeding and re-queening bees. He used lantern slides to highlight points during his talk.

He joined the SBA executive in 1923 and was a very active member. Mr Howie went on to be President of the Scottish Beekeepers Association from 1928 to 1930. He travelled extensively over Europe and was to be president of the SBA again in 1934. Unfortunately, he died suddenly on 31st January 1934, just before he could start his second term of office. In *The Scottish Beekeeper* it said that he had done much to make the SBA Insurance Scheme a success. In a talk in 1939 Dr J Anderson called him one of the beekeeping giants of the GDBKA along with Mr Baillie and Dr Tennent.

At the AGM in 1925 the association decided to give lessons to boy scouts and girl guides. Mr Malcolm McCallum was elected president. In June 1925 Mr and Mrs Pratt's apiary was visited and in July it was the turn of the president to host a visit to his apiary at Mearns Road, Clarkston. He demonstrated wax rendering and honey extraction. At the meeting on 16 January 1925 the Association had its biggest turnout for that time to hear a lecture from Dr John Anderson. At the meeting on 12 November 1925 another eminent beekeeper spoke – Mr John J Walker of Kilmaurs on the Stewarton Hive. He talked about the need to study bees more from the outside than the inside of the hive. It was honey produced in Stewarton Hives that ran off with the prizes at the Show at Crystal Palace in 1874.

The meeting on 25 November was a talk by J Todd a member of GDBKA on – "Anatomy of bee, Parts and Uses".

In 1926, GDBKA had a very successful Honey Show where Mr Walker of Kilmaurs ensured there was a selection of appliances including a Stewarton Hive, a hive feeder, foundation mould and wax on show. In 1926 1lb of heather honey sold for 3/6, a section for 3/6 and 1lb of clear honey 2/4.

At the AGM in 1927 Miss Adam became president and Jean Hepburn the Secretary of GDBKA. Miss Nicolson, also a member, was an SBA certificate examiner during this period.

In January 1928 Major Yuille from Kilmarnock, another eminent beekeeper, gave a talk. Miss Adams continued as president and apiary visits that year included going to Kilmaurs and visiting JJ Walker and John Dickson at Brig O

Doon, Ayr.

At the 1929, AGM Mr David Barnett became president of GDBKA. It was also reported that the heather site for GDBKA at Stockiemuir was no longer yielding and that a new site was required.

The president in 1930 was Mr P Bebbington who was a founder member and the first Secretary of the Association. He had met Mr Herrod-Hempsall demonstrating at the Royal Agricultural Society of England Show at Carlisle which sparked an interest and he started beekeeping thereafter. It is said he was devoted to queen rearing rather than producing large crops of honey. He resigned from the position of President in December 1930.

Mr George Gadsby became President in 1931, (Mrs Shepherd spoke very highly of his practical beekeeping skills). The membership had risen to 216 paid members, the highest recorded.

In 1932 the January lecture was given by Mr A. Limond a very well know beekeeper of that time on the importance of the Drone in Queen Rearing.

In 1933 George Gadsby was still president. That year the association visited the apiary of Mr J J Walker and then Mr Steven's apiary in Kilmaurs. At a similar Apiary visit in 1927 Mr W Shepherd got the idea of his swarm control board – (an adaption of a board he saw Mr Steven use) and wrote about in his article in *The Scottish Beekeeper* November 1937.

Mr Gadsby was again president in 1934 and 1935. Mrs Shepherd of Greenfield, Newton Mearns became secretary to the association and Dr Tennent became the association's representative at SBA meetings and Mr T B Baillie became secretary of the SBA in 1935.

At the AGM in 1936 Mr Hodge of Langbank became president. The schedule for the association for 1936 was a full one with a number of Apiary visits – Mr Gracie at Crossford, the Brown Brothers at Gargunnock and on the 4 July - 107 members visited Mr W Shepherd's apiary in Newton Mearns where he explained how he had been experimenting with the Snelgrove system of swarm control, explaining some of the modifications he had made. Dr J. N. Tennent of Clairmont Gardens, Glasgow was Vice President of the SBA and Mr T. B. Baillie of South Brae Drive, Jordanhill, continued as secretary of the SBA.

In 1937 Miss Allan became President, Mrs Shepherd as Secretary with Mr Shepherd becoming Treasurer that year. The talks that year included JJ Walker on "My Life among the Bees" when he explained the differences between a Stewarton and a Renfrew Stewarton Hive. Miss Allan who had lived in Holland for many years spoke on "Bees in Holland." On the postcard sent out to members promoting the talk, it stated that Miss Allan was going to speak about Beers in Holland. Miss Allan told the members this was a subject she knew nothing about and poked fun at the secretary Mrs Shepherd for her error before speaking about Bees in Holland.

In 1938 Miss Allan was president, the association had 104 members and 4 life members. Talks in 1938 included Robert Black, Dr Tennent on Heather Honey and Dr Anderson who talked about the 4 main factors for honey production. In 1938 The Empire Exhibition was held in Bellahouston Park, Glasgow where the SBA had a prominent pavilion. The event was seen as a great success and part of this was attributed to the organisational skills of Mr Baillie who was in charge of the SBA Pavilion. It is at this event that the Shepherds' got the idea for the Beekeepers' Club that was to be set up in 1942 and existed until the end of 1975, around a year after the death of Mrs Shepherd, who died in October 1974. Anyone who has been a steward at the RHS or any big event will have enjoyed meeting other beekeepers and having a cup of tea or coffee, something to eat and a chat about bees. This was the essence of the Beekeepers' Club.

At the 1939 AGM Mr Black became President. A special class of enthusiasts had fortnightly meetings to study for the expert Beemaster certificate. This was organised by Mrs Shepherd. Mr Percy Thomson presented a cup for the Honey Show which was to be called the Shepherds' cup in recognition of their good work. This cup is still awarded at GDBKA Honey Shows.

At the 1941 and 1942 AGM, Mr Black was re-elected as president. At the 1942 AGM the idea of opening a Beekeepers' Club in Glasgow was put to the members for consideration, who agreed it was a good idea. On Friday 22 May the Club opened at 213 Buchanan Street by Mr Struthers when 40 members attended. The club was open to all beekeepers – membership fee was 5 shillings entry fee and 1 shilling per annum subscription. Initially the club was open on Wednesdays, 3 to 9pm later it was 3 to 8pm and tea served from 3 to 6.30pm. Members were welcomed by Mr and Mrs Shepherd and a reference library and beekeeping journals were available to read and spare beekeeping equipment from members was advertised for sale.

On the 22nd October, 1942, twenty two people started a beginners class in beekeeping with the GDBKA. That year the Association visited Mr Shepherd's apiary in Newton Mearns. Due to the war sugar was rationed. It had been negotiated by Mr Baillie the Secretary of the SBA, The UK Government Agency, that 5lbs of sugar per hive allowance to feed the bees would be made available to improve the bees' chances of surviving the winter period.

In 1943 there were talks from Mr Black, Mr High and Mr Smith. Mr Smith spoke on how he built up his colonies in spring. He told the members who were present the secret of relieving congestion was to draw the bees upward in the hive and if done properly it reduced swarming to a minimum. He clipped his queens in April/May and did inspections every 9 days.

In 1943 Mr Tinsley, Chief Lecturer on Beekeeping at the West of Scotland Agricultural College, Auchincruive, spoke to GDBKA. At this meeting he told the group of his interest in what happened at GDBKA and that he was in good part responsible for its inauguration.

At the 1944 AGM Mr High became president of GDBKA. On the 27th May, Mr Hamilton's apiary was visited at Auchenairn with over 100 members and friends visiting. The work of opening and going through the hive was undertaken by Mr Shepherd. He inspected the brood, clipped and marked the queen with nail varnish. The association visited an apiary in Helensburgh on 17 June, on this occasion Mr Shepherd demonstrated how to make up a nucleus and introduced a mated queen.

A case of American Foul Brood (AFB) was discovered in Giffnock, Glasgow and a special meeting was held on the 21st of September 1944 with the largest attendance ever. All beekeepers were invited whether members or not. It was held to consider methods of stopping the spread of the disease. It was agreed to re-stock the unfortunate beekeeper's apiary the following year. All beekeepers gave consent to their hives being examined. The executive committee thought that this was the first case in the South West of Scotland in thirty years however, it was found that this was not the case as all cases at that time were not being reported and the Colleges and Rothamsted needed to be approached to get more accurate information.

In 1945, Mr Arthur West became president. The big issue this year was around AFB and the need for compulsory notification and the need for compulsory powers in its control.

Mr J. Airth became president in 1946. The Beekeepers' Club now at 104 Renfield Street, had an outing on the 29th June to Dumfries and the estate of Countess Liverpool. The demonstration as usual was carried out by Mr W. Shepherd.

Mr Airth was re-elected as president in 1947.
The talk in October was by Mr W Smith.

Dr Tennent did a talk on Research in Beekeeping and the need for a Scottish

Centre for this.

In 1948 Mr Charles McDonald became president. The Secretary and treasurer was now Mr J. Airth. There is no mention of Mr or Mrs Shepherd being on the committee. The eminent speaker was Robert Skilling who talked on honey production and successfully wintering.

In 1949 Mr Charles McDonald was re-elected as president. The association decided to relinquish the sub tenancy of the ground at Newton Mearns for the Association Apiary.

In 1949 the association had an apiary visit in Haddington.

Mr Baillie resigned from his position as secretary of the SBA. On the 31st July 1950 the SBA presented him with a gold watch as a tangible token of its appreciation for the service rendered. Under his stewardship the SBA expanded with *The Scottish Beekeeper* circulation quadrupling in that period. He was instrumental in securing the allowance for winter feeding during the war period and during a period of price control for honey obtained a more advantageous price for beekeepers.

Dr Tennent was now writing regularly in *The Scottish Beekeeper* on "Glimpses of Beekeeping Abroad."

At the AGM in 1950 Mr Ireland was elected president. The membership for 1949 had increased from 175 to 200 members. Some people believed that the growth of beekeeping from 1940 to 1953 was partially due to the rationing of sugar and the sugar allowances for winter feeding and rearing queen honey bees. Sir Hector McNeill the ex-Lord Provost opened the Honey Show at the Christian Institute that year.

In 1952 a large contingency of Glasgow and District BKA members friends and family visited Buckfast Abbey.

The first Scottish National Honey Show and Exhibition of Apiculture was held on 17 and 18th October 1952 at the McLellan Galleries, Sauchiehall Street, Glasgow. It was officially opened by Mr T A Kerr, the Lord Provost of Glasgow. it was heralded as an all round success. The Glasgow Association members under the leadership of Charles McDonald, were responsible for all local arrangements and the result was a triumph of planning and of teamwork in the execution.

1953, Mrs Shepherd was 70 and the Beekeepers' Club presented her with a box containing 70 Coronation shillings as an appreciation of the work she had done.

In 1954 Mr DC Barnett became president of GDBKA – and a series of talks under the heading "Focus on…" were run. The first was "Focus on Queens" and the second was "Focus on Increase". Later that year GDBKA members met to hear Father Les Smith of Buckfast Apiary talk about "All Aspects of Beekeeping". Buckfast Abbey at that time had 320 colonies kept in 9 apiaries in Modified Dadant hives.

Mr Savage of the Agricultural college at Auchincruive also gave a talk on beekeeping that year. Mr Savage had been a prisoner of war in Germany during the war. During this period, he had been allowed to keep bees and was involved in introducing and training other POWs in the art of beekeeping.

In 1955 Dr Tennent was still writing in *The Scottish Beekeeper*– this year he was writing about American Beekeeping. Andrew Dick of Glasgow wrote an article on his beekeeping system in *The Scottish Beekeeper*.

Mr Thomas Smith became the president in 1955 with Mr N. J. Hill as Vice president. Mr Charles McDonald past president died this year. In his obituary it stated that he had achieved his finest hour as Chief Steward at the Scottish National Honey Show of 1952. The outstanding success of this venture was in no small measure due to his efforts. Charlie as he was better known, kept his bees in a small back garden, in a built up area. He qualified as an Expert Beemaster and Honey Judge and was an authoritative speaker on beekeeping matters.

Norman Hill Vice president and former secretary of GDBKA emigrated to Canada. Before leaving he was made an Hon. Life Member of the GDBKA. Even after emigrating he continued to contribute articles to *The Scottish Beekeeper* magazine.

Mr Andrew Smith became Vice President and in December Mr W Smith talked to the Association. A wonderful quote from Mr Smith was "Never bother about the honey or the money, look after the bees and they will give you the return."

The president for 1956 and 57 was Mr T Smith. In 1957 a new Association out apiary was opened for the benefit of beginners and to provide a place where members could meet and spend a pleasant afternoon. Mr T Smith also wrote an article for *The Scottish Beekeeper* on Beekeeping within a City.

In 1958 the Beekeepers' Club, to advertise beekeeping and to improve beekeeping interest in the community, arranged window displays in shops which sold honey, where an observation hive was put in the window. Three hives were necessary for a one-week display, with the hive being changed every two days. The Club was also at this time looking for a site for a beekeeping museum which was to include an observation hive. James Burns became secretary for GDBKA in 1958.

Mr Shepherd was made Hon. President of The Beekeepers' Club. Mr Robb of Dundee was commissioned to make an observation hive after a visit to Dundee Museum by members of the Beekeepers' Club, in March 1958. The hive was installed at Kelvingrove Museum and Art Gallery on the 23 May 1959 and the Beekeepers' Club undertook to take care and restock the hive ongoing.

In 1961 the Association visited one of Hugh Howatson's apiaries at Garlieston.

In 1964 there was a talk from Robert Couston on Spring Management and the president for that year was Mr Oddy.

In 1965 there were talks from Dr Tennent and Dr Butler. Thirty people turned up to hear Dr Butler – not a high number for such an eminent speaker. Mr Burns the secretary said that the turnout was poor and not good enough.

Apiary visits that year were to Captain Thake's apiary, and the apiary of James Burns (secretary) in Thornliebank.

In 1966 there was a talk from Mr Savage. Mr Oddy was re-elected as president and his report stated that enthusiasm was greater in 1918 than in 1966 and that he felt that access to suitable apiary sites was a great hindrance to beekeeping. Speakers this year included Mr W Smith who talked about his Smith Hive and Mr A S Deans from Aberdeen. There was an apiary visit and demonstrations at Mr Burn's apiary at 17 Woodlands Road, Thornliebank, a regular fixture for many years.

The Beekeepers' Club, to try and arouse interest in beekeeping arranged with the BBC the filming of the transfer of bees from the winter box into the observation hive at Kelvingrove Museum and Art Gallery. The film was broadcast in the programme "A quick look round."

In 1967 Mr Morgan was elected president. In 1967 Dr J N Tennent was awarded the SBA's Dr John Anderson Memorial Award.

In April 1967 Mr W Shepherd died, he had been treasurer of GDBKA from the mid 1930s to mid 1940s. He demonstrated beekeeping manipulations at apiary visits, he was the mentor of other beekeepers, a main stay at the Beekeepers' Club, and the inventor of the Shepherd tube which was part of his swarm control system.

Dr J. N. Tennent died on the 28th April at the age of 69. He had been president of SBA from 1938 to 1940, and Convener of the Moir library, doing much to make it one of the finest beekeeping libraries in the world. He was a great contributor to *The Scottish Beekeeper* magazine. He was known as Scotland's Beekeeper travelling ambassador, he was well known internationally. He had been president of the SBA when the great Empire Exhibition had been held in Glasgow in 1938 and had been involved in the running of the SBA pavilion. He had been a member of the SBA executive committee for almost 30 years.

In November 1967 W S Robson gave a talk to the association which was a follow up to the association's summer outing to the East of Scotland Agricultural College apiary at Lauderhill.

At the 1968 AGM all the officials were retained in office. The summer outing was to Culzean.

In 1969, Mr J Morgan was the president. From an Association survey the average number of colonies per member was 5 with an average harvest per colony of 100lbs for 1968.

Neil Anderson wrote in *The Scottish Beekeeper* that the bees in the observation hive at Kelvingrove were still alive and kicking under the care of Dr Burkel of the Natural History Department. It was hoped that with a bit of luck that the 2 frame stock would over winter successfully. Mr Anderson commented, that on two occasions there had been two queens seen side by side for a few days.

The March Association Meeting was a talk from Mr A S Deans, Head of Beekeeping at the North of Scotland Agricultural College which was well received. He said that the best strain of bees for an area was the ones that we had and that it was futile in a City like Glasgow to try and keep a pure strain of bees as drones can gather from up to 7 miles away. The outing was to Garlieston and to visit one of Hugh Howatson's apiaries at the heather.

1969 was a swarmy year and on the 11th June it was recorded in *The Scottish Beekeeper* that at 8 pm in the evening, a swarm at 4, Petershill Road, Glasgow was taken off a tenement wall one floor up with the help of a ladder from a fireman. A policeman, a fireman, an army of reporters, and a group of children watched the swarm being taken.

A presentation was made to Mrs Shepherd at the Jubilee Celebration for the association that year.

At the 1970 AGM it was reported that the yield per hive had gone down from 100lbs to 46lbs of honey for 1969. Mr Philip McBarron was elected president. That year there was the customary visit to Mr Burns' apiary in Thornliebank. Mr Burns did a talk on his swarm control method and Mr John Harris did a talk in December on wax rendering. The average yield per hive for 1970 was 61lbs of honey per hive.

In 1971 the officers and committee were re-elected. That year there were talks from Robert Hammond NDB, Mr Peter Morton SBA president and Mr Robert Skilling. That year Mr Eric McArthur, Member of GDBKA wrote a letter for beekeepers to stand up and be counted with regards to the "denuding of the beekeeping advisory educational and watchdog services" due to Mr J Smith NDB, the West of Scotland College Beekeeper Advisor, not being replaced. The average yield of honey per hive from the association survey went down slightly to 58lbs per hive.

In 1972 the average yield of honey per hive from the association survey went down to 41lbs per hive. Eric McArthur under the nom de plume of Apis Fanatica, wrote about the fight to overcome apathy to unite and take control over insecticide usage.

The Association outing and apiary visit was to Rothesay and to the apiary at Ascog where 3 working Stewarton hives were seen.

Ian Craig gave a talk on – "My method of beekeeping" that year. Philip McBarron did the December talk. The Association hive and bees had been destroyed – they had been kept in the Garscube estate.

In 1973 Hector McPherson was president

Mr A Stirrat, of Auchincruive gave a talk on 'The Bee Year'. The usual visit to the secretary's apiary in Thornliebank, was held on 17 May giving beginners the opportunity to see and handle live bees.

The death of Mr D C Barnett a past president of GDKA was reported.

Mrs Shepherd was made an Hon. Vice President of the SBA in recognition of her long and active association with beekeeping.

At the 1974 S.B.A. AGM, Mrs Shepherd got a special welcome by the president Mr Andrew Smith.

In October 1974 the death of Mrs Shepherd was reported. In her obituary it said she would be remembered for her helpfulness, friendliness and great love of bees and beekeeping; Beekeepers were welcomed to her home and Apiary in Newton Mearns; the Shepherds were founder members of The Beekeepers' Club in Glasgow (the Club was the brain child of the Shepherds). Mrs Shepherd was one of the first recipients of the SBA's Dr Anderson Memorial Award for Services in Beekeeping in 1945.

In 1975 The Beekeepers' Club in Glasgow felt unable to carry on and requested and was merged with Glasgow District Beekeepers' Association.

Mr S Doak, a GDBKA member gave a talk on making mead.

In 1976 John Harris became president. Allan McCorquodale of Clarkston died – Allan was an experienced and well known beekeeper who kept bees from the South Western Highlands to East Kilbride. In 1973 he lost eight hives which were destroyed by fire caused by vandals.

Mr Eric McArthur was the judge at the association Honey Show that year.

In 1977 John Harris was president again. The winter session had talks by Mr R Skilling "Beekeeping for Beginners" and Mr Ian Craig on "Presentation of Honey".

In 1978 Mr John Harris was still president. On 2nd October the Association had a film show at Eastwood recreation theatre to which other associations were invited. Mr Thomson of Callander gave a bee talk on "60 years of Beekeeping." Mr Eric McArthur a member of GDBKA, started to write articles for *The Scottish Beekeeper* under the name of Rambler.

In 1979 Dr D Andrews became President of GDBKA. The Association had talks from Mr James Smith NDB and Mr Ken Steven who was the SBA tour speaker. He talked on "False Economies in Beekeeping and Mistakes in beekeeping." Suzanne Ullman wrote an article in *The Scottish Beekeeper* – "Beginners luck – but surely no substitute for experience."

In 1980, 1981, 1983.... Eric McArthur wrote prolifically under his own name, as well as that of Rambler and Apis Fanatica in *The Scottish Beekeeper* on all aspects of beekeeping.

In 1983 Mr Hugh Gilles of Westerton, Bearsden was president of GDBKA. The March talk was by Tom Dunlop and Bill Henderson of D & H Apiaries, Bishopbriggs, on queen rearing.

Mr Eric McArthur published his book - "Milestones in bee-keeping and the swarm trigger."

In 1984 Clive De Bruyn talked to the association on Oil Seed Rape.

In 1985 the Association wrote to all local association secretaries in Scotland, asking for signatures for a total ban on the importation of queen honey bees. GDBKA set up an anti-varroa lobby group led by Mr Eric McArthur.

In 1986 Mr Charles Irwin took on the role of secretary of GDBKA, he was to retain this role for the next 20 years.

In 1987, Mr McArthur wrote a letter regarding the Beekeeping Advisory service and the SBA subsidising this service. He wrote in 1988 that we were living in dangerous beekeeping times with varroa on the horizon and the advisory service disappearing.

Mr Gordon Stewart was the president in 1988. At the November meeting Mr Charles Irwin gave a talk on "Wax Moth Damage and how to Prevent It." After the tea break he did a talk on queen rearing.

At the 1989 AGM Mr John Mitchell became president. The Association outing was to the Stoakleys' apiary. Ron Brown spoke to three Associations from Glasgow and the surrounding area, which included GDBKA on the 19th October the subject was "Bees in Winter."

In 1989, 1990 and 1991 Mr Mitchell continued as president. On 14 February 1990 Willie Taylor of Kilbarchan did a slide talk on the identification of pollen in honey. Mr Taylor returned in November to be judge at the association Honey Show. Mr R Skilling spoke in March 1990 about Spring Management.

In March 1991 Mr Brown from Strathblane did a talk on beekeeping in former time. Dr Christison did a talk on microscopes and their use in identifying bee diseases. The Association visited the apiary of Mr Peter Aird, Dalry.

On Tuesday 23 April 1991, representatives of the 5 associations in the greater Glasgow area agreed to the formation of a new group which was called the Clyde Area Beekeepers' Association. The Aim of the association being to improve communication between the associations, share the expense of invited speakers and to arrange travel facilities to meetings held in places more distant than association members would normally travel. The first CABA meeting was on 10 October 1991 when Margaret Thomas was the speaker on swarm control methods.

In 1992 Gordon Smith became President of GDBKA. On the 12th of February the Association talk was by Willie Taylor of Kilbarchan on queen rearing. On the 8th of April the talk was on "Varroa and how to treat it" by Mr Eric McArthur this was around the time of the discovery of Varroa in Britain. The September talk was on preparing bees for winter and the speaker was Archie Ferguson. In October Norman Stark spoke about his efforts on setting up an apiary on Arran and Gordon Stewart spoke about his system of management.

In 1993 Gordon Smith remained president.

James Burns past secretary died 18th February 1993. His success with his bees he attributed to his friend and mentor Mr W Shepherd and he was a great advocate for Mr Stepherd's system of swarm control.

Speakers this year were Mr Gilchrist, Mr Garrow, Mr and Mrs Stoakley, the SBA president Mr Iain Steven and Mr Blair of Kilbarchan. At the April meeting Mr Gordon Stewart brought along the Robb Hive that had been on display at Kelvingrove Museum. The hive had been returned, by the museum, to GDBKA and was no longer being used. This hive was later to be given to Willie Robson at Chainbridge. At the October meeting Mr Ian Craig spoke on the criteria for a perfect apiary site.

At the 1994 AGM Mr Smith was elected president again. Speakers this year included R. Brown of Dumfries on Spring Management, Mr Charles Irwin on making wax foundation, Mr Ian Morrison on Operation Sunflower. Mr Hugh Gilles had an article in *The Scottish Beekeeper* – 'Another Swarmy Tale'. Ian Craig was the judge at the Honey Show this year.

1995, Mr J Morgan was elected president. Speakers this year were Mr Gordon Stewart, Dr Moody, Mr Les Webster and Mr Cowle of Eastwood. *The Scottish Beekeeper* magazine ran a series of articles by Ian Craig on his double brood chamber method of management. George Duncan from Ayr was the judge at the Association honey show.

At the 1996 AGM Mr John Morgan was re-elected for a second year as president. Speakers this year included Mr George Hood of Ormiston, East Lothian. Mr Hood had over 600 colonies and had been a friend of Mr Willie Smith. He explained how he managed his bees and the use of his equipment. In March Dr Chard spoke about statutory bee diseases. The CABA outing that year was to Denrosa Apiaries where there was a talk from Mr Murray McGregor, a tour of the honey house and storage facilities, before several apiary sites were visited. Mrs Thomas was the CABA/ SBA tour speaker that year. Mr Eric McArthur became the editor at *The Scottish Beekeeper,* in April 1996, a position he would hold until August 2005.

Mr John Morgan was again GDBKA president in 1997. At the AGM Hector McPherson vacated the post of treasurer. Hector had held this position for an amazing 50 years, later that year Hector McPherson died on 28th September. He had been one of the founder members of the Beekeepers' Club in Glasgow, had held the offices of President, Secretary and Treasurer for both the Glasgow and Paisley Associations. He was made a life member of GDBKA in 1988 and had received the SBA's John Anderson Memorial Award in 1989. He was a quiet unassuming man, well known and respected through the length and breadth of Scotland for his services to beekeeping. Mr Eric McArthur became publicity Officer and Mr Charles Irwin continued as secretary. At the May meeting Dr Suzanne Ullman a member of the association spoke about pheromones and bee communication. Ian Craig a member of GDBKA became president of the SBA. In *The Scottish Beekeeper* that year there was an article about a fire in an apiary where an old caravan and two huts which were being destroyed by burning them which got out of control setting several hives on fire. In the article by Gordon Stewart he records how Charlie Irwin saved the hives and bees from the fire. Ian Craig wrote about his method of rearing queens in *The Scottish Beekeeper*. On 4th September 1997 Ian Craig as president of the SBA was notified that Varroa had been found in seven out of ten apiaries around Canonbie in Dumfries and Galloway and thus began the spread of varroa and its effects through Scotland.

1998, Mr Norman Stark was elected president of the GBKA, Mr Phil. McBarron was elected Vice President. In February Dr Chard spoke about Statutory Bee Diseases. Ian Craig wrote an article in *The Scottish Beekeeper* magazine, on detecting and monitoring varroa mites. In April Mr R Simpson (Flowers of May Aparies, Fife.) spoke about the season ahead. In the summer, the Association apiary was used to hold demonstrations and there was an Association outing to "Flower of May Apiary" hosted by Bob and Joan Simpson, Cupar Fife. The September and October speakers were Kevin Cowle and Willie Taylor respectively. The judge at the Honey Show was Colin Watson.

At the 1999 AGM Mr Norman Stark was re-elected as president. At the AGM it stated that the finances of the Association were in good shape due mainly to the funds from the Higgins' Trust. Frank Higgins had been a long-time member of GDBKA who on his death, had bequeathed 14 colonies to the Association. Some of the hives were sold and the remainder were managed for the benefit of the association, the money going to subsidise speakers, buy bee keeping equipment and books for the Association.

At the February meeting Ian Craig spoke about his method of beekeeping. The Association promoted all aspects of beekeeping at the Drymen Agricultural Show, Pollock Park, the open day held by friends of the River Kelvin. The Association apiary was used to hold demonstrations for beginners. On Sunday 16th May the CABA outing was to the apiary of Ian Kirkwood of Heather Hills Honey Farm at Bridge of Cally. The weather for the outing was a mixture of cold and wet, surprisingly a number of hives were opened. In October there were talks from Norman Stark and Charles Irwin on preparation of honey and wax for show. Eric McArthur was the judge at the Honey Show this year. Mr Charles Irwin was judge at the Royal Highland Show this year.

In 2000, Mr Norman Stark was re-elected as President. The Association talk in February 2000 was "Beekeeping in India" by GDBKA member Suzanne Ullman. An article was published in the June 2000 Scottish Beekeeper. Morna Stoakley spoke about Bits of Bees at the April meeting. Colin Watson spoke about "Preparation of Honey for Show" in September and Diseases disorders and pests was the subject for October, the speaker was John Goodman. Enid Brown was the judge at the Honey Show. Dr Christison was awarded the SBA's Dr John Anderson Memorial Award.

At the AGM in 2001 George Morrison was elected president. On the 14th of February Ian Craig spoke about his method of management. Speakers this year included Willie Taylor – mini nucs and queen rearing in March; Enid Brown – her method of beekeeping was her talk in April. A microscopy night was held in May and the Association visited the apiaries of Eric McArthur in Dalmuir with a focus on queen rearing and then Charles Irwin and Gordon Stewart at Torrance. Charles and Gordon did a demonstration on swarm control. The SBA tour speaker for the CABA meeting this year was Albert Knight. Leslie Webster the SBA president did a talk on "Life of a Beekeeper" and the October talk was Bob Simpson on "Beekeeping as a Business." Ian Craig was the judge at the GDBKA Honey Show this year. Ian Craig was awarded the SBA's Dr John Anderson Memorial Award.

In 2002, George Morrison continued as President. The talk in February was from George Hood on "Maximising your Honey Crop." The speaker for both March and December was Ian Craig; John Goodman spoke in April. In May, Willie Taylor spoke about the "Powerful Drive in a Swarm." The June visit to George Morrison's apiary included a demonstration on swarm control. It was the turn of Dr John Durkacz to speak of "Beekeeping in the North of Scotland" at the September Association meeting.

On the 25th November Hugh Gilles a past president died. He was a keen and successful practical beekeeper. At apiary visits to Hugh's home apiary, demonstration on swarm control and queen rearing were normally carried out.

George Morrison was re-elected as President in 2003. Speakers this year included Andrew Abrahams whose talk was on "beekeeping on Colonsay"; Suzanne Ullman and Mike Thornley spoke on "Beekeeping in Tobago" – an article was published in *The Scottish Beekeeper*, Archie Ferguson with a talk on Summer Management with Varroa in mind and Dr Flora Isles talking about

Bee Diseases. George Morrison's apiary was visited in early May with queen marking demonstrated. Later in the month the apiary of Mike Thornley at Rhu was visited. Unfortunately, two of the hives due to be inspected, swarmed an hour before the event. *I have heard Mike speak about this visit on a couple of occasions and I believe Eric McArthur returned after the visit, with much needed help and equipment as well as to check that Mike was okay.*

At the AGM January 2004 John Miller was elected president. In 2004 Mr Eric McArthur was presented with the SBA's Dr John Anderson Memorial award.

Speakers this year included: Ian Craig, "Late winter into Spring Management"; in May it was Les Webster. Ian McLean was the tour speaker and spoke at the CABA meeting on "Handling Bees, Art and Science."

In *The Scottish Beekeeper* Mr Eric McArthur wrote articles on oxalic acid sublimation as a treatment for varroa infestation.

In 2005, John Miller was re-elected as President. John Taylor was speaker for the March meeting and then the association visited his apiary in Dunkeld in June. In May, Mrs Susan Irvine of Science and Advice in Scottish Agriculture (SASA), gave an "Update on Bee Diseases", the talk in October was by Alan Riach on "Beehives through the Ages." And the honey judge at the Association Show was Enid Brown. Apimondia 2005 was in Dublin and the SBA had an exhibition stand which was awarded the bronze award. The funds to have this stand in Dublin was supported by donations made in part from the Association and individually by members of GDBKA such as Ian Craig, Charles Irwin and Eric McArthur.

At the AGM in 2006, Mr Charles Irwin was elected as President. The May speaker was Ian Craig his topic was on "Spring Build up."

In 2007, Charles Irwin was again President of GDBKA. Ian Craig wrote his series of monthly of articles on his "Beekeeping Year." The SBA Tour speaker was Bryan Hateley who spoke to CABA about recovering beeswax and solar wax extractors. The Summer outing with CABA was to Auchincruive. The December speaker was Dr Peter Stromberg on "All about Beeswax." Dr David Christison died October 2007. He was a member of both GDBKA and Kelvin Valley BKA. He was born in 1915 and brought up on the Marquis of Bute's estate at Auchinleck where his father was a game keeper. He won a busary to allow him to do medicine at Glasgow University. After he was called up during WW2 he was sent to Singapore in 1942 two weeks before the Japanese arrived. He was one of the camp doctors on the River Kwai where he was responsible in saving many lives. He ended the war as a prisoner of war in Japan. On return to the UK he became an ophthalmic surgeon and then consultant at the Glasgow Eye Infirmary. He retired at 67 and then took up beekeeping. He was an active member of both GDBKA and Kelvin Valley. For a time he was SBA Area representative for the West. He ran 3 apiaries and was keen to teach and help other beekeepers wherever possible. He was awarded the SBA's Dr John Anderson Memorial Award in 2000.

In 2008 Charles Irwin was re-elected as President. The February speaker was Eric McArthur on the "Bee Breeding Project and update." March was Alan Teale on "Integrated Pest Management." In April, Phil McAnespie spoke about "How we conduct a Bee keeping Demonstration." In May, Graeme Sharpe spoke on Swarm Control. Ian Craig wrote two articles in *The Scottish beekeeper* on Swarm Control. The summer outing under the CABA umbrella was to John Mellis's Bee Farm near Dumfries. The SBA tour speaker was Philip McCabe from Ireland, he spoke to CABA on "Spring and Summer Management."

At the 2009 AGM George Morrison was elected President. The April speaker was Jeanne Robinson on Bumble Bees in Spring. This was followed by a talk by "Swarms and Supersedure" by Graeme Sharpe in May. The SBA tour speaker was Clive De Bruyn who spoke to CABA about "What can the Hobby Beekeeper learn from the Commercial and Professional Beekeeper." The CABA summer outing was to Coupar Angus and Andrew Stirret's Apriary. The December meeting had Dr Peter Stromberg speaking on "Bee Stings and Stinging." Eric McArthur wrote an article in the December *Scottish Beekeeper* on the CABA Apiary project.

2010, saw George Morrison re-elected as President. In February, Ian Ferguson spoke on his system of beekeeping. Jim McCulloch spoke about Bee Breeding in March, and in April, Alan Riach spoke about the "History of Beehives." Graeme Sharpe spoke about foulbrood in May, and July in conjunction with CABA the summer outing was to Coupar Angus and Murray McGregor's Bee Farm. Dinah Sweet was the SBA tour speaker. She spoke about "Swarm Management for the Hobbyist Beekeeper." The November talk was by Steve Sunderland (Scottish Government Lead Bee Inspector) on "Beebase and the present position of Beekeeping in Scotland." Dr Peter Stromberg spoke about "Honeybee Navigation" in December.

At the AGM in 2011 George Morrison was re-elected as President. Charles Irwin spoke in February on "The Use of Nuclei"; in March John Durkacz spoke about the "Black Bee in Scotland." Jim McCulloch spoke about "Bee Breeding" in April, then Gavin Ramsay on "Mechanisms of Resistance in Varroa Honeybees."

In *The Scottish Beekeeper* magazine under the heading of "Spotlight on a Beekeeper", articles were published about Eric McArthur (May), Charles Irwin (July) and Ian Craig (November). These articles gave information on how they got started and memorable things that had happened to them over the years. The summer outing under the auspices of CABA was to Andrew Scarlett's apiary at Longley's Farm Meigle, Blairgowrie.

2012 At the AGM Ian Craig was elected President of the Association and Eric McArthur elected as Vice President. The Summer outing under the CABA umbrella was to Auchincruive. The SBA tour speaker was Richard Ball who spoke to CABA members on "Integrated Pest Management." In October there was a video night organised by Peter Stromberg and then in November he gave a talk on Chalkbrood. In December Mike Thornley spoke about "What Makes a good Association."

2013 At the AGM Ian Craig was re-elected as President of the Association. In the New Year's honours list GDBKA President Ian Craig was made a Member of the British Empire (MBE). An article was written in the February *Scottish Beekeeper* congratulating him and listing his many achievements. Eric McArthur in a letter published in *The Scottish Beekeeper* magazine highlighted the significance of Ian Craig getting an MBE for services to beekeeping, as he was the first Scot to be recognised in this way. Eric put it so well when he wrote "that the Association was extremely proud to be associated with a man who has done so much and given so selflessly of his time, effort and experience over more years than he cares to remember for the future of the honey bee."

At the AGM in January a motion was laid before the members;

"That the Glasgow and District BKA show its concern for the health and safety of the honey bee and other beneficial invertebrates in the light of the current mounting scientific evidence against the neonicotinoid pesticides by endorsing the need to invoke the Precautionary Principle for amoratorium on the use of these substances until they are proved to be harmless to these creatures."

In March, Steve Sunderland spoke about "The foulbroods and their Current Position". Dave Goulson from Stirling University spoke in April on the "Ecology and Conservation of Bumblebees". In May Ian Craig covered "Swarm Control" in his talk. The summer outing organised through CABA was to the Apiary of Enid Brown.

2014 Ian Craig was re-elected as president.

In February Ian Craig wrote about "Observations at the Hive Crownboard and Entrance" and in July about his "Beekeeping Recollections", both were published in *The Scottish Beekeeper* magazine. The March CABA speaker was Celia Davies. The SBA tour speaker for 2014 was Pam Hunter. Ian Craig spoke in June to the Association on the "Variable Brood Chamber." The Summer CABA outing was to John Mellis's apiaries, GDBKA also visited to the Edinburgh and Midlothian Beekeepers' Association (EMBKA) apiary at Gogar. In September the Association had a discussion evening and in October a Practical Microscopy evening was held. In November the talk was on "GM Crops the pros and cons" with Peter Dominy.

In April 2014, Dr Suzanne Ullman died, aged 78. She was born in Budapest and had lived in the UK from the end of WW2. Her parents who were Jewish had been visiting Britain at the time of the outbreak of the war, so Suzanne was left with her twin brothers in the care of her grandmother. During the war they had a harrowing time evading Nazi agents. It was not until 1946, when aged 10 years she was re-united with her parents. Suzanne was a Zoologist and long-serving lecturer in zoology at the University of Glasgow, taking early retirement in 1992. She was a fluent speaker of Hungarian, German, English and Italian and enjoyed travel, in particular travel associated with her beekeeping activities in her later years. Her visits were usually to beekeeping meetings followed by field trips to local beekeepers, this included Argentina, Turkey and Ukraine.

At the 2015 AGM Kathy Friend was elected president of the Association. George Morrison retired from the role of treasurer and was made an Hon. Member of GDBKA for his work for the Association over the years. In February the Association had a talk on Wing Morphology; by Jim McCulloch in March Enid Brown spoke about her "System of Beekeeping." The CABA summer outing was to Coupar Angus and the apiaries of Murray McGregor. The SBA tour speaker was Graham Royle who gave his lecture "Apis Through the Looking Glass" at the CABA meeting in September. Ian Craig spoke on "Preparation of Honey for Showing." The December speaker was Neil Sandison of Helenburgh speaking about "Setting up an Association Apiary."

In 2016 Kathy Friend was re-elected President of the Association. Alan Riach spoke to the Association in February which was on "an Introduction to Microscopy", March was on "Beekeeping using a Single Brood Chamber" by John Coyle and in April, Christine Matthews spoke about "Useful Products Made from Hive Products". In May Ian Craig spoke about "Local Nectar and Pollen Sources" and "Identification of Honey." In October Ed O'brien and Kathy Friend spoke about Urban Beekeeping. The CABA summer outing was to Chainbridge and a talk from Willie Robson. The SBA tour speaker for the year was Dan

Basterfield who spoke at the CABA meeting in October. Eric wrote Monthly actions under the name of Rambler ll in *The Scottish Beekeeper* magazine.

In 2017 Kathy Friend still president of the Association. Mhairi Neill at the February meeting spoke about Manual Handling, Health and Safety, in March Graeme Sharpe spoke about "Common Diseases" and in May Gavin Ramsay spoke about "Bee Farming." The CABA outing was to Abernethy and the apiaries of Gavin Ramsay. In September Charles Irwin spoke about "Observation Hives;" in October Julian Stanley spoke about "Digestion, Trophalaxis and Bee Communication" and Enid Brown was the judge at the Honey Show in November. Sharon Dennett wrote an article in *The Scottish Beekeeper* magazine on the Soap and Cosmetics Workshop at the recent Ayr and SBA Convention and followed up by demonstrating her skills at the Association craft night in December.

2018 – At the AGM, Dr Taylor Hood was elected President. Eric McArthur, Charles Irwin and Ian Craig were made Hon. Members of GDBKA for the work and support they had given the Association and its members over the years. Mr Alasdair Gray the Scottish Author and Artist was also made an Hon. Member of the Association for a drawing of a bee that he donated to the Association for use on the Convention programme and Centenary Book cover.

From January to May the Association had a display at Kelvingrove Museum. The display was organised by Susan Fotheringham and a piece along with photographs was written about it in the *Scottish Daily Record*.

On 9th February 2018 a Civic Reception/ dinner was held by Glasgow City Council to celebrate the Centenary of GDBKA . This reception was attended by members of GDBKA and speeches were made by the Depute Lord Provost, Mr Braat, the President of GDBKA Dr Taylor Hood and the President of the SBA Mr Alan Riach. At this reception GDBKA also twinned with Les Amis des Abeilles du Val-D'Oise, Jim McBeath and Christophe Woirgard had spent two years forging close links between the two groups which led to the formal twinning and declaration of friendship and co-operation. In May a picnic at Kelvingrove Museum and Art Galleries has been arranged. In September the Scottish Beekeepers' autumn Convention is being arranged in conjunction with GDBKA. The Convention is to be at the Royal College of Physicians and Surgeons in Glasgow and the speakers, Clive De Bruyn, Margaret Murdin and Margaret Lear.

For 2019 refurbishment of the observation hive at the Museum has been planned employing the expertise of the sculpture Alan Kain.

Much of the success of these recent activities is due to the enthusiasm and energies of Mhairi Neill (Secretary), Susan Fotheringham (Committee member and past president of Lanarkshire Beekeepers' Association), Sharon Dennett (Committee member) and Kathy Friend (Past President of GDBKA)

3. Ian Craig, MBE

Recollections

I was brought up on a sheep farm in the moors of Wigtownshire where I began to keep bees in 1950. It was only five years since the end of World War II and sugar was still rationed. Many of the farms and cottages had beehives, not all of them had bees in them but their owner still claimed a sugar ration for every hive. I don't recall anyone ever checking. My father did not keep bees, but I was mildly interested. One summer evening an old farmer, a few miles away, arrived with an ancient WBC hive plus bees strapped to the carrier at the rear of his ancient Singer motor car. I was presented with the hive and a flame of interest was kindled in me which is still burning to this day. The bees were British Blacks and how they managed to survive into the first winter is still a mystery as I peered into them almost every day. I do not recollect the bees being bad tempered, which is just as well because the beesuit had not been invented. The usual attire of the apiarist was an old pair of trousers smelling of sheep's wool, a sports jacket, hat and a veil resembling a present day midge hood which slipped over the hat and tucked into the neck of the jacket. Gloves were not worn. The bees were descendants of black bees given to my benefactor by Joseph Tinsley, then head of beekeeping at the West of Scotland Agricultural College at Auchincruive. The college ran an experimental permanent out apiary on the heather site on the old farmer's land. The site is described in chapter 12 of "Beekeeping Up to Date" by Joseph Tinsley. The saying that "the grass seems to be greener on the other side of the fence" applies especially to beekeepers. I have bought French Black bees from Steele and Brodie; they were good honey producers but they could sting through my inadequate veil. I was prone to swelling in those days and many a time I came home with wrists twice their normal size and lips resembling those of a platypus. I bought Caucasian bees from Mountain Grey Apiaries in Yorkshire; they were quiet and excellent honey gatherers, but their first cross with my Mongrel bees was unmanageable. I bought Buckfast bees on three separate occasions all from different sources in England. Only one of these, from Birdwood Apiaries in Somerset was to my liking. I still have a yellow band from these bees appearing in some of my present day colonies from time to time. Next, I purchased two Amm queens which were from Bernhard Mobus's Maud strain. They were nice bees to handle, had a compact brood nest and lovely white honey cappings. The problem was that they were bred for the heather and did not build up sufficiently until the beginning of July, whereas I wanted strong colonies for the Sycamore in early May. I learned the hard way that breeding local bees is best. In my early beekeeping days, I was a member of the SBA and of Western Galloway Local Association (LA). The Agricultural Colleges must have been funded differently in those days as there were into teens of Advisors in Beekeeping attached to the then four colleges in Scotland. The advisor in Western Galloway was Ian Maxwell who was based in, I think, Stranraer. The highlight of my beekeeping year was the frequent summer visit to apiaries in the county; the excellent clover fields near Stranraer, Hugh Howatson's bee farm at Garlieston and Willie Paterson's heather site west of Kirkcowan. In those days, the colleges took turns at running an annual beekeeping weekend consisting of lectures and apiary visits. As time passed, the beekeeping departments in the Colleges gradually closed as lecturers retired and funds were reallocated. Beekeeping was well down the list of Scottish Government (SG) priorities. As readers will be aware, the increased awareness of environmental issues together with the outbreak of AFB and EFB stimulated the SG and SASA into providing what is now an excellent service to the beekeeping sector. In the fifties and sixties the weather patterns were different to those of today. We had our share of inclement weather, but there were long spells of fine beekeeping conditions. Hay was cut from mid-July, allowing the meadows to be full of wildflowers which attracted a myriad of bees, butterflies and other insects. Nowadays the meadows are planted with ryegrass which is cut for silage in May and again in July and are consequently devoid of what farmers now refer to as 'weeds'. Wildflowers were also prevalent amongst the corn. Sprays were not much used and were less lethal to insects in those days. The Newton Stewart to New Galloway road led to excellent sites for both Bell and Ling Heather, sadly now gone thanks to tree planting by the Forestry Commission. In the fifties and sixties, Western Galloway had over sixty members, which dwindled to a handful in the nineties. This LA has since been rejuvenated and is strong and forward looking once again. A similar situation developed in Kilbarchan BKA. In the mid-sixties the Association had dwindled to only three members. I wonder if the President at that time, Matthew Muir was a relative of Don Muir whom Alan Riach met on his recent trip to Australia, described on page 164 of the June 2014 *Scottish Beekeeper*? At that time a number of keen beekeepers including myself, joined the Association and breathed new life into it. In my early years, beekeeping was mainly a countryman and villager pursuit. The forage was diverse and plentiful; winters were colder and sunnier; we did not seem to have such long spells of poor weather and we did not have spraying problems. Nowadays many more people from all sections of society are becoming interested in beekeeping and the environment. Agricultural practices are different, but the new generation of beekeepers have not known anything else. The government and many businesses are aware of the value of bees to the environment and give encouragement to beekeepers. The number of beekeepers in the country is steadily increasing. Bees and beekeeping have a future.

Observations at the Entrance and Crownboard by Ian Craig MBE

Bees can be stressed by too frequent disturbance by the beekeeper during the year and especially in the winter. A great deal of information can be gleaned by observation, without opening the hive. Hives showing some sort of abnormality should be noted and be subject to a closer examination when weather permits.

Small particles of wax

When bees are confined to their hive for long periods of winter, the appearance of small dust particles of wax at the hive entrance is a sign that the bees are uncapping and eating into their stores and that all is well. Hives should be hefted from the rear in order to ensure that the colonies have enough food.

Medium particles of wax

Particularly in autumn, wax particles up to 2.5mm suggests robbing by other bees or wasps. The robbers being in a hurry to tear down cappings and make their escape.

Large pieces of wax

In winter, if wax pieces up to 12 mm are seen at the entrance it suggests that a mouse is in residence. Mice can nest in the hive, feed on honey and urinate on combs. Combs soiled by mice will never be used by the surviving bees and should be disposed of and a clean floorboard given.

Small wax platlets

Prior to the start of the summer, small platelets of wax can sometimes be seen at the hive entrance. This indicates that there is a surplus of young wax secreting bees. A frame of new foundation should be given.

Pollen

Small pollen pellets on workers' corbiculae early in the season is an indication that small amounts of pollen are available. At other times large pellets indicate a laying queen, with plenty of hungry larvae to feed.

Hard grey pollen pellets

These are often seen on the alighting board when bees are expanding their brood nest in spring. They are usually the size of a worker cell, resemble Chalk Brood but are brittle if crushed, breaking up into layers. They are a sign that some pollen store in the autumn had not been covered with honey and sealed. A late autumn syrup may have prevented this loss of protein.

Combs with large patches of hard pollen are usually eaten down to the mid-rib by the bees in an attempt to remove the pollen. Combs like this should be removed by the beekeeper at the first opportunity.

Chalk Brood mummies

These are flatter than mouldy pollen and are like poorly developed pupae and do not crumble into layers. They can be whitish or almost black with fungal fruiting spores.

White grubs

If the colony is on the verge of starvation they eject first drone brood pupae, then worker pupae. This is an indication that immediate feeding is required. Occasionally long, thin, caterpillar-like grubs are ejected. These are an indication of the presence of wax moths. It also is an indication that some very old abandoned combs are in the hive.

Few bees flying in early spring

If fewer bees are flying from a hive than others in the apiary, especially in winter or early spring, it might be an indication that the strain of bee does not fly in cold weather, or that the hive is in the shade or it might indicate a weak or diseased colony. Such colonies should not be united to another until the cause of the weakness is ascertained.

Drones Flying

In April, this is a sign of an early colony build up. In late autumn it is a sign that the bees are unhappy with their queen. She may be unmated or failing.

Drones being ejected

This is a normal occurrence in early autumn and is an indication that all is well. In summer it is a sign of starvation or that the hive has been satisfactorily requeened by supersedure or by the beekeeper and has no further need for drones.

Small drones

If these are seen flying or on the alighting board of a weak hive, they are the sign that the colony has laying workers or a drone laying queen. It is futile trying to requeen such colonies.

Dead bees on ground at entrance

During the year bees are continually removing the dead bees and dropping them some distance from the hive. During a mild day in early spring, bees in a healthy colony can be often be seen clearing out bees which have dropped off the winter cluster. Such bees usually form a small cluster at the entrance.

Crawling bees on the hive front and on grass stems Bees with blackish, swollen abdomens, fluttering 'K-wings,' walking on grass in front of the hive and cannot fly are probably suffering from Acarine and Chronic Bee Paralysis Virus.

Chemical poisoning

In good weather when bees are foraging, if the number of flying bees is less than usual and large numbers of dead bees, usually with their proboscis extended, are in a pile on the ground outside the hive entrance, that is a sign of chemical poisoning. There will probably be fighting because poisoned bees will not be allowed back into the hive where they could come into contact with the larvae. Dying bees are likely to be crawling, trembling, falling over and spinning round on their sides similar to bees with Chronic Bee Paralysis Virus. The colony is likely to be aggressive and throwing out infected bees which have managed to avoid the guards. Since returning foragers are not allowed to enter the hive, stored honey will usually be unaffected by chemicals.

Bees clustering in front of the hive

Brown smears on the hive front

This could be a sign of overcrowding, lack of ventilation or a surplus of young bees due to a reduction in the queen's egg laying in autumn or during a honey flow. It is also a common occurrence when supers have been removed from a single brood chamber colony at the end of the honey flow.

This can be caused by a number of factors. The colony could have gone into winter with some unsealed stores of honey or sugar syrup which had fermented because of their intake of dampness, causing dysentery in early spring. Spring dysentery is also, but not necessarily, a sign of Nosema.

Collecting a sample of bees to test for disease

On a cold spring day when nurse bees are eating large quantities of stored pollen their rectums will rapidly fill with pollen husks which they struggle to get rid of in poor flying conditions, forcing them to defacate on or near the hive.

Honey Flow

During a honey flow, bees are flying in and out of the entrance in vast numbers. They are in a good mood, intent on harvesting as much as possible. In the evening, there will be a strong scent of nectar flavoured water being evaporated and condensed moisture will be seen on the alighting board.

Fanning bees

Bees fanning at the entrance can be a normal process of ventilation especially in hot weather or the evaporation of excess moisture when ripening honey.

Robbing

At first glance robber bees resemble returning foragers during a honey flow, but on closer inspection they usually hover momentarily with their legs hanging down, before deciding to try and enter the hive. The robbers' abdomens may be shiny due to the nibbling of guard bees and they will never be carrying pollen loads. Cluster of guards bees will be confronting them at the entrance and fighting will also occur.

Slugs

Slugs are sometimes to be seen on the hive floorboard or on the alighting board in the evening. They feed on the hive debris. Bees cannot deal with them but they are fairly harmless.

Wasps

Wasps forage on small insects and grubs until the supply dries up about the end of July when they start to look for other sources of food. Strong hives which have their entrances reduced can fight them off but they can be observed freely entering weak hives and small nuclei which can be robbed out.

Bumble bees

In spring large queen Bumble Bees can be seen scouting for a nesting site. They are usually escorted off by hive guards without any recourse to fighting.

Heat detectable on crownboard and feedholes

This is a welcome sign in late winter and early spring that breeding has commenced.

Smells at the feedhole

Especially in early spring, smells at the feedhole can give valuable clues as to conditions in the colony. A pleasant yeasty smell means that the queen is laying and open brood is present. A smell of urine or ammonia indicates that mice are present. A damp, musty smell suggests that the colony may be dead

Noise from the feedhole

In early spring, if you put your ear to the feedhole and give the side of the hive a few taps a queen right colony will emit a gentle hiss which quickly subsides. A queenless colony will emit a characteristic roar. A cracking sound may indicate fidgeting due to the effects of Acarine.

Ants on the crownboard

During the summer, on lifting the roof off, hundreds of ants, trying to scurry off with their white eggs or pupae, can be exposed on the top of the crownboard. Ants can be observed going in and out of the hive entrance where they are too small to be a target for bee stings. Generally, ants are not a problem unless the apiary is close to ant-hills.

AIDS TO COLONY MANIPULATION

By IAN CRAIG, MBE

At the request of the Editor, I have agreed to write an article, commencing next month, describing my method of beekeeping using double brood chambers.

As a prelude to that article, I should like to describe three pieces of equipment which I consider to be essential for my method of management I feel sure that they would be useful to other beekeepers. These items can be made now, in preparation for the coming season. The dimensions given are to suit National hives.

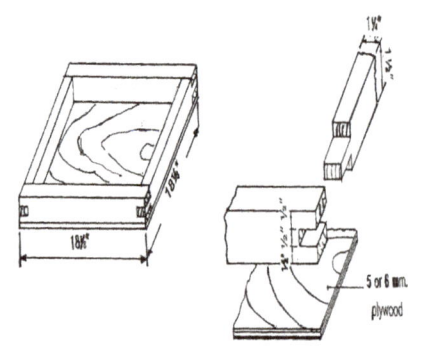

MANIPULATING BOARD (Figure 1)

I use two of these manipulating boards. When you remove a super or a top brood chamber during a hive inspection, where do you put it? It is considered bad practice to lay the super on the ground. Some beekeepers rest it on top of a hive roof. This imprisons the bees but it also crushes many which are clustered on the bottom bars of the frames, especially if you use a top bee space hive like the Langstroth or Smith. Most beekeepers set the removed super diagonally on an upturned roof. This does not crush bees but they are free to escape and add to the numbers in the brood chamber being examined or to sting the beekeeper. By using a manipulating board, or boards, the bees are safely imprisoned in the super by the manipulating board below and a manipulating cloth above. Their presence can then be discounted during the manipulation of the remainder of the hive. The board has other uses, such as acting as a temporary roof or as a tray for transporting small pieces of equipment or jars of honey.

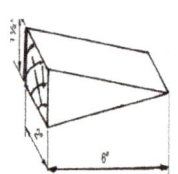

HARDWOOD WEDGE (Figure 2)

HARD WOOD WEDGE

I have found this to be indispensible for use in conjunction with the hive tool to lever two chambers apart. I frequently find the top bars of frames in the bottom brood chamber sticking to the bottom bars in the top chamber. The wedge can be inserted to keep the chambers from going back together and crushing bees while the hive tool is being used in another position to prise apart any other sticking frame.

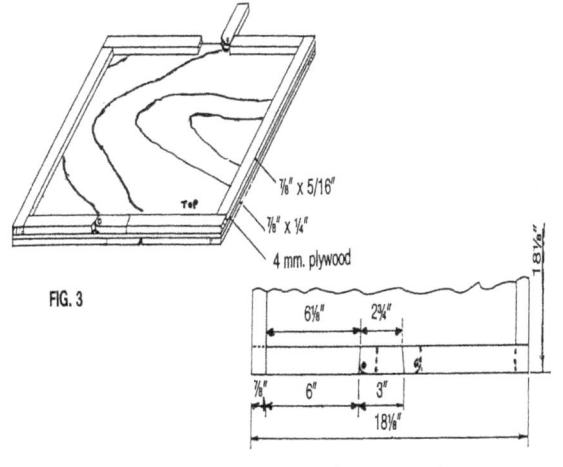

FIG. 3

SWARM BOARD (Figure 3)

This is similar to a Snelgrove Board but less complicated. I have one of these swarm boards for each colony. The swarm board is the only extra piece of equipment necessary for my swarm control system. No extra floorboards, roofs, crown boards or brood chambers are required.

Is there adequate room in a single National brood chamber for the requirements of a good queen? By Ian Craig. MBE

Readers to this magazine will be aware that I am of the opinion that a queen should have the use of more than eleven British Standard combs when she is in full lay and also during the autumn when space is required for winter stores of honey/sugar syrup/pollen, rearing "winter" bees and for the bees themselves. The only time in which I consider eleven BS combs to be desirable is during the summer when the need for bulk brood rearing has passed. I reduce my colonies to a single brood chamber about 20th June because eggs laid after that time will not produce foragers for another six weeks, which will be into the first week of August. If colonies are maintained on large brood chambers during the main honey flow and the weather is poor, unnecessary brood will continue to be reared and any surplus honey will be stored in the brood chamber. Access to a single brood chamber will force the bees to cut back on their brood rearing and store any surplus honey in the supers. This works for me because I only take a few stocks to the heather. The heather stocks require to be boosted by uniting them to a large nucleus containing a young queen and a force of young, fresh bees before going to the moor. Queens usually reach their peak of egg production in May/June during the spring honey flow. If they are confined to a single brood chamber in which to breed, the queen will be restricted in her egg laying, the nest will be congested which will prevent queen substance being adequately distributed to all bees and swarming will be the inevitable result. Let us consider the following facts and assumptions.

If we count the number of complete cells visible on a single side of a sheet of foundation after it has been inserted into a DN1 British Standard frame we will see that there are 41 rows containing 56 complete cells. This multiplies out at 2,296 (say 2,300) cells on each side of a newly drawn, perfect, comb. We all know that combs do not remain perfect for too long and if they are drawn out in a single brood chamber there is often a space between the comb and the bottom bar of the frame. In a national hive, if we ignore the outside of the two outside frames which seldom contain brood, we have 20 sides available which equates to a total of 46,000 cells.

Now consider the assumptions:

Assume per side	Total Cells
25 drone cells (x 20 sides)	500
400 honey cells	8,000
400 pollen cells	8,000
125 holes; mis-shapen cells; Comb bottoms; etc.	2,500
100 cells being cleaned and prepared	2,000
Grand Total	**21,000**

Therefore our initial 46,000 less 21,000 (as detailed above) leaves 25,000 cells available for the production of worker brood. The books tell us that a good queen at her peak is capable of laying 1,500 to 2,000 eggs per day. If we take the lower figure, this equates to 31,500 over a 21 day period. This means that she is 10,500 cells short; the equivalent of 4.6 comb sides or say 3 frames. Remembering that in a second brood chamber the outside of the two outside combs would be unlikely to contain brood, this would indicate that 4 extra frames would be required, making a total of 15. That is for a queen laying 1,500 eggs per day. She may be capable of laying more. From the above I would argue that two BS brood chambers, each containing eight combs flanked by dummies would be required. In practice, I winter my colonies on 16 combs, expand them to 18 during the spring honey flow, then some to 20 about the beginning of June, before reducing to 11 in late June, and returning them to 16 when I remove the summer honey crop.

Double Brood Chamber Management by Ian Craig MBE

In this three part article I shall describe my method of beekeeping in the West of Scotland, where I have kept bees for over forty years. My beekeeping experience has all been in Western Galloway and Renfrewshire where the climate is mild, and the annual rainfall varies between 45 inches in Renfrew and 65inches 10 miles away in Kilmacolm. I am well aware that weather patterns and honey sources in other parts of Scotland are different and consequently methods of beekeeping may have to be varied to suit the conditions.

I try to run as many colonies as possible on double brood chambers, but flexibility is the key. With this method the number of British Standard brood combs can vary between twenty in May/ June and eleven during July/August. Also it would be imprudent to winter colonies being built up from summer nuclei on more than eleven combs in fact, they are frequently wintered on eight or nine. Such colonies usually build up rapidly in spring, show little tendency to swarm and progress unchecked into the main honey flow.

I have found that my method of management gives a satisfactory crop of honey, virtually complete control of swarming and provision for re-queening. It is vital for honey production to have colonies as strong as possible from May to September. The flow can come at any time, depending on the weather and it may be brief or prolonged.

At that time the hives consists of a floorboard; entrance block; two brood chambers each containing eight combs flanked by two dummy frames; crown board with piece of glass covering the feed hole; a two-inch thick sheet of expanded polystyrene and a deep metal covered roof. (See Fig.1).

Advantages:

1. Adjustable - 11 combs (or fewer) to 22 combs.
2. Good air circulation and ventilation.
3. Easy to manipulate – no combs propped outside the hive.
4. Easy to check for swarm preparations.
5. Room available when most required – for expansion in spring and to accommodate bees, brood and stores in autumn.

Fig. 1

6. Frames are interchangeable between the two chambers.
7. Foundation can be drawn in the warmth of the top brood chamber.
8. Safety valve in case of late supering – this should not be allowed to happen.

Disadvantages:
1. Two brood chambers are required.
2. Four dummy frames are required.

No through ventilation via the feed hole is employed. Under natural conditions, swarms occupy cavities in trees etc. and usually go to the top of the cavity where they start to build their combs. In this situation there is unlikely to be a through draft of air. If we put on perforated zinc over the feed hole during autumn, the bees will fill the perforations with propolis. Why do we think that we know the bees needs better than they know themselves?

At the beginning of November, match sticks are inserted under the two rear corners of the crown board. This allows the escape of moisture-laden air containing carbon dioxide, thus providing dryer wintering conditions. Dr David Raven, writing in Beecraft, in September and October 1989, states that a level of 1% to 2% carbon dioxide is beneficial to keeping bees in a state of inactivity during winter. This is quite a high concentration, considering that the carbon dioxide in normal atmosphere is only 0.03%. Central ventilation through the feed hole acts against this. Since I have started to use polystyrene over the crown board in winter there has been no sign of drops of condensation under the glass covering the feed hole and the bees cluster right up to the glass, especially when brood rearing commences. In my case this is not an indication that the bees are short of food, copious quantities of sugar syrup having been given during September. From November until early February the hives are given a visual check once a fortnight to ensure that no storm or animal damage has occurred. From January onwards they are "hefted" (lifted gently, first one side then the other) to check that their stores are plentiful. In early February, the match sticks are removed from under the crown board. This conserves heat during early brood rearing and any condensation produced can be used by the colony to dilute their stores, thus reducing their need to fly in adverse conditions with a consequent loss of life. During the latter half of March, depending on the weather the floor boards are cleaned and the debris sent to East Craigs for a varroa check. At this time potential oilseed rape colonies which have an abundance of stores can be stimulated to produce more brood by feeding them "sweetened water" (1kg sugar to $1^{1}/_{2}$ litres water), this further reduces the flying losses. Because of cool weather conditions at this time of year, feeding is best accomplished using a "contact" feeder. Colonies which are short of food should be fed stronger syrup. Colonies which have ample stores and are not destined for early flows from rape or sycamore are better not fed, as this might produce a large force of under-employed bees with proportionately few losses that are likely to make swarming preparations at the start of the honey flow. Weak colonies which have adequate stores are better not fed. Feeding such colonies encourage them to fly in adverse conditions which hastens the death rate and also causes an expansion of the brood nest beyond which the colony can cover adequately during periods of cold spring weather. If the "signs at the hive entrance" are favourable and heat is detected through the crown board, all should be well until flowering currant time. At this time, which for me is usually late April, the first inspection is carried out on a fine warm day. This is the best time to mark and clip all queens which have been reared the previous season. Colony populations are at their lowest at this time of year and there are few, if any, drones in the hive, therefore the queen is much easier to find. She is the only large bee and providing you have not used excess smoke, will usually be found on a frame of eggs in the top brood chamber. I have stopped marking or clipping queens during the summer as I have sustained a few losses at the time and none in the spring. When the queen has been found, a circular queen marking cage is positioned on top of her thus pressing her gently against the face of the comb. In her struggles, she usually pushes one pair of wings through the mesh of the cage where they can be clipped. The large wing is clipped just far enough down so that 2 mm is removed from the tip of the smaller wing on the same side at the same time. The queen is then marked on the centre of the thorax and the pressure of the cage on her back is relieved. The newly clipped and marked queen is allowed to run about in the cage until the marking paint has hardened and until other hive manipulations are complete. In this way the whereabouts of the queen is known. By using a cage for clipping and marking, the beekeeper has both hands free and does not transfer any human odours to the queen. If you are concerned about the odour of the marking paint, a drop of honey can be placed on top of it, but I do not bother with this.

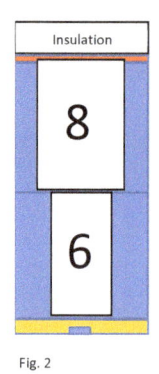

During the first inspection a check is made for the presence of any brood diseases and the two outer combs in the bottom brood chamber are removed. These two combs are probably old, damaged and due for replacement. No reorganisation and consequent disruption of the brood nest is carried out on this occasion, the brood chamber being left with eight + six combs as in Figure 2.

I do not believe in reversing the brood chamber at this time as the brood nest is then split, as in Figure 3.

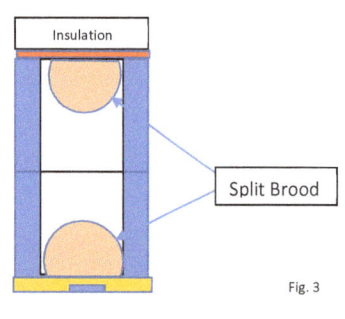

If all the brood was in the upper chamber and the chambers reverse, you would certainly induce the queen to lay rapidly in the warmth of the new top brood chamber and get a larger force of bees in late May and June, but I have found that this retards the entry of bees into the first super during the first two weeks of the spring honey flow which might be the only flow there is. The colonies are better to be left to expand downwards, naturally, at their own pace if you want early honey. Fortunately none of my apiaries are within reach of the oil seed rape. I can therefore control the number of colonies foraging on rape and keep the others on the sycamore. I move over half of my colonies to the rape and two-thirds of these I reduce to single brood chambers during the first inspection, thus concentrating the brood and inducing early entry into the first super. Early swarming, due to congestion, is not a problem with rape colonies as loss of foragers is heavy and a colony "balance" is maintained.

I firmly believe in disturbing the colony rhythm as seldom as possible during the year. A great deal of information can be gleaned from observing the hive entrance. The second inspection of the season is carried out when the sycamore flowers are beginning to open. A quick check is made in the top brood chamber for the presence of eggs and sealed brood, indicating that the queen has not suffered any ill effects from her clipping and marking. There is no need to find her at this time. A check should also be made to ensure that combs of overwintered stores are not preventing expansion of the brood nest. If they are, they should either be uncapped or removed to a beeproof store. A first super of DRAWN comb should be added above the queen excluder. Bees occupy drawn comb much more readily than foundation at this time, thus relieving the start of congestion and allowing early honey to be stored. It is futile to give the bees foundation to draw too early. Under natural conditions they only require to make new combs during a honey flow or after swarming.

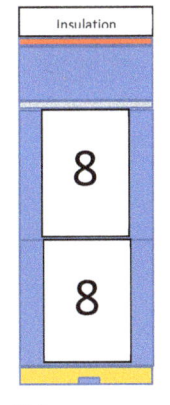

After the first week of the sycamore flow, re-arrange the brood chamber. Move two frames of brood from the top brood chamber to the bottom one and add two frames, fitted with foundation, to the top. The foundation should be placed at the edge of the brood nest. It will be drawn perfectly, right down to the bottom of the frame, when placed in the top brood chamber above the warm brood nest. See Figure 4.

Fig. 4

One week later, move a further comb of brood from the top to the bottom brood chamber and add two further frames of foundation to the top brood chamber. This allows for the expansion of the brood nest, keeps the wax makers gainfully employed and reduces the chances of swarm preparations being made. See Figure 5.

Fig. 5

At this time a second honey super can be added if the first super is occupied. Do not delay in providing abundant super room at this time of the year. If the bees sense the impending congestion, the seeds of future swarming will have been sown. If this second super is of foundation, I place it below the first one, directly above the brood chamber. This has the added advantage that progress in sealing the honey in the first super can be observed by only having to raise the crown board. If the colony is expanding rapidly and the flow is good, more foundation can be added to the brood chamber, giving ten + ten combs. As the season progresses, the colonies are checked weekly to see that there is enough super room and that

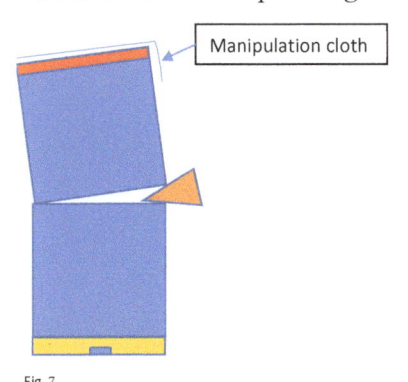

Fig. 7

swarming preparations are not being made. My method of checking for queen cells is to remove the supers onto a manipulating board, cover the queen excluder with a manipulation cloth, lever the two brood chambers apart with the hive tool and insert a wooden wedge. See Figure 6.

The top chamber is then pulled forward 20mm to ensure that it won't slide off the bottom chamber when it is tilted. Smoke is puffed into the opening created by the wedge and the top chamber "hinged" upwards to allow a quick inspection to be made of the centre of the brood nest. By peering upwards and downwards between the frames, the presence of sealed brood and embryo queen cells can be determined. The whole operation takes a couple of minutes and a minimum of disturbance is created. The odd queen cell could remain undetected using this method, but it is a great time saver and, in any case, the queen is clipped.

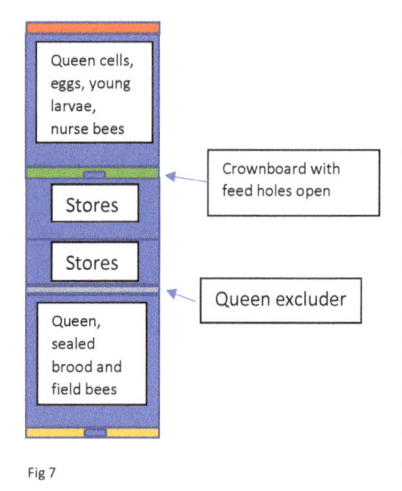

I shall describe the treatment of the colony should queen cells be discovered, later in the article. In the meantime let us assume that the colony has a satisfactorily laying queen, plenty of room has been given for expansion of the brood nest and the storage of surplus honey and that no swarming preparations have been made. Around 20th June I reduce all colonies to single brood chambers. This is not so necessary if the summer turns out to be good, but if it is average or poor, you will get little or no honey when using double brood chambers. The colony will overbreed if this is not prevented, honey in the supers will be "chimneyed" (stored in the centres of the supers) and too much honey will be stored in the brood chamber which should not be extracted. Re-arrange the brood chambers by putting the queen, sealed brood and older larvae in the bottom chamber, then the queen excluder, honey supers, crown board with feed holes OPEN, top brood chamber with eggs and younger larvae and, finally the swarm board or spare crown board on top. See Figure 7.

The bees will re-arrange themselves, field bees will gravitate to the bottom brood chamber and the nurse bees will move up to take care of the young brood. This will relieve congestion in what is now the brood chamber and, contrary to popular belief, will actually reduce the tendency to swarm. The reason that the sealed brood is put into the bottom brood chamber is that it will be the first to hatch, thus creating room for the queen to continue laying albeit at a reduced rate now that the need to sustain a huge birth rate is past.

The top chamber should be checked after one week as the bees, now far removed from the influence of queen substance, will likely have drawn emergency queen cells. As no special feeding arrangements have been made, these cells are likely to produce inferior queens and should be removed. No check need be made on the lower brood chamber for at least two weeks, as one of the methods of swarm control, i.e. the removal of bees, has been carried out. One potential problem which could occur is that the drones which hatch in the top chamber will pass through the open feed hole, into the supers and clog the queen excluder in an attempt to escape. This can be prevented by pinning a piece of zinc queen excluder over the feed hole. The crown board can be removed on a fine day thus allowing the trapped drones to make a hasty and noisy escape. The top chamber can remain in place until all the brood has hatched. It can then either be cleared of bees and removed to the honey house, or allowed to remain above the open feed hole where the bees will look after it. Alternatively, at the stage when the queen cells were removed, two or three nuclei could be made from the top chamber and given desirable queen cells from another source. During July and early August the whole massive force of bees will be caring for a relative small brood nest, thus releasing a large force of foragers to fill the supers with honey. With this system, very little honey is stored in the brood chamber. It sounds good if you can do it!!

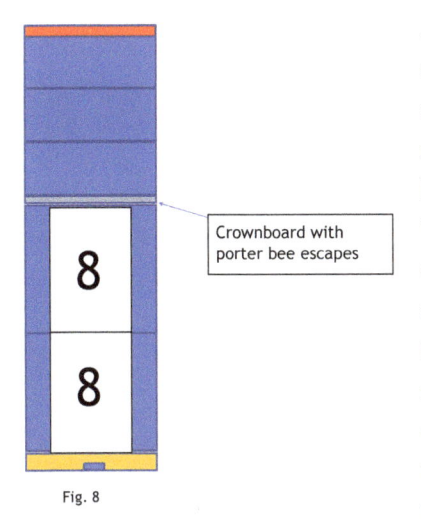

Fig. 8 — Crownboard with porter bee escapes

At the end of the summer honey flow the bees can either be taken to the heather, which is another story, or prepared for the winter. This is done by putting the colony back to a double brood chamber and removing the honey. Three combs are taken from the bottom brood chamber, the remaining eight are centralised and flanked by dummy frames, making sure that the two poorest brood combs are placed next to the dummies, from where they can be easily be removed during the first manipulation the following spring. The three spare combs, with bees, are placed in the top brood chamber, which was stored above the supers. Combs will have to be removed from this chamber so that it only contains eight combs, flanked with two dummies. The crown board, with Porter escape(s) now fitted, is put on top of the brood chambers and the honey supers placed on top of that. See Figure 8.

In my opinion, a double brood chamber is essential at this time to accommodate a large colony. It provides ample room for winter stores, for autumn breeding, (so essential for spring build up) and for the bees themselves. A Miller feeder is placed on the colony and unless there was a late flow from ragwort etc., one and possibly two 28lb honey tins of autumn syrup (2kg sugar to 1 litre water) is fed. I also add a little thymol to the feed. This inhibits fungal growth in feeders and in any unsealed syrup stored in the brood chamber. A strong colony can empty a Miller feeder in twenty-four to forty-eight hours.

I have explained my method of beekeeping with a well behaved colony. I shall now describe how to treat a colony when queen cells are found during a weekly inspection in late May or early June.

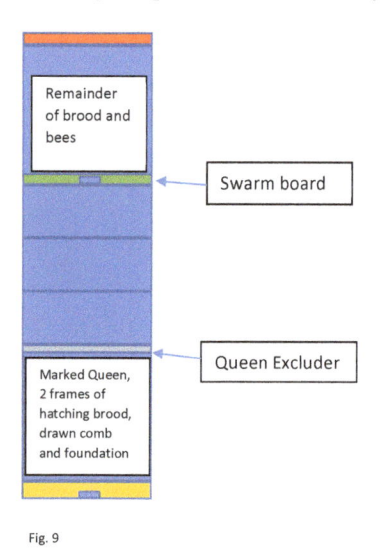

Fig. 9 — Remainder of brood and bees / Swarm board / Queen Excluder / Marked Queen, 2 frames of hatching brood, drawn comb and foundation

Queen cups are usually present in most colonies during the period when swarming can be expected. These should be ignored. Don't bother to cut them out, the bees will only waste time building them again. Action is only required if the cups contain larvae which are being fed royal jelly. On the discovery of partly complete queen cells, action should be taken. It is a waste of time cutting out these cells as they will be rebuilt by the next visit and the bees might have attempted to swarm. In any case, if a colony shows signs of swarming, despite all the preventative measures which have been taken, I would prefer it to do so sooner rather than later. Measures can then be taken to satisfy the swarming urge and get the colony settle in time for the main honey flow. Cutting out queen cells on a weekly basis cannot be relied upon to prevent swarming and the morale of the colony will be affected. Depending on weather conditions, the bees could eventually build inferior queen cells on three day old larvae (i.e. on day six after the egg is laid),which could lead to a swarm three days after the last inspection. At the first sign of larvae in queen cells being fed, I artificially swarm the colony by rearranging the brood chambers: find the marked queen, put her in the bottom brood chamber with two frames of hatching brood and no queen cells, fill up this brood chamber with empty drawn comb if it is available or with frames and foundation, queen excluder, honey supers, crown board with feed hole open, top brood chamber with the remainder of the brood and bees, finally with the swarm board on top. By the following day the field bees will have moved to the bottom brood chamber with the queen, and the nurse bees will have moved to the top chamber to take care of the brood. In this way, an artificial swarm has been created which will satisfy the instinct of the bees. At this stage the swarm board, with its top entrance open, is placed under the top brood chamber. This allows bees flying from the top entrance to return to the original entrance thereby reinforcing the "swarm." See Figure 9.

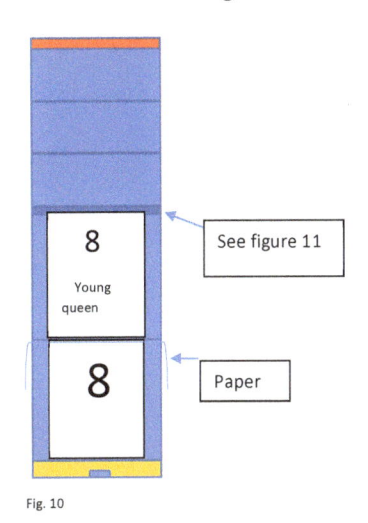

I keep detailed hive records and have placed the queens heading all of my colonies into one of three categories: (i) breeder queens, (ii) queens from which I would be happy to breed, (iii) queens with undesirable characteristics which must be eliminated. If the queen is in category (ii), I will put a feeder on top of the upper brood chamber and keep up to six of the original queen cells. In my opinion, these cells have been reared under optimum conditions when the queen was reducing her rate of laying, prior to swarming. It is essential to feed the nurse bees in the top brood box, to ensure that the queen cells are adequately nourished, because the field bees have been forced to join those in the bottom brood chamber. A frame with pollen is also essential in the top box for feeding to older worker Larvae. After four days, more bees are diverted to the bottom brood chamber by using the swarm board: open the top back entrance, close the top front entrance and open the bottom front entrance. Before the first queen cell in the top brood chamber is due to hatch, the cells must be reduced to one. Alternatively, the top chamber can be split into two nuclei and the remainder of the cells dispersed to other nuclei. When the worker brood in the top chamber has hatched, the virgin queen in what is virtually the nucleus will hopefully mate and commence laying. The bottom brood chamber can be checked two weeks after the original manipulation and, if all is well, left for the remainder of the honey season. The performance of the young queen in the top box should be monitored. This nucleus, flying from the upper rear entrance, should be left to build up. At the end of the season, I kill the old queen, remove the honey supers and re-queen the colony all in one operation. See Figure 10.

Fig. 10

A special frame, with a small rear entrance, can be

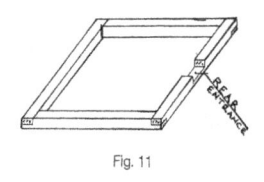

inserted under the crown board to allow the field bees from the nucleus access to the hive. See Figure 11.

Fig. 11

The supers are removed and after two days the special frame is also taken away. The sixteen comb colony with its young queen is now fed and prepared for the winter as described earlier in the article. As I stated in my introduction to this article, I have used this method of management for many years and it works for me. Why not give it a try on a couple of colonies as an experiment?

Queen Excluders, Supers and Supering by Ian Craig. MBE

The modern parallel-wire queen excluder has sturdy spacer straps, is framed and has a bee space on one side only. If you use a bottom bee space hive, like the National or WBC, the excluder bee space must be underneath and if you use a top bee space hive, like the Smith or Langstroth the excluder bee space must be on top. It is more expensive to buy than the traditional "sheet" type but with reasonable care it will last a lifetime. When considering which type of excluder to use, the restriction of hive ventilation must be considered. When nectar is first stored in the supers, it has a high moisture content, which must be reduced during the process of converting it into honey. This process becomes more difficult if the excluder forms a partial barrier to the free flow of air. The slotted zinc or slotted plastic excluder has many drawbacks. Beekeepers often use it by laying it directly on top of the frames of the brood chamber, where it has to be peeled off during inspections, with the constant risk of distortion. This is not quite so bad if using a bottom bee space hive, where the frame tops are flush with the top edge of the brood chamber, with a top bee space hive the excluder tends to sag causing brace comb to be constructed both above and below it. This can be avoided to some extent by framing the excluder but a frame consisting of four edges still allows sagging. At least two additional cross pieces are required to reduce this. If the excluder is merely lying on top of the frames it forms a barrier, both to the bees and the circulation of air, because bees and air can only pass through the slots which lie between the frame tops. The slots which cover the top bars might as well not be there. In addition, only the slots which have their short dimensions fully clear of frames will allow worker bees to pass through. To reduce the effects of this the excluder should be placed with the slots at right angles to the frames, if the brood chamber dimensions allows this flexibility, i.e. some brood chambers are not square. Sheet excluders can be purchased "short slot" or "long slot." The long slot marginally increases access and ventilation but it does so at the expense of reliability in excluding queens and drones. Overcrowding of bees is recognised to be a barrier to ventilation and to the distribution of queen substance, thus leading to the onset of swarming. Not every beekeeper can recognise crowding in a colony. If bees are occupying all the space available to them, then they are overcrowded even although they seem to have enough room to move about. They do not need to be thick on the comb in the brood chamber or supers to be overcrowded. Since temperature, the need for space to "hang nectar out to dry" and other considerations determine the number of uncrowded bees on a comb face it is not feasible to use the number of bees per unit area as a measure of crowding. Severe overcrowding is obvious but the uncrowded colony can only be distinguished by the presence within it of some unoccupied comb space. Bees will occupy more comb space in hot weather than in cool weather so it is better to give them more room than seems necessary if they are examined on a cool day. When a single brood chamber is crammed wall to wall with bees, it is overcrowded. It requires part of another brood chamber or, at the very least, a super to be added. The number one purpose of a super is for accommodating bees. Supers are needed whether or not there is any surplus honey. The bees are programmed to appreciate if extra space is available should it be required. If that extra space is not available the seeds of swarming will be sown.

Honey supers can consist of boxes having the same sized frames as those in the brood chamber or, more usually, shallower frames. In the use of National or Smith hives, these shallow supers are capable of holding 11-13kg of honey and are much easier to lift than deep supers. I would recommend the use of Hoffman self spacing frames in the brood chamber, but these frames in a super are a waste of money, the spacing projections on the sides of the frames are a hindrance when uncapping and they do not easily allow a wider spacing to be employed. I would also recommend that brood chamber frames are of the DN5 type consisting of a 26mm wide top bar which reduces the incidence of brace comb being built between the frames. I use the excellent SN1 shallow frames, which are made of 22.5mm wide wood, in my supers. The frames, when newly fitted with wired wax foundation, are set apart using narrow plastic spacers 36.5mm wide. A National super holds eleven of these frames. If frames with foundation are spaced more than 41mm apart, there is a likelihood that the bees will draw combs between some of the sheets and make a hopeless mess. When correctly spaced, the bees will draw the foundation into perfect combs which will, hopefully be filled with honey and capped. These frames are then extracted, given back to the bees to clean up and stored in their supers for future use. In subsequent years the plastic spacers can be removed and the frames put on to "wide spacing" using 9 slot castellated pacers. Castellated spacers should never be used in the brood chamber as they prevent frames being slid along the hive runners. Full supers containing nine combs will contain more honey than those containing eleven because they are only ten passage ways between the combs instead of twelve. The combs will be fatter, heavier and easier to uncap. If you have 100 supers to extract, you will only have to handle 900 frames instead of 1,100, which is a saving in cost and time for beekeepers with a large number of hives. The first super should be added, above a queen excluder, when the bees are starting to occupy the inside face of the outside combs in the brood chamber. When a wire excluder is in use and the first super consists of drawn comb, I have not experienced any difficulty in getting bees to enter the super. When the bees (not necessarily honey) are occupying two thirds of the first super, a second should be given. This second super can be of foundation. If it is, it should be placed below the first where the bees are forced to pass into it to reach the super in which they are already working. Also, the direct heat from the brood chamber will assist in getting the foundation drawn. One of the aims of supering is to relieve congestion in the brood nest. Bottom supering helps to do that because bees will occupy the new super faster than they might if it had been placed on top. Remember that bees only draw foundation when there is a honey flow. If it is given at other times they will tend to chew and make holes in it. The foundation must be fresh from a sealed packet. If it has been on a colony the previous year and stored undrawn, try heating it gently with a hair dryer to remove "blooming" and raise its aroma.

Bees tend to seal honey from the top downwards. This, together with bottom supering means that the sealed honey will be found in the top super where it can be inspected and removed more easily. It should be remembered that a super full of sealed honey has less room for bees than an empty one, because the inter-comb space has been reduced. If such supers are at the top of the pile, the state of crowding may not be apparent until these supers are taken off as there may only be guard bees in them. If there are a lot of bees in a full and sealed super, the probable reason is that they are overcrowded.

In light of the preceding discussions, plenty of super room should be available in late spring and early summer to reduce the pressure of swarming. Towards the end of the honey season surplus super room is less desirable and should be reduced in order to try to encourage the bees to concentrate, ripen and seal the honey which they have collected. Partially completed frames or supers can be given to stronger hives to complete. If the beekeeper is likely to be on holiday or at work for an extended period in late spring to early summer, two shallow supers can be given at the same time. If they are given as top up supers, a sheet of polythene which has a 30mm space cut all round can be placed between the supers. This will conserve heat and the bees will go around the edges of the polythene if they need to enter the second super. The polythene should be removed at a convenient later date.

Beekeepers Year by Ian Craig. MBE

The SBA has recruited a fair number of new members during the past few years. I have been prevailed upon to produce a series of twelve monthly articles on practical beekeeping, with these new members in mind. I hope the series will also be thought-provoking for all beekeepers as it is useful to pause for a while and consider if the methods we are using are the most suitable for our individual conditions. I am writing these notes taking into account the flora and weather conditions in Mid-Renfrewshire. Readers will have to make allowances depending on which part of the country their apiaries are situated. I am giving my opinions based on fifty-five years beekeeping experience beginning in Wigtownshire and now in Renfrewshire. Everyone has their own ideas on colony management which may not coincide with mine so, hopefully, this series of articles might generate some response in the 'letters' page from experienced beekeepers in other parts of Scotland.

JANUARY

At this time of year bees are in their winter cluster and are better left undisturbed except for visual checks every two weeks or after gale force winds. It is also beneficial to clear any dead bees from the hive entrance using a piece of bent wire. I never remove snow from hive entrances because bees will be attracted out by any sunshine, with the possibility of becoming chilled and unable to return to the cluster. There is no danger of suffocation as some top ventilation has been provided. Food stores should also be checked by gently 'hefting' the rear of the hive. Alternatively, heft both sides of a single brood chamber hive as the stores may be concentrated on one side only. If stores are considered to be running short, the beekeeper has been remiss with the autumn preparations. Remove the crown board, place an 'eke' or empty shallow super on the exposed brood chamber, place a block of fondant or four damp 1kg sugar bags over the cluster, cover with an old blanket or sack (these could become damp and, if so, should be changed occasionally during the remainder of the winter), replace the crown board and roof. My colonies are housed in National hives, consisting of a floorboard; two brood chambers each containing eight combs flanked by two dummy frames; crown board with pieces of glass covering the two feed/Porter holes and matches under the two rear corners to allow excess moisture to escape from the winter cluster; a 50mm thick sheet of polystyrene insulation and a deep, metal covered, ventilated roof. (See Fig.1).

The majority of the roofs are deep enough to almost cover the top brood chamber and they never blow off. Roofs which I have bought in the past fifteen years are of the Fig. 1 shallow design, as the deep ones are no longer obtainable. These require to be 'weighed-down' with a brick to prevent them from being blown off. Beginners to beekeeping are advised to keep to a single brood chamber, containing eleven combs, until they gain some confidence in handling their bees, otherwise the set-up is the same as I have just described. I use wooden entrance blocks which have two 140mm x 6mm slots along the bottom to allow for access by the bees, prevent entry by mice and provide extra ventilation. Some beekeepers advocate facing the slots upwards arguing that in that position they are less liable to become clogged with dead bees. I do not agree with this. Placing the slots downwards allows condensation to run out of the entrance. Many beekeepers don't use entrance blocks, instead they prevent mice from gaining access to the hive by covering the entrance with perforated zinc in which a slot of height 6mm has been cut. I dislike the commercially available mouse guards made of sheet metal in which a series of holes have been punched. This type is more prone to becoming clogged with dead bees and it has a tendency to strip valuable pollen loads from the bees' legs in springtime. My hives are situated on concrete slabs on which two 30mm x 15mm wooden strips have been laid in order to keep the floorboard clear of the slab, thus allowing a free passage of air under the floorboard to alleviate dampness as much as possible. The front strip should be positioned so that laden foraging bees returning to the front of the hive cannot mistakenly fly underneath the floor, thus wasting valuably energy. Slabs containing hives consisting of a solid floor should be tilted forward, enough to allow excess moisture to drain out of the entrance. If solid floors have their original entrance rotated to the rear and a 'varroa floor' (which is different from an 'open-mesh' floor) placed on top, then the slabs should be tilted towards the rear. I do not believe in using hive stands with their resultant maintenance problems. I examine the brood chambers from a kneeling position. In any case, during the summer when, hopefully, up to four or more supers and a top brood chamber are on the hives, lifting heavy crates at eye-level is too much of an effort. For beginners, and others, this is a good time to purchase hives and other equipment. Some of the suppliers offer much reduced prices during January. I would always advise beekeepers to purchase new (first or second quality) hives or very good second hand hives. Hives must be bee-tight, water-tight and all parts interchangeable. Hives which I purchased new fifty-odd years ago and kept properly maintained will be good for another fifty years. Beginners who are considering the purchase of second-hand equipment should seek the advice of an experienced beekeeper (not the seller). Beware of a hive in which the bees have died out unless you are sure that their demise was not due to disease. January is a suitable time to prepare a new hive record book. I keep my hive records in a small loose-leaf folder which can be carried in a pocket of my bee suit. I don't keep my records pinned underneath the hive roofs as my sixty colonies are in eight out-apiaries, it is therefore important to be able to refer to the record book at home prior to visiting the apiaries to plan what is to be done and what equipment might be required. A page is allotted to each colony, in which is recorded: year, colony number, year in which the queen was born, strain, if the queen is marked and clipped, if she is identified as a future breeder, record of manipulations, previous honey record, disease record (eg. chalk brood), feeding record and nuclei.

FEBRUARY

By February colonies should be showing signs of activity. Queens should be laying patches of eggs in the warmth of the cluster and workers will be taking cleansing flights and bringing in pollen on mild days from snowdrops, crocuses and winter heaths. Around the middle of the month, in order to assist brood rearing by reducing heat loss, the matches should be removed from the rear of the crown board thus cutting off top ventilation and the varroa trays should be inserted under open-mesh floors to cut off the cold air. Every three or four years, during a spell of cold weather, if the outside of the hives are dry, they can be given a coat of liquid insecticide-free preservative. Choose a cold, windy day when few, if any, bees are flying. Paint should not be used on wooden, single-walled hives as it prevents the wood from 'breathing' thus causing the hive walls to become saturated during winter, leading to a hive which has the comfort of a refrigerator. When the hive walls start to dry out, the paint will blister and flake off. Fortnightly checks of apiaries should be carried out to see that surrounding fences are stock-proof, no vandalism has taken place and there is no woodpecker damage to the hives. Each hive should be hefted to check on the quantity of food stores remaining. Check the signs at the hive entrance. There should be a considerable amount of fine particles of wax from the uncapping of stores. If large pieces of wax are on the alighting board, you have a lodger in the form of Mr and/or Mrs Mouse. Spots of faeces may be evident on the front of the brood chamber, caused by the bees' over long confinement due to severe weather; or dysentery caused by fermenting stores or Nosema disease. Bees crawling, with fluttering partially spread wings (K-wing), clinging to plant stems and unable to fly suggests Chronic Bee Paralysis Virus which is common in bees suffering from Acarine Disease or one of the viral diseases which are associated with varroa and are becoming more common. If bees are flying freely and bringing in large pollen loads and, by placing the back of your hand against the crown board, you can detect heat then all should be well. If very small pollen loads are being taken into a particular hive there is either a shortage of plants yielding pollen or the queen may not be laying for some reason. If there are fewer bees flying from this hive compared to others in the apiary or no flying bees at all, a quick check can be made by raising the crown board. If the colony is dead, try to establish the cause and close-up the hive in order to prevent robber bees gaining access to any remaining stores which may be harbouring some infection. By the end of the month, new hives which were purchased in January should have been assembled and given a heavy coat of preservative. Repairs should have been carried out on any spare equipment damaged during the previous season. New frames should be assembled and nailed. I have encountered a surprising number of beginners who have never been told to nail their brood and super frames, leading to disastrous consequences. I would advise beginners to purchase Hoffman self-spacing brood frames having 27mm (1 $^{1}/_{16}$") wide top bars and 35mm (1 $^{3}/_{8}$") wide side bars with internal grooves. With top bars 27mm wide, very little brace comb will be built between the frames. The grooves in the side bars act as a location for the wax foundation, leading to nice straight combs. Eleven frames will be required for a National or Smith brood chamber. Never use castellated spacers in the brood chamber as they make it impossible to slide frames along the runners. For shallow honey-super frames, I consider Hoffman spacing to be a waste of money and a nuisance when uncapping. The Hoffman side bars get in the way of the uncapping knife. It is far better to purchase the cheaper 22mm ($^{7}/_{8}$") top and side bar frame. When using this frame with foundation in the super, it must be spaced using narrow plastic ends (35mm) during the first year until the foundation is drawn out, filled with honey, sealed and then extracted. The comb is then stored for use during the following year, when the narrow plastic ends can be replaced by wide plastic ends (47mm) or nine-slot castellated spacers. If wide spacing is used with foundation, the bees will probably draw extra combs in the space between the adjacent sheets of foundation. Eleven frames are required using narrow spacing, these will reduce to nine on wide or castellated spacing. In a good honey year, if you have two hundred supers to extract, a great deal of time is saved by using 22mm flat frames on castellated spacing giving nine frames per super. Only nine frames have to be uncapped and extracted instead of eleven and you get a heavier, more easily uncapped frame. Also, you get a heavier super because the nine-frame super has only ten air-spaces between the combs whereas the eleven-frame super has twelve air-spaces. An alternative is to use supers each containing nine Manley frames which have a fixed spacing of 41mm (1 $^5/_8$"). That is the absolute maximum spacing if extra combs are not going to be drawn between adjacent sheets of foundation. I do not like Manley frames because the long shoulders get fixed together with propolis in my district, making them very difficult to remove for extraction. I also find them slower to uncap compared to the ordinary frame described in the previous paragraph.

MARCH

Brood rearing should now be in full swing, depending on the bees having access to fresh or stored pollen and sealed stores. The top of the crown board should be warm and large pellets of pollen should be going into the hives from early sources such as heaths, willows, hazel and gorse. Hive floor 'inserts' should be checked for the presence of varroa, pending possible treatment early next month. Hives should be lifted off their floorboards and placed diagonally across an upturned roof, without disturbing the brood chamber(s) or crown board. This will allow the floorboards to be scraped free of winter debris. If the floorboards are reversible, this should be done, or better still give a clean recently preservative-treated floorboard. It is a real boost to colonies to get a clean, dry floor at this time. In order to conserve heat, an entrance block with a narrow entrance should then be put in place and the varroa-sampling drawers should be inserted into open mesh floors. If the bees have been wintered on a single brood chamber there is a likelihood that their food stocks will be getting low. Bees which are short of food must be given spring syrup made by mixing 1kg of white sugar (not brown or castor sugar) with 1litre of water and fed using a contact feeder. Miller or Ashforth feeders are unsuitable for early feeding because the bees will not go up over the cold 'weir' to reach the syrup. I don't bother with so-called spring stimulation, but some beekeepers in early districts do try to push their colonies to take advantage of spring rape. Colonies with ample stores do not need feeding. Weak colonies with adequate stores should not be fed. All that will happen in this case is the queen will be encouraged to lay in the periphery of the cluster and if a period of cold weather follows, the cluster will contract, leaving these eggs to become chilled and 'bee energy' will have been expended in vain. Feeding also causes bees to fly during inclement weather, leading to a further loss of field bees. I don't believe in feeding sugar bags at this time because bee energy will be required to convert the hard sugar into a usable form. Bee life is extremely precious at this time of year when beekeepers are striving to build up their colonies to take advantage of early honey flows from oilseed rape or sycamore. Weak hives should not be united until you have established why they are weak, as there is a danger of spreading disease. If the queen is thought to be at fault she is certainly not worth saving. The importance of pollen is often not appreciated by the beginner. Bees cannot rear brood without pollen. If the weather is cold and foraging restricted, colony development will be severely curtailed, no matter how much syrup is fed.

Any hive where the bees have died should be taken home for renovation. The inside of the hive and the top and bottom bars of the frames should be scraped clean of brace comb and any inferior combs cut out of their frames and stored in a sealed container, to avoid the attention of wax moths, until they can be rendered into blocks of wax. Actually I dispose of my inferior brood combs by burning. Thus getting rid of disease spores and pyrethroid residues. Very little wax is obtained from old brood combs in any case. A stack of hive parts together with serviceable combs, some of which might contain stores, should be made in the open air. The stack should be fumigated using absorbent pads, placed between each box and soaked with 80% glacial acetic acid. The stack should be closed off and left for a week, then dismantled and the boxes aired in a bee-proof shed and left until required later in the season. Fumigation should be carried out using goggles, face-mask and rubber gloves because acetic acid is corrosive to flesh, metalwork and concrete. If the brood chamber is some years old it will probably have tinplate frame runners. These should be smeared with Vaseline to reduce the effects of corrosion.

Now is the time to finally settle on the location of new permanent apiary sites. When making your choice the following should be considered:

(i) Availability of forage - nectar, pollen and water;

(ii) Colony density. Dr Colin Butler stated that in order to obtain a worthwhile honey crop, colony density should not exceed one colony per ten square kilometres. If my mathematics are correct: 3mile flying range = 4.8km Area = pi x rad. squared = pi x 4.8 squared = 72 sq.km i.e. giving a colony density of 72/10 = 7 colonies per apiary. When calculating colony density in a district, colonies owned by other beekeepers must be taken into account. The above calculation is not valid when considering colony density for intensive crops such as oilseed rape and heather.

(iii) Danger - to other humans, to animals, from animals;

(iv) Elements - wind, air-drainage, frost pocket, too hot, too shaded, winter sunshine, dripping trees, proximity of power lines, danger of flooding;

(v) Access;

(vi) Out of public view.

Once you have settled on a site, its layout should be considered:

(i) Allow plenty of room for manipulation of colonies;

(ii) Consider hive orientation - irregular pattern, direction faced;

(iii) Install hive bases.

APRIL

This month bees will be collecting nectar and pollen from flowering currant, dandelion, willow, cherry, gorse and blackthorn, leading to rapid colony build-up. Continue to monitor the food supply and the number of varroa on the hive inserts. On the alighting board, hard grey old pollen pellets the size of a cell may be seen. If crushed between the fingers they will break up and show layers and sometimes a trace of colour might still be visible. The appearance of old pollen pellets is a good indication that the bees are expanding their brood nest. Chalk Brood mummies can also be observed at the hive entrance. These do not crumble into layers, they are usually smaller and flatter than pollen pellets and are often recognisable as poorly developed pupae. Chalk brood can appear if bees have wintered in damp conditions and can be seen in nuclei which are short of bees. In other words, where the bees have been under stress. Some strains of bees are more prone to it than others. If widespread, chalk brood can hinder colony build up. There is no 100% cure, but I have found that Apiguard encourages hygienic conditions within the hive and reduces the occurrence of chalk brood. In severe cases the colony should be re-queened from a different strain and any badly affected combs replaced with acetic acid treated combs. In mild cases I sprinkle table salt on top of the brood frames to encourage hygienic conditions. This is repeated twice at weekly intervals.

If you require to treat varroa in spring, it should be carried out before the hives have honey supers added. I do not use Apiguard or any other thymol-based treatment prior to the honey flow as its odour can remain in honey for some time. Not long ago, a honey judge at one of the major shows rejected some exhibits because he could detect a slight odour of thymol in the honey.

Towards the end of April, when the sun is warm and there is little wind, a first inspection can be made. The hive entrance should be smoked and the beekeeper should wait for two minutes before opening the hive in order to allow time for the soothing effect of the smoke to take effect. Gently lever up the crown board and give it a sharp shake above the open hive to dislodge the queen back into the brood chamber if perchance she has been driven on to the crown board by the smoke. Two manipulating cloths are laid on top of the brood chamber such that only the particular comb which you are going to remove is exposed. This assists in controlling the bees and prevents brood becoming chilled. The first thing I do at this first inspection is to find the queen. She is more easily spotted at this time because there are fewer bees in the colony and as there are no, or few, drones, she is the largest bee. She is likely to be found on a frame containing eggs in the top brood chamber. She might not be the queen which you are expecting to find. The queen that you saw during your last autumn inspection may have been marked and/or clipped, but a younger unmarked queen may now be in the hive, the old one having been superseded in late autumn. April is the best time of the year to mark and clip the queen. I do not mark and clip queens during the summer in which they were born because there is a danger that the bees will detect your odour or that of the paint and supersede your new queen. Whereas if the marking is done in April, before the drones are flying and fertile, the bees know that she cannot be replaced and there is little likelihood that she will be killed. As a further safeguard, I never handle a queen. When found, I use a 'press-on' type queen cage to first clip and then mark her. Only about a quarter of one pair of wings need be clipped. When marking, ensure that you allow the paint to reach the hard surface of the queen's thorax. If you only paint the thorax hairs, the paint will very soon wear off. I keep queens for no more than two full seasons, therefore I only use yellow or white marking paint, which are more easily seen in a populous colony. Once marked, I ease the pressure of the cage on the queen and keep her in the cage until the paint has dried. When looking for a queen, concentrate solely on that task. When found, clipped and marked, she should remain in the cage until you have carried out other tasks, so that you know her whereabouts. The remainder of the brood chambers can now be checked quickly. It is important that you memorise the appearance of healthy, sealed and unsealed brood so that any abnormality can be given a closer examination. Check that there are nice areas of eggs and larvae also slabs of sealed brood containing few 'missed' cells. Check for signs of American Foul Brood, European Foul Brood and Chalk Brood. If AFB or EFB are suspected, consult an experienced beekeeper. Continue to monitor varroa numbers by checking the hive inserts and/or uncapping drone brood with an uncapping fork. Be careful not to kill too much drone brood or you may seriously reduce the drone gene-pool which will be available for queen mating later in the season. Check that the colony has enough food to last it until your next visit.

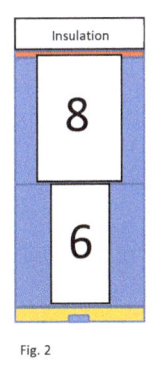

If the colony is on double brood chambers, remove the two outer combs from the bottom box. These should be the poorest combs and as such should have been deliberately put on the flanks during the previous autumn so that they could be removed when empty, in spring. The opportunity should also be taken to remove any other misshapen combs or combs clogged with old hard pollen and replace them with drawn combs which you had stored from the previous autumn. The renewal of three or four brood combs per year helps to rid the brood chamber of the spores of bee disease pathogens and residues of pyrethroid varroa treatment. The hive will now consist of eight combs in the top brood chamber and six in the bottom (See Fig. 2).

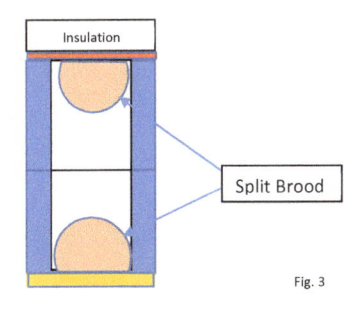

Fig. 2

Finally, don't forget to release the queen before closing the hive. Do not reverse the brood chamber at this, or any other, time as you will split the brood nest and give the colony a severe setback. Let the brood expand downwards naturally as nature intended. (See Fig.3).

Fig. 3

Now that I know the age and pedigree of my queens I can decide which I am going to deploy as breeders. By consulting my previous hive records I divide my queens into three categories: (i) breeder queens from which I am going to actively raise queen cells, (ii) queens which are satisfactory and from which I will use queen cells if they are produced and (iii) queens with undesirable characteristics which must be eliminated. I do not believe in selecting a so-called 'best colony' and rearing all my queens from it, thus losing the vast gene pool of all the others. In practice, I rear queens from the best forty of my sixty colonies and exchange young queens between my eight apiaries to reduce the chances of in-breeding. In-breeding will become more of a problem now that the feral colonies have been more or less wiped out due to varroa. I shall describe my method of queen rearing in my June article. Some beekeepers raise queens in May and re-queen their colonies as early as possible in the knowledge that queens of the current year rarely swarm. It is also true that the most prolific colonies are the first to raise drones. So the beekeeper who can re-queen in May/June with current years queens shouldn't have swarming worries and drones should be from good stock. I prefer to concentrate on maximum colony strength, with the potential to produce a large spring honey crop, by not using bees and brood to make nuclei at this stage and rely on controlling swarming if and when it comes.

MAY

May brings the start of the spring honey flow from sources such as oilseed rape, sycamore, horse chestnut, bluebell, top fruit, etc. and in order to derive maximum benefit from these the colonies must be strong. Indeed, the beekeeper's skill is in keeping colonies as strong as possible from May until August. We have no way of predicting what the weather is going to be like or how long good foraging conditions will last, therefore we must endeavour to maintain our colonies as strong as possible so that they can make the most of any flow. This means that we must deter or delay swarm preparations at least until the spring flow is over and exercise effective control over swarming if or when it comes. Never open a hive unless you have a reason for doing so and have a plan worked out in advance for what you are going to do and what equipment you require to have on hand. In early May the colonies should be checked in order to see that the queen has not suffered any ill effects as a result of clipping and marking. There is no need to find the queen, just check for the presence of patches of eggs. There should now be drones in the hive. Bees work best when they have some drones, but you don't want too many. Incidentally, drones do not become fertile until they are twelve to thirteen days old.

Check that the queen has plenty of space in which to lay. In a double brood chamber you might have to remove surplus stores of sealed honey and replace them with drawn comb. Do not give frames with foundation at this time unless you are prepared to feed syrup to simulate a nectar flow. Always try to do what nature intended. When do bees in the wild require to draw new comb? During a nectar flow or after they have swarmed. If you don't have a supply of drawn comb the alternative is to scarify the cappings on combs of stores with the corner of your hive tool to encourage the bees to move the stores, thus creating room on the periphery of the present brood nest into which the queen can expand. You must be able to 'read' a colony just like you would read a book. Understand what is normal, then the abnormal will stand out. Answer the following questions: Do you see the queen? Is there sealed brood? Is it in solid slabs with few missed cells? Are the cappings worker or drone? Are there eggs? Are they in slabs? Are the eggs two or more per cell? Are they fixed to the bottom or the sides of the cell? Are there any signs of disease in sealed or open brood? Has the colony got adequate food supplies? Is it as strong as its neighbours? Has it got adequate room? Keep a watchful eye on the sycamore (or oilseed rape), a week before it is likely to flower the brood chamber is checked and two combs from the top brood chamber are put into the bottom brood chamber (remember that the two outside combs were removed from the bottom brood chamber last month) and one deep frame of foundation and one shallow frame of foundation are added to the centre of the top brood chamber. The shallow frame is fitted with worker foundation and this will be drawn out with worker comb. The space underneath the shallow frame will have natural, drone comb built in it. This 'sacrificial' drone comb will be preferred by female varroa mites and when it is sealed it must be cut out and destroyed, (or given to a friendly fisherman) thus removing a large number of varroa mites from the hive without the use of chemicals. The frame, still containing its worker brood is then returned to the hive where the bees will repeat the process. Be careful to cut out this drone brood before it hatches or you will be breeding varroa. A queen excluder and the first honey super are put on top. This super should be of drawn comb so that when the flow starts, honey will be stored in the super and not allowed to restrict the space in the brood chamber. (See Fig.4).

Fig. 4

If you have been working on the single brood chamber system you might want to experiment with the sixteen comb doubling system which I advocate. If so, this is the time to double your colony. In the unlikely event that you have spare drawn combs, these should be used, otherwise you will have to use five frames containing foundation. Put the foundation in the top brood chamber and feed three litres of spring syrup to simulate a nectar flow in order to get the foundation drawn. Don't super the hive until the feeder is removed or syrup may be stored in it. Be aware that on the year that you first double a colony it will be at the expense of the spring honey crop unless it is a very good year, but you should reap the benefits during the summer and in subsequent years. The advantages of the double brood chamber system are:

1. Adjustable --- 11 combs (or fewer) to 22 combs.
2. Good air circulation and ventilation.
3. Easy to manipulate – no combs propped outside the hive.
4. Easy to check for swarm preparations.
5. Room available when most required – for expansion in spring and to accommodate bees, brood and stores in autumn.
6. Frames are interchangeable between the two chambers.
7. Foundation can be drawn in the warmth of the top brood chamber.
8. Safety valve in case of late supering – this should not be allowed to happen. After a week of the honey flow more breeding and super room should be provided. Fig. 4 Move one comb of brood from the top brood chamber into the bottom brood chamber and add two frames of foundation in the centre of the top brood chamber (See Fig.5).

Fig. 5

If the foundation is given on the flanks of the top brood chamber instead of in the centre, there is a tendency for the bees to draw part of it in drone cells. I have found that another advantage of the shallow frame drone trap which I have described in the penultimate paragraph (see explanation for figure 4) is that the bees have less need to draw drone comb elsewhere. Add a second super at this time if the bees have occupied the second-outside frames of the first. Even if the weather is wet, provide extra super room for the bees or you will get congestion. This second super could be of foundation which will employ any idle wax makers and take them up out of the brood chamber thus further alleviating congestion. If the second super is of foundation it should be placed below the first, to take full advantage of heat from the brood chamber. As the season progresses, the colonies are checked weekly to see that there is enough super room and that swarming preparations are not being made. My method of checking for queen cells is to remove the supers on to a manipulating board, (see Fig. 6)

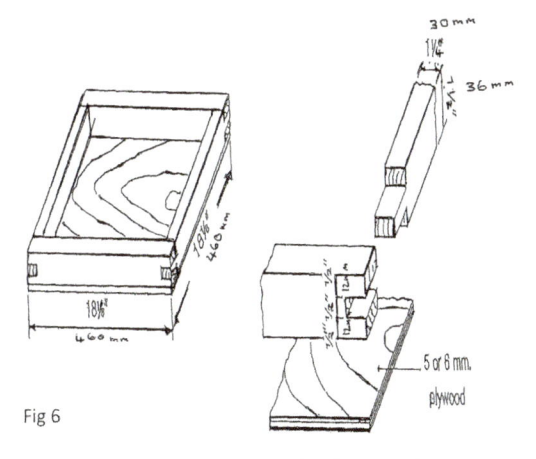

Fig 6

cover the queen excluder with a manipulating cloth, lever the two brood chambers apart with the hive tool and insert a wooden wedge (See Fig. 7).

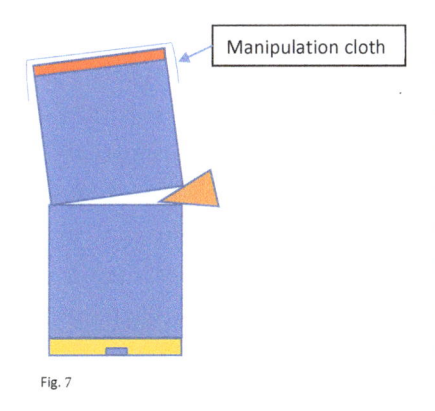

Fig. 7

The top chamber is then pulled forward 20mm to ensure that it won't slip off when the top chamber is tilted. Smoke is puffed into the opening held by the wedge, then the top chamber is hinged upwards to allow a quick inspection to be made of the centre of the brood nest. By using a little smoke and peering upwards and downwards between the frames, the presence of sealed brood and queen cells can be determined. The whole operation only takes a couple of minutes. When queen cells are built they are usually to be found along the lower edge of the combs in the top brood chamber which, of course, is the centre of the brood nest. The odd queen cell can be missed using this method, but it is a great time saver and causes minimal disruption to the hive rhythm. In any case, the queen has been clipped. Ensure that there is enough super room for final storage of honey and for the bees to 'hang nectar out to ripen' in spare cells. I tend to over-super in spring and consolidate in late summer. If queen cell cups, with nothing in them are found, this is quite normal and it is a waste of time removing them. If the cups are being extended and contain an egg, or more critically, a larva being fed, then you must assume that swarming will take place soon after the queen cells are sealed on the ninth day after the egg was laid. I do not believe in cutting queen cells as a method of swarm control as it cannot be relied upon. If cells are removed they will probably be rebuilt the following week and if cut again the bees will sulk, become fed-up and build queen cells on larvae up to three days old, ie. five or six days after egg laying. The cells will be sealed three days later and a swarm is likely to be lost long before your next visit. These cells will produce queens which are inferior to those produced from newly hatched larvae which have been copiously nourished from birth. The queen cells which were found first of all will contain larvae properly fed from birth therefore they will produce the best queens and should be utilized if the strain is thought to be suitable. In order to go as near as possible to what would happen in nature, I artificially swarm the colony. This prevents the loss of a swarm, (and your future honey crop), satisfies the bees' instincts and provides for re-queening later in the year. First set an empty brood chamber on a manipulating board beside the hive. Find the queen, (she is marked and clipped), and put her together with the comb and adhering bees on which she is found in the centre of the new brood chamber. Put another frame of sealed brood and bees beside the frame containing the queen. Cut out any queen cells which are on these two frames. Without shaking any bees, reassemble the hive as follows: floorboard; new brood chamber with queen and two frames plus seven empty drawn combs, if available, plus the two outside combs from the original bottom brood chamber put on the flanks, (if no drawn comb is available, foundation will have to be used); queen excluder; the original two or three supers; crown board with feed holes open; the original brood chamber(s) with the queen cells; second crown board. This operation can also be used with single brood chamber management. If you are using double brood chambers, two may be required on top. In practice, these can be reduced to one the following week after some of the sealed brood has hatched, the surplus combs are then used when artificially swarming other hives.

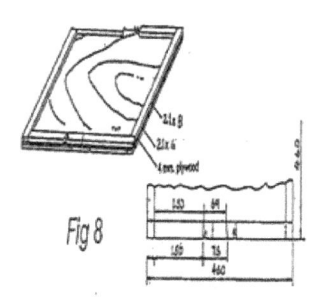

If the queen cells are hanging down from what was the original top brood chamber they are likely to be damaged against the crownboard in the rearrangement just described. In which case a small eke, (the one I use in autumn to accommodate Apiguard containers), should be used. After twenty four hours the colony will have rearranged itself. All the field bees will have joined the queen in the lower part of the hive and most of the nurse bees will be in the top brood chamber(s). The open crown board is now replaced with a solid swarm

Fig 8

board, (See Fig. 8), with an entrance open to the front of the hive, (see Fig.9).

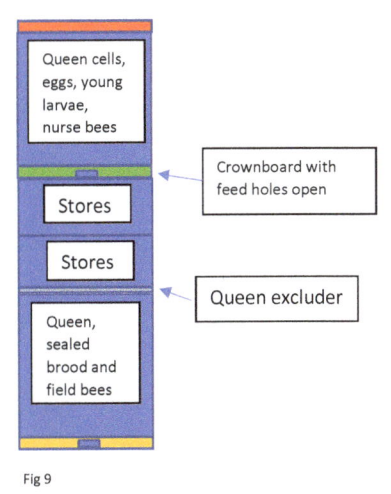

If the queen cells in the top brood chamber(s) are unsealed a contact feeder containing spring syrup should be placed above the feed hole because the top colony, which is now a big nucleus, will have lost all its field bees. Five days later the top front upper entrance is closed, the top front lower entrance opened and the top back upper entrance opened. This will further divert new field bees to join the queen below thus adding to the artificial swarm. The new field bees from the considerably weakened nucleus will fly from the back. The queen cells in the top brood chamber can be reduced to one, which is really unnecessary because the nucleus is too weak to throw a cast. Or they can be distributed to other similar nuclei which are of an undesirable strain. A virgin queen will fly and eventually mate from the rear of the top brood chamber, where her brood and the temperament of her bees can be assessed during the summer. If these are deemed to be satisfactory the top brood chamber with its new queen will be used for requeening the original colony or another colony on another site. Ten days after the original splitting of the colony, the bottom brood chamber should be checked to see that the colony has settled down and no further queen cells have been drawn. If the original queen cells were being drawn because the bees realised that their queen was failing, they will likely still be trying to replace her and any queen cells will be few in number as they will be supersedure cells. Colonies treated as described usually settle down before the main nectar flow is due and behave like a prime swarm, hopefully producing lots of honey. If, as in 2006, some colonies still persist in drawing queen cells, you have the option of cutting them out or killing the queen and reducing the queen cells to one. Bear in mind that colonies work better if they have a queen. Also, a virgin queen hatching in a strong colony will not usually mate and commence laying until most of the sealed brood from the previous queen has hatched. This causes a gap in brood rearing which weakens the colony prior to the main honey flow. What I would recommend, if you find more than two queen cells seven days after the original split, would be to cut them all out and check again in another seven days. If queen cells are still being drawn, kill the queen and unite one of your top brood chamber nuclei (you should have some young queens laying by now) using the newspaper method. In this case two sheets of newspaper, one below and one above the young

Fig 9

queen's brood chamber, should be used. After a further seven days, reduce the brood chambers to one.

JUNE

Sometime in the first half of June, the spring honey flow will finish. At the end of the month the main flow from sources such as white clover, bell heather, field beans and raspberries will commence. Before the start of the main flow there will be a period referred to as the 'June gap' when, if the weather is good, the bees will just be able to support themselves. If the weather is poor or if you have removed all of the spring honey crop you will have to monitor the food stores and be prepared to feed. Feeding at this time is a problem. I don't feed syrup when supers are on the hives in case it ends up in the honey. I don't want to take the supers off when feeding as that would cause congestion. My solution is to feed with one or two damp sugar bags laid over the feed holes in the crown board and remove them as soon as fresh nectar is coming in from the start of the summer flow. About the middle of the month, I remove any full or nearly full supers of spring honey, leaving unsealed honey on the hives. There is no oilseed rape grown in Renfrewshire. I extract the honey when it is still warm, strain it as it is taken from the extractor, run it into honey buckets and put into storage. I shall explain this process more thoroughly in the August magazine. Some beekeepers do not believe in removing any honey at this time, arguing that it should be left to tide the colony over the June gap. My position is that honey is more expensive than sugar. Therefore I am going to remove the honey and am prepared to feed the bees if necessary. In any case, if the summer turns out to be poor, that honey is all I might get. Continue to monitor the natural mite drop on the varroa floors and estimate the number of mites in the hive to make sure that they don't reach anywhere near to the danger level before late August, when the supers are removed and they can be treated. Another quick method of checking for mites is to uncap drone brood where varroa are easily visible on pupae at the 'purple eye' stage. Read the free booklet entitled 'Managing Varroa' published by CSL/DEFRA in June 2005. Your Local Association Secretary should have copies. As the spring honey flow is nearing its end, bees often start to make swarm preparations, especially if they are short of super room, or the brood chamber is congested with semi-idle workers cooped up in a single brood chamber, leading to many being starved of queen substance. Last month I explained how I dealt with swarming as soon as queen cells were being built and how I raised a new queen from these cells provided the queen heading the colony was classed as category (i) or (ii) as described at the end of my April article. I utilize queen cells from the first batch to be sealed, as these are the best cells, made in the bees' own time, from properly fertilised eggs and copiously fed with royal jelly from the moment they hatched. If you break down these cells and use the second batch they are likely to be inferior because they will have been made in more of a hurry and could even have been drawn from old larvae thus producing 'scrub' queens. J Woyke demonstrated that queens produced from eggs were 10% heavier than those produced from 24 hour old larvae, also they had 8% more capacity in their spermatheca and 3% more ovarioles. Any colonies which draw queen cells before 20 June will be artificially swarmed as I described in my May article. From these, NATURAL queen rearing will be carried out, provided the original queens were graded category (i) or (ii). Swarming is the natural method by which bees reproduce themselves and therefore cannot be eliminated. Some strains are more prone to swarming than others. As far as I am concerned, if colonies are of a mind to swarm the earlier they show signs of it the better. They can be dealt with as described, the swarming urge satisfied and the bees settled prior to the main honey flow. All colonies which have not shown signs of swarming by 20 June are reduced to single brood chambers containing eleven combs in order to prevent overbreeding. Bees produced from eggs laid after that time will take no part in the main honey flow. If the summer turns out to be poor or moderately poor you will get little or no surplus honey from colonies on large brood chambers. They will store any surplus in the brood chambers and the supers will be empty. Colonies treated as described will be short of young bees for the ling heather. The treatment of these will be discussed in my August article. Proceed as follows: find the queen and put her on the comb on which she was found in the centre of the bottom brood chamber; make the chamber up to eleven combs with sealed brood because this will soon hatch making room for the queen to lay; queen excluder; supers; crown board with feed hole open; and the remainder of the mainly unsealed brood on top. The bees in the top brood chamber are remote from the influence of queen substance and are likely to produce queen cells. Check after a week and remove them because the bees will not have been fed therefore any queens produced will be scrubbers. Check the bottom brood chamber at the same time to see that queen cells have not been started there. Contrary to what some beekeepers believe, reducing to a single brood chamber at this time does not induce swarming because you have removed both bees and brood from the brood chamber. The brood in the top chamber should be allowed to hatch. Most of the bees will migrate through the open feed holes to join those below. Some will remain in the top box and will 'look after' the combs during the summer. If the summer nectar flow is good they will even store some honey in the top box which will form part of their winter food when the colony is returned to a double brood chamber in August. Three weeks after the original manipulation, the rear of the top brood chamber should be propped up with a 6mm piece of wood for a day or two in order to allow the trapped drones to escape thus preventing them from clogging the queen excluder. If the colony has been graded category (i), then they should be INDUCED to produce queen cells in the top brood chamber. The procedure is exactly the same as I described in the May magazine for artificial swarming, except that the top brood chamber must be fed with spring syrup in order to simulate a nectar flow and frame containing plenty of pollen must be inserted beside a marked (with a drawing pin) frame containing newly laid eggs in the centre of the chamber. (See Fig. 10).

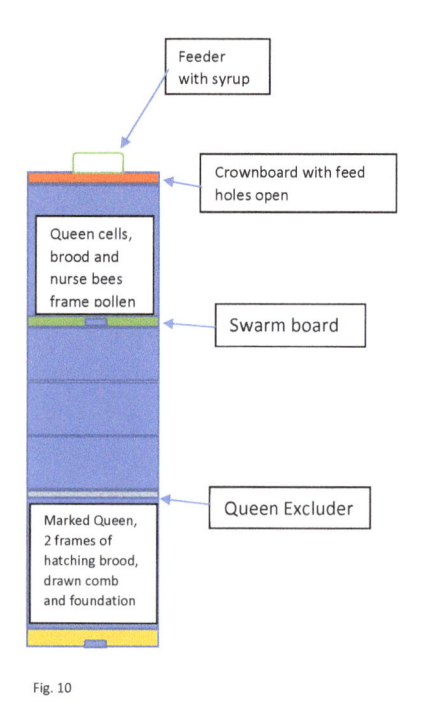

Fig. 10

If you don't feed the top brood chamber, which has lost its field bees, inferior queen cells will be produced. The top brood chamber must be checked three days after the artificial swarm was created and any sealed queen cells removed because they will have been drawn from older larvae and will produce inferior queens. When the queen cells are within a couple of days from hatching, they should be reduced to one. The others should be cut out, using a sharp knife, well clear of the cell to avoid damaging it and distributed to other nuclei which are of an undesirable, [(category (iii)], strain. For more details read my articles on Queen Rearing in the 'ature ofScottish Beekeeper' of April 1997 pages 401-402 and May 1997 pages 431-434. (Pages 89-92 and 92-96).

JULY

This is the month when we will find out if our colony management is going to bear fruit. Given good weather, the bees will be foraging at the main honey flow on plants such as white clover, bramble, bell heather, lime and willow herb. Varroa should continue to be monitored. During the first fortnight, check to see if some colonies are still attempting to swarm, although most if not all will have passed that stage and the brood chamber need not be opened. Make sure that the hives have adequate super room and you can go on holiday during the second half of the month. Unless you have re-queened your colonies in May or June (which I do not) you must check your colonies every week from mid-May until mid- July in order to control swarming thus giving yourself the chance to obtain a good honey crop. Early in the month, move the shallow frame drone traps described in my May article to the outside of their respective brood chambers or into top brood chamber nuclei. When the worker brood which they contain has hatched, the frames should be removed and stored in a bee proof and wax moth proof location for deployment during the following May. This will save the bees having to draw the worker foundation from scratch next spring. Check the top brood chambers, or nuclei, deployed in queen rearing to see if the virgin queens are mated and laying and that the brood pattern is satisfactory. Sealed brood should be in complete slabs with few 'missed' cells. (except, perhaps, along the lines of the foundation wires). The proportion of empty cells in the sealed brood area is an indication of the extent of in-breeding. In the honeybee a fertilized egg is diploid, having 32 chromosomes and will develop into a female caste; an unfertilised egg is haploid, having 16 chromosomes and will develop into a male caste. With multiple matings of the queen with, say, up to ten drones diploid females are usually produced. If in-breeding occurs some of the eggs would produce diploid drones. Diploid drones are detected by nurse bees whenever the egg hatches. They are not seen in the colony because they are not tolerated and are eaten by the worker bees leaving a 'pepper pot' appearance to slabs of brood. In-breeding can occur in isolated apiaries affecting up to 50% of the brood. However, with the demise of the feral colonies of honeybees, due to the ravages of varroa, in-breeding is becoming more likely in the more heavily bee-populated areas of the country. Colonies where in-breeding has taken place fail to build up to full strength. In-breeding should not be confused with any of the brood diseases. Two further checks should be made on Queen mating nuclei, Check if the Queen has been slow to mate because of bad weather and has become a drone breeder. If you find more than one egg in each cell, the young queen is either just starting to lay or there are laying workers in the hive. If it is laying workers, the eggs will have been deposited on the cell walls because the worker's abdomen is too short to reach the bottom of the cell. If for example the virgin queen has been lost during her mating flight, the bees have no means of raising a replacement. Some workers will develop the ability to lay unfertilised eggs which will only produce drones. By the time laying workers appear in a colony, it is hardly worth trying to save it because the bees will all be nearing the end of their lifespan. In any case, colonies containing laying workers are almost impossible to re-queen because the beekeeper has no way of recognising which workers are laying and the bees assume that they are queen-rite. If I find a colony with laying workers I carry it one hundred metres from its site, shake all the bees on to the ground and return the hive to its original position. The idea being that the laying workers will be too heavy to fly and the others, all being field bees, will fly back home. You can try to unite a queen-rite nucleus to the colony twelve hours later or set it above the open feed hole of another colony. If it is a good honey year, like 2005 and 2006, early nuclei in top brood chambers, whose queens have been laying since June may be becoming congested. This is an excellent opportunity to add frames with foundation. The developing colony will consist of surplus bees of wax-making age and they will readily draw foundation with few if any drone cells, because they have a young mated queen and drones are no longer required. A larger back entrance may have to be given in hot weather. This is easily achieved by propping up the rear of the top brood chamber with a 6mm piece of wood. When adding foundation to an expanding brood nest during a honey flow, it should be put two positions in. It is a mistake to add it to the flanks as the bees will draw it and quickly fill it with honey before the queen has time to lay in it. New foundation should not be added to weak nuclei if the one previously added has not been drawn. Some beekeepers like to produce some comb honey from the summer and ling flows. Spring honey is unsuitable for cut-comb or sections because oilseed rape will granulate in the comb and sycamore is too strong in flavour for most palates. In the past I have produced cut-comb in supers containing nine Manley frames. These frames are spaced at 41mm, (1 5/8in). That is the absolute maximum distance between sheets of foundation if you are going to prevent the bees drawing extra combs between those that you want. I now space cut-comb supers using narrow (34mm/1 3/8in) plastic ends because the resulting sealed combs are a little bit thinner and they fit better under the lid of the cut comb containers which are sold by the suppliers. Another problem I have experienced with the Manley's is that the bees heavily propolise the frame shoulders, making the frames difficult to remove from the crate. This could be overcome by smearing the shoulders with 'vaseline'. When working for cut comb I use thin unwired foundation cut into four horizontal strips, instead of full sheets which tend to buckle when being drawn. The bees will draw the strip of starter foundation with worker comb and the space below with drone comb. I have not experienced any difficulty in getting bees to enter cut-comb supers, they can be used in conjunction with ordinary supers where required. It is preferable not to place a cut-comb super next to the brood chamber as the bees might store some pollen in it, also the cappings soon become 'travel stained' by the bees moving to supers which are above. I produce square sections in hanging section holders each containing three sections. Eight section holders fit into a shallow super allowing twenty-four sections per crate. Metal 'separators' must be suspended between each section holder otherwise the bees are likely to spoil many of the sections by building comb where the beekeeper does not want it. Bees do not like being restricted in little sections and are often reluctant to enter them. The queen is unlikely to enter, so no queen excluder should be used. Full squares of thin fresh foundation should be used when working for sections. Special crates, holding thirty-two sections can be purchased from the suppliers but bees do not like them. Round sections are preferred by some producers but I have no experience of these. Strong colonies are required when working for sections and bees must have a laying queen which was reared during the current year otherwise they are likely to swarm due to congestion and their reluctance to enter the 'little boxes'. Colonies chosen for section production should have a section super put on top of the current supers one week before the start of the targeted honey flow so that the bees can become accustomed to it. At the start of the flow the section super should be placed on top of the brood chamber with the queen excluder removed and the original supers put above the clearer board. These supers, cleared of bees, can be given to other colonies to complete. To further entice the bees into the section super a partly completed section, saved for the purpose from the previous year, should be placed in the middle of the super. Do not add a second super until the bees have started to draw comb in the first or they may sulk and start to 'loaf'. The second should be added below the first and if the colony is strong and the flow is good a third super should be placed on top of the first. If the flow continues the third super can be placed below and so on. In this way the bees are given tasks a little at a time, whereas if faced with a huge space containing innumerable 'little boxes' they may give up and sulk, or swarm especially if the queen is of the previous year. Towards the end of the month I put cut-comb supers, intended for going to the heather, on top of the supers on strong honey producing colonies in the hope of getting them drawn out. This will give the bees a quick start at the heather.

AUGUST

The summer honey flow should be in full swing at the beginning of the month and beekeepers must ensure that the bees have ample super room in order to take full advantage of nectar from the willow herb and Himalayan balsam.

During the early part of the month, a check of the supers should be made with a view to removing any which have not been occupied and concentrating partially completed combs on the strongest hives in an attempt to get as much of the honey ripened and sealed as possible by the end of the flow. Flower honey keeps best if its moisture content is in the range of 17-19%. Above 20% there is a danger that the honey will ferment sooner or later depending on storage conditions. The most accurate method of determining the moisture content of honey is to use a refractometer. If you do not possess such an instrument, it is safer to extract only honey which has been sealed. If there is still a combination of sealed and unsealed honey in the supers a week after the end of the flow, select each frame in turn, hold it over the hive and give it a good shake. If no unripe honey splashes out, it could be extracted, but beware that its moisture content could be near to the critical limit. It should be used as soon as possible and not stored for any length of time. With the exception of the colonies which have been moved to the heather earlier in August, I remove my honey crop starting about the middle of the month, put the colonies back on double brood chambers and re-queen the colonies, where required, all in one operation. (See Fig. 11).

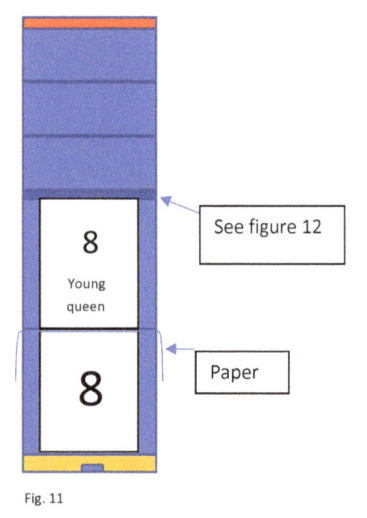

Fig. 11

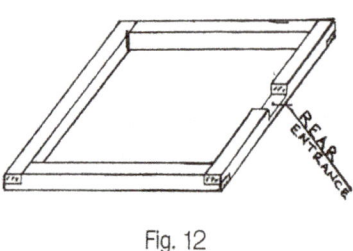

The old queen in the bottom brood chamber is removed, the combs reduced to eight flanked by dummies, a sheet of newspaper spread over the top of this chamber and the top chamber containing eight combs and the young laying queen placed on the newspaper. A special frame with a small rear entrance (See Fig, 12) is placed on top of the second brood chamber to allow the field bees from what was the top chamber access to the hive and a clearer board is put on top. Any surplus brood combs are distributed to other hives which are being doubled and don't have a nucleus on top. Great care must be taken to ensure that the stack of supers is bee proof or, as they become cleared of their own bees, they will be robbed out if bees can gain access from the outside. The supers are removed as darkness begins to descend the following evening. Guard bees do not fly in the dark. The special frame (Fig. 12) is removed after a few days, when the bees using the back entrance will find their way round the front and be accepted.

Fig. 12

The honey supers are taken home, kept in a warm room and extracted that night or the following day. Honey is taken from the extractor and strained, first through a stainless steel sieve and then a 200 micron nylon conical bag suspended in a stainless steel storage tank. A length of cotton tape should have been sewn on to the apex of the cone and be long enough to reach the bottom of the tank. This allows the strained honey to run down the tape, reducing the amount of air gathered in the strained honey. After settling in the tank for a few hours to allow more air bubbles to rise, the honey is run into plastic pails and put into storage. At this time of year when ambient temperatures are fairly high honey can be extracted and strained with a minimum of effort. Warm honey does not incorporate as much air during the extraction and straining process as it does when it is cold and thick. Shortly after the removal of surplus honey the colony, now requeened and on sixteen combs, must be treated for varroa. Bayvarol or Apistan or Apiguard or some other effective and suitable varroacide should be used. Remember, if you use either Bayvarol or Apistan (both are pyrethroids) year after year, the pyrethroid will build up in the wax combs, thus subjecting the varroa to continuous small doses of the chemical which will lead to the development of pyrethroid resistant mites. It is essential that varroa numbers are reduced to a minimum in order to prevent damage to the young bees which are going to have to carry the colony through the winter months ahead. I have been treating with Apiguard for the past three autumns and have found it to be effective if applied early enough when the ambient temperature is in excess of 15deg.C. In order to conserve heat, I place a 50mm thick sheet of expanded polystyrene on top of the crown board. A 20mm deep eke is placed under the crown board to accommodate the Apiguard. I have also found that Apiguard considerably reduces the incidence of chalk brood. One snag with it is that 50% of colonies refuse to take liquid feeding while the Apiguard is in place.

The ling heather will start to bloom about the tenth of the month. Heather yields more nectar in the first half of its blooming period, weather permitting. This gives me a dilemma because the willow herb and balsam are in full bloom. Consequently, I only move eight colonies to the heather and leave the others on their permanent sites as described above. This means that I can be well on with my autumn work by the end of August and only have the heather colonies to deal with in mid September. At the beginning of August, I select eight strong colonies to go to the heather. These colonies will have a large force of field bees, but many will be old and would not last long in the harsher conditions often experienced on the moors. It is also important to try to keep the queen laying while at the heather to produce 'winter bees'. Therefore a queen reared during the current year is essential. I treat these colonies destined for the heather almost as I have described in paragraph two above. (again see Fig. 11). The only difference is that when I remove the old queen from the bottom brood chamber, I leave it on the original eleven combs. I also put two cut comb supers above the queen excluder and the clearer board on top. Five days later, when the honey supers have been removed and the young queen has been accepted and laying, the hive is reduced to a single brood chamber; queen excluder and two cut comb supers. As much of the brood as possible is concentrated in the single brood chamber and surplus combs are used to strengthen other nuclei. In rearranging the brood chamber, combs containing mostly sealed brood are placed in the centre and those containing mostly eggs and young larvae are put in the flanks. The sealed brood in the centre will hatch first leaving room for the queen to continue laying. If the brood distribution had been the other way round, space would have been created in the flanks which would have been filled with incoming nectar, leading to a severe reduction in space for egg-laying and less honey in the supers. At this time the crown board is replaced by a ventilation screen and the hive is secured by means of a nylon strap, ready for removal to the heather. About the tenth of August the hives are moved to the heather. In the evening prior to moving, when flying has ceased, a piece of foam rubber is pushed into the hive entrance. The hives can be moved during the night or first thing in the morning. On arrival at the moor, the hive straps are kept in place, a sheet of insulation is placed above the ventilation screen, the roof replaced and the bees released.

SEPTEMBER

The main nectar flows are now over except for the end of the heather and Himalayan balsam. As I stated in my notes for August, the honey from all colonies which have not been taken to the heather has been removed and the hives put on to sixteen British Standards (BS) combs. This arrangement gives plenty of room for autumn laying, for winter stores and for the bees themselves; none of which is available in an eleven BS comb single brood chamber. The bees will continue to store honey from late sources, which I leave for the bees themselves and any shortfall is made up to about 25kg by feeding sugar syrup (1kg white sugar to 550ml water) or ready made liquid feeding purchased commercially. I use Miller feeders for autumn feeding which allows the colonies to be fed in a very short time, certainly in less than a week. When feeders are first put on the hives they should be 'primed' by running some syrup down through the feed hole to entice the bees up into the feeder. I try to have all my colonies fed by mid September. Some beekeepers are of the opinion that if bees are fed too early they will use some of the food for breeding and will have to be given more food later. I do not agree with this. What better preparation can there be for next spring than to have colonies going into the winter with an abundance of young bees, even if it means having to top up their food stocks during the last week of September. Bees are thrifty insects, they do not waste food. Some beekeepers make the mistake of feeding syrup far too late in the autumn, causing the bees to become exhausted trying to reduce its moisture content and trying to get it sealed before the onset of cold weather. After honey has been extracted I usually put the wet supers over the open feed holes of hives which are lightest in stores. When they have been cleared of honey the dry supers are taken home, sprayed with 'Certan' as a preventative against wax moths and stored in my bee shed for the winter. The supers are stacked, with a sheet of newspaper between each, in piles which are made rodent-proof by placing a swarm-board top and bottom. The use of Paradichlorobenzene for preventing wax moth attack is now illegal. Another, much quicker, method of deterring wax moths is to store the supers 'wet' as they come from extraction. This is successful, but it produces a terrible sticky mess in the bee shed the following spring. One advantage is that bees enter the supers very rapidly in spring to mop up the honey which has more than likely fermented due to dampness over the winter. I am always afraid that this fermented honey will taint the future honey crop especially if it is oilseed rape. It is claimed that wax moths only attack combs in which brood has been reared. That being the case, stored spare brood combs are particularly vulnerable and should be sprayed with 'Certan' and made rodent-proof.

Around the middle of the month the heather hives should be brought home. Their honey supers must be removed, the colonies put on sixteen combs, treated for varroa and fed. The air temperature is too low for Apiguard to be used, so some other method of treating varroa must be used. I try to top up the winter stores by feeding autumn syrup. In order to prevent robbing, all colonies needing food in an apiary should be fed at the same time. Feeding should be done in the evening and care must be taken to avoid spilling syrup in the vicinity of the hives and entrances should be reduced to 25mm to deter robbers and wasps. Beekeepers who have many colonies at the heather usually do not have time in September for liquid feeding. They winter their bees on fondant or dampened sugar bags, neither of which encourage robbing. Cut comb and sections can be prepared for market as soon as they are removed from the hive. Bees should be removed from the frames using some type of clearer board, never by shaking the combs as they are easily damaged. The use of smoke should be severely curtailed as it can be detected on the comb for weeks. When preparing cut comb for exhibition it should be cut to the exact shape of the plastic container in which it is going to be exhibited and put on a wire baking tray to allow exposed honey to drain out before it is put in the container. If I am preparing cut comb for the market it is cut out using a 'Price' comb cutter and not drained. I have never had any complaints about liquid honey being in the bottom of the opaque container. It is all good stuff! This does not apply to beekeepers who individually wrap their comb in transparent cellophane bags or sheets. In that case the liquid honey would appear at the bottom of the pack and detract from its appearance. When using a 'Price' comb cutter, I cut close to the bottom of the frame, this leaves 12mm of comb at the top of the frame which will act as a 'starter' for the following year and the frame wont require to be re-waxed. I put the cut comb supers above the feed hole back on the hives to let the bees clean up the surplus honey remaining in the starter comb. All honey extraction equipment should be cleaned and stored until required the following season. The extractor, metal strainers and tanks should be washed with cold water to remove most of the stickiness, followed by hot water to which washing soda has been added in order to remove any dirt or wax particles. They should then be rinsed in hot water, thoroughly dried, wrapped in polythene and stored in a dry place. I still use two Porter bee escapes fitted in the crown board when clearing bees from supers. Some beekeepers prefer to use Canadian clearer boards, claiming that Porters get gummed-up and cease to function. In my opinion Porters only cease to function because of lack of maintenance. They should not be left on the hives after the supers have been removed otherwise the bees will propolise the springs to close off the air space. At this time all Porters should be submerged in a pot of water to which a handful of washing soda has been added and heated to a temperature above the melting point of wax. Before storing, the distance between the tips of the springs should be checked and adjusted, if required, so that they are about 3mm apart.

OCTOBER

At this time, weather permitting, the bees could be topping-up their reserves of both nectar and pollen from the ivy. This is the last important bee plant of the year. At the beginning of the month, winter entrance blocks must be put on the hives to keep out rodents and to allow winter ventilation. Read again my notes for January. Strips or empty trays used in controlling varroa during late August or September must now be removed after they have been deployed for six weeks. Bayvarol or Apistan strips must not remain in hives for more than this time because the pyrethroid will be losing its strength and varroa coming into contact with it will not be killed. The varroa will adapt to be able to withstand the effects of the weakened pyrethroid, leading to the evolution of resistant mites. Very soon neither of these pyrethroids will be effective against varroa and alternatives such as Apiguard or Oxalic Acid will have to be deployed, or you will lose your bees. I place a sheet of 50mm thick expanded polystyrene insulation on top of the crown boards and weigh down the shallow roofs with a brick. Apiary hedges can be cut and trees which are shading the apiary should be pruned. Fences and gates enclosing the apiary must be made stock-proof. Hives should be checked to see that they are watertight.

Drones should have been expelled from the hives by October. Any hives still having drones flying during the second half of the month should be investigated, as there could be something wrong with the queen. If the colony is queenless, it should be united to a nucleus or to another colony.

During the active season I store wax, obtained from cappings, pieces of broken comb and frame scrapings gathered during the manipulation of colonies, in large polythene buckets. This wax is now melted down into cakes and stored until the following spring, then I make and wire my own foundation. Making your own foundation or making candles is the most cost-effective way to utilize the wax crop. I make no effort to extract wax from old brood combs as it is a messy job and for all the wax obtained it is not worth the bother. In any case, if pyrethroids have been used in the brood chambers to treat varroa, residues will have built up in the wax. This also applies to wax purchased from the equipment suppliers. As I stated in my March article, I burn all my old brood combs thus getting rid of pyrethroid residues and disease spores.

I will probably have already marketed some of my spring and summer honey but the bulk of it will be in store. I like also to have in stock a surplus from the previous year in case of a poor current season. In my work for August I described how my honey was extracted and strained without the need to apply heat and stored in sealed buckets to granulate naturally. Now is the time to start to get stored honey ready for the market. If honey has been allowed to crystallize in a container it cannot be jarred without the application of heat. Great care must be taken not to overheat honey. If it is overheated, especially with the top removed from the storage container, there will be a loss of flavour and aroma by volatile oils and other substances of plant origin being driven off. Hydroxymethylfurfural (HMF) is a substance produced by the chemical breakdown of fructose in the presence of free acids, a process which is occurring in honey all the time. The rate of production of HMF is dependent upon the temperature to which the honey is subjected. The higher the temperature and the longer the storage time the more HMF the honey will contain. Honey extracted and strained without the application of heat will contain very little HMF. The present EU Honey Regulations state that the HMF of honey must be no more than 40mg/kg at the point of sale. Therefore care must be taken during processing not to heat honey at too high a temperature and/or for too long a time. If honey is jarred immediately after extraction and straining there is no guarantee how long it will take to set. Some honeys are slower to set than others. Some set with a soft texture and fine grain, others with a hard texture and coarse grain, others with coarse grain in the bottom of a jar and liquid honey at the top. None of the honey just described must go anywhere near the shops. In general, if the honey has a high glucose content it will granulate quickly producing a fine crystal structure and if it has a high fructose content granulation will be slow and the crystal structure will be coarse. To produce clear honey, the bucket of stored honey has to be gently heated to completely re-liquify it. If any small amount of crystals remain, the honey will become dull in appearance and will reset. The honey should be heated in a thermostatically controlled air heating cabinet at an absolute maximum temperature of 50°C for 36 to 48hr. It should then be re-strained into a honey tank and allowed to cool rapidly in a cold place until its temperature falls to 32°C, then run into clean jars which have also been heated to 32°C. The jars should remain in a warm room overnight to allow air bubbles to rise. They should then be stored in a cool room until required for sale. It will have a shelf life of a few months before it starts to granulate again. Some authorities claim that, if the shelf life is to be extended to six months or more, the hot newly jarred honey with the lids screwed on should immediately be further heated in a thermostatically controlled water bath to a temperature of 60°C for 3/4hr to kill off the yeasts and then cooled rapidly.

To produce soft set honey, first of all check the texture of the stored honey. If it has set with a fine texture and a soft grain which will spread like soft margarine, all that is required is to heat the bucket in the air heating cabinet at a temperature of 32°C until the honey has a consistency of thick porridge, then poured into a honey tank. In this form it should be stirred to ensure even mixing, left to settle to let air bubbles escape and run into clean jars preheated to 32°C. The jars are left in a warm room overnight to allow more air bubbles to rise and then stored at 12°C to re-set before being put on the market. Alternatively, if the stored honey has set with a hard (spoon bending) texture and coarse grain it will have to be completely melted and 'seeded' with a fine textured soft set honey like clover or oilseed rape. The stored honey is heated in an air heating cabinet at no more than 50°C for 24 to 36hr to completely re-melt. It is then poured into the honey tank and allowed to cool rapidly to 32°C. At the same time 'seed' honey is heated to 32°C and stirred. Then pour the seed into the melted honey in the proportion of 1kg of seed honey to 10kg melted honey. The greater the proportion of seed the faster the mixture will set. The mixture is then stirred with a wooden stick until an even mixing has been attained, left to settle to let the air bubbles escape and run into clean jars preheated to 32°C. The jars are left in a warm room overnight to allow more bubbles to rise and then stored at 12°C to reset before it is put on the market. Honey treated as described above, paying particular attention to the temperatures at which blending and jarring take place, should set as a smooth textured product which should be free from shrinkage at the edges of the jar and free from 'frosting' which has the appearance of white streaks or patches in the honey. In both cases there is nothing wrong with the quality of the honey, only its sales appeal will have been reduced.

NOVEMBER

November is the start of a period when colonies should be disturbed as little as possible. At the beginning of the month I insert the hive tool under the rear of the crown board to crack open the propolis seal and place a match under both rear corners of the crown board. This will create a small air-gap which will allow moisture-laden air, from the winter cluster, to escape. I don't believe in leaving the feed hole open. If this operation is carried out in October, or earlier, there is a likelihood that the bees will propolise the gap again. Many beekeepers believe in giving more top ventilation than I do, also many are now using open-mesh floorboards which allow the winds to sweep upwards through the hives in the belief that this will lead to better wintering conditions. Undoubtedly these floors give a drier hive in winter and it is claimed that 20% of varroa mites will be lost through an open-mesh floor. As far as I am concerned, the 'jury is still out' on the deployment of open mesh floors. I am experimenting at present, with ten of my colonies on these floors and over the past three years I have lost a slightly higher proportion of them, compared to those on solid floors and they definitely have been slower to build up in spring although by mid-summer there was little difference. I also think that there is a slightly reduced incidence of queen cell production, but this could be because they had not attained full strength by the swarming period in June. Mid-Renfrewshire is an early district, where strong colonies are required for the sycamore flow, therefore a rapid spring build-up is essential not only for the prospect of obtaining early honey but also for the advancement of queen cell production and the consequent satisfying of the swarming impulse prior to the main honey flow. Dr David Raven, writing in *Beecraft* in September 1989, stated that a level of 1% to 2% carbon dioxide is beneficial in keeping bees in a state of inactivity during the winter. The concentration of CO_2 in normal atmosphere is only 0.03%. In 1960, the Russian researchers Taranov and Michailov found that a fairly high concentration of CO_2 in the centre of the winter cluster was essential, as it slowed up metabolism and reduced the consumption of stores. Although bees could regulate the concentration of CO_2 and the relative humidity in freely ventilated hives it was attained at the expense of heavy energy expenditure, excessive use of stores and worn-out bees. The effect on wintering colonies of these findings is contained in the researchers' excerpts:- "If winter colony death rate is taken as a criterion, better results are obtained from colonies wintering under conditions of good ventilation. However, the real loss in bees can be seen only some time after the spring flight when colonies lose their enfeebled winter bees. An estimation of the number of full frames occupied by bees in autumn and spring has shown that bees are in better condition when wintered under conditions of poorer ventilation. These experiments have shown that increasing hive ventilation considerably weakens the bees and reduces their ability to rear larvae in spring. With a small degree of ventilation in the hive, the concentration of CO_2 at the centre of the cluster is increased, this reduces the consumption of food, conserves more energy and ensures more ability to rear brood by the over-wintered bees. The data obtained suggests that it may be necessary to revise ideas on hive ventilation and the arrangement of hive entrances over the winter period. Reduced air exchange in the hive can mean better wintering of bee colonies and more intensive brood rearing in spring." For this reason I insert the underfloor varroa tray from 1 st February until 1st May to conserve heat. November is a good time to study hive records from the year just past and select possible breeder queens for the following season. I don't believe in breeding solely from my so-called best colony because there is no guarantee that the new queen's progeny will be as good as their mother and now that we have varroa, the feral colonies will soon be wiped out, leading to the increased chance of in-breeding. By breeding from one best colony a vast gene pool of desirable qualities will be lost. I have eight out-apiaries and try to cut down the risk of in-breeding by requeening one apiary with queens cells from another, or by exchanging nuclei between apiaries. In doing this, care should be taken not to interchange bees from one part of the country with those from another as it can take years for bees to acclimatise to a new district. Beekeepers who purchase bees please note. It is a fact that queens reared as early in the season as possible and introduced into their source colonies do not, as a rule, (except for the 2005 season) show any tendency to swarm during the remainder of the current year. However, by re-queening all colonies every year you might have cured the swarming problem but that system does not allow for good queen selection. A queen is judged by the performance of her offspring. Therefore you cannot claim to have a worthwhile breeder queen until the qualities of her daughters can be assessed. This might not be possible until the following year. Therefore, queens, thought to be suitable breeders have to be overwintered for use as breeders in the following season. Good queens in one generation should mean good drones in the next. If we keep on culling unsatisfactory queens we will soon get rid of the poor drones. I don't make any effort to have dedicated drone-producing colonies because the drone will pass on the qualities of his mother (he has no father), whereas the qualities of the new queen and her workers will depend on those of both their father and mother. If a colony is bad-tempered, it must be requeened from a better strain, unless you try to breed out the bad temper by having its virgin queens mated with drones in a different apiary. Over time, it is possible to obtain a desirable strain by crossbreeding and to fix it by line-breeding. The beginner will have observed that the colours of all the workers in a colony are sometimes not the same. That is because their mother will have mated with many drones, some of which could have been dark and others yellowish.

DECEMBER

December is a quiet month in the apiary except for an occasional visit to ensure that all is well. It is also an opportunity to conduct a preliminary search for a suitable apiary site. The sun is low in the sky at this time of the year so you will be able to observe if your likely site gets some winter sunshine and is not a frost pocket. At this time of year there should be little or no sealed brood in the hives, (although I do support the idea about winter breeding put forward by Eric McArthur on. p. 330 of the December 2005 *Scottish Beekeeper* Magazine) consequently the vast majority of varroa mites will be on the adult bees thus exposing them to treatment. All colonies should have been treated for varroa during autumn, after the honey flow. There is a chance that the autumn treatment was ineffective or that there has been a reinvasion of varroa mites from collapsing feral colonies or from the colonies of the 'feral beekeeper'. When the ambient temperature is about 6°C, all colonies should be treated with Oxalic Acid by either the trickle or sublimation methods, the application of which have been adequately described in this magazine in previous years.

During a frosty spell when wax is brittle and easily removed, the opportunity should be taken to clean all queen excluders. I use the wired type which is framed and has a bee-space on one side. The excluder is placed flat on the workbench and then a 20mm x 20mm piece of wood is slipped under the far side thus raising the excluder clear of the bench. The bulk of the wax is removed from both sides in turn using the hive tool in a 45deg. cutting motion, taking great care not to displace any of the wires. The remainder of the wax is then removed by means of brushing parallel to the slots with a wire brush.

If you are planning to start beekeeping during the coming season or you are thinking of adopting a different hive pattern you should carefully consider which pattern you are going to adopt, prior to placing an order with the supplier early in the new year. You should then ensure that any subsequent equipment which is purchased is fully interchangeable, and that there is only one hive pattern in your apiaries. My advice would be to decide whether you are going to adopt top or bottom bee-space. If you are going to adopt top bee-space I suggest that you should adopt the Smith hive or if bottom bee-space the National hive. My preference is for the National because the brood chamber is of a suitable capacity in summer and for the heather, it is not too heavy to lift, it takes British Standard frames with long lugs for ease of lifting out, it has excellent hand-holds and, if you are buying or selling hives or bees, the majority of beekeepers have this hive pattern. Whether you adopt top or bottom bee-space is a matter of personal preference so long as you do not mix the two in one hive. Some beekeepers argue that top bee-space is better because you can set a second brood chamber or super on top of one already there and slide it into position without crushing bees which are on the frame-tops of the lower box. But what about the bees which are hanging from the underside of the frames in the box which is being slid over? They are much more likely to be decapitated because you cannot see to smoke them, whereas those at the top of a bottom bee-space box are easy to see and be driven down with a puff of smoke. It is my opinion that the WBC hive is fine if you are only having a few of them and are not going to practice migratory beekeeping. They are far too much work for the many-hive beekeeper as well as being far too expensive to purchase. The Langstroth hive is probably the most common worldwide but I consider its brood chamber to be too large in summer. In the west of Scotland, and I suspect in other areas as well, a brood chamber containing eleven BS frames is adequate during the summer and can be increased by adding a second brood chamber at other times of the year. In summer, the use of a large brood chamber, or double National or Smith brood chamber, will give an excellent honey crop in a good season but little or no honey in a middling or poor season. All that you will do then is rear large numbers of bees which will spend most of their time in brood rearing and store any honey they gather in the brood chamber. The Modified Dadant hive has far too large a brood chamber for Scottish conditions and is far too heavy to lift. This concludes my series on practical beekeeping throughout the year. I hope that beginners have been assisted in planning their work during each month and that more experienced beekeepers have been stimulated to review their current methods to ensure that what they have been doing is best for their particular conditions. One of the many joys of beekeeping is that there are numerous way of achieving whatever you desire from the craft. Happy beekeeping!

Swarm Prevention by Ian Craig. MBE

During 2007 I wrote a series of monthly articles describing my methods of beekeeping throughout the year. Readers will recall that I try to reduce the commencement of swarming preparations by giving plenty of room for bees, brood and stores. I only practice the "artificial swarm" method of swarm control on the colonies which produce queen cells. My method of beekeeping involves having 16 to 20 combs in the brood chamber until the 20th June then reducing to a single, 11 comb, brood chamber for the summer before returning to 16 combs in the autumn.

For beekeepers who prefer to restrict colonies to single brood chambers, the artificial swarm method of swarm control can still be used. There are however two methods of swarm prevention which I have used in the past and which are worthy of consideration. They are the Demaree System and Worthwhile system.

Demaree System

This is a method which is described in many books on beekeeping but the following description was given to me many years ago at a course conducted by the late Andrew McClymont while he was a lecturer at the then West of Scotland College of Agriculture at Auchincruive and to the best of my knowledge it, or a variation of it, is still practiced by Graeme Sharpe at Scottish Agricultural College (SAC) to this day.

This system of swarm prevention is generally applied to colonies which have come through the winter in the strongest possible condition and are headed by queens not more than two years old which are capable of expanding their egg laying to make use of the extra capacity in an additional brood chamber which is provided. It should be applied before colonies make swarming preparations.

1. The colony is allowed to build up until there are nine or ten frames with brood and plenty of bees to fill the brood chamber from side to side. Depending on seasonal variations, colonies should reach this strength during May or early June. There is likely to be one or two honey supers on the hive.

2. A second brood chamber is prepared with at least two empty drawn combs, the remainder being frames with foundation. Giving more than two drawn combs is an advantage as the queen is not inhibited in her egg laying.

3. The colony is examined to find the queen. She is caged on the frame on which she is found and placed in an empty nucleus hive until she is required.

4. The original brood chamber is then removed to one side and on the floorboard is placed the new brood chamber filled with foundation towards the outside and drawn comb in the centre, leaving space for one or two frames of hatching brood to accompany the queen.

5. The queen is then transferred into this new brood chamber and the queen excluder, honey supers, original brood chamber and finally the crown board is placed on top.

The hive is then closed and can now be left for 7 days. The top brood chamber should then be checked to see that no emergency queen cells have been built. The emergency cells will possibly have been built over older larvae and if so, will produce inferior queens. During this examination any drones trapped above the queen excluder can be released by inserting a 6mm piece of wood under the front of the top brood chamber for a week. The bottom brood chamber should be examined to check that the queen is laying and she has plenty of spare room in which to lay. If she is likely to be short of room before the next inspection, exchange frames of undrawn foundation from the outside of the bottom brood chamber with drawn combs from the top brood chamber which by now is unbrooded. This exchange of frames can be repeated in subsequent weeks until the new bottom brood chamber contains nothing but drawn comb. The foundation which is placed in the top brood chamber is better to be placed in the centre as this will allow the bees to cluster conveniently around it in order to conserve warmth for comb building. By this method, beautiful combs can be drawn in the top brood chamber which is now used as a super and if the weather is kind, will be filled with honey which can be extracted. The extracted combs can subsequently be used to replace any old or damaged combs in the bottom brood chamber or stored for future use. In many cases queen cells will not be constructed in the top brood chamber, especially if only one super is separating the two brood chambers. However, it is very important that a check should be made, otherwise the system will collapse if queens are produced and allowed to hatch. If this happens, there is a strong possibility that the old queen, plus a high percentage of foraging bees will swarm and leave the young queen trapped above the excluder and unable to fly from the hive to be mated and will eventually become a drone layer. Later in the season it may be necessary to repeat the operation. Re-Demareeing is quite common if the weather in spring and early summer has been exceptionally favourable, but in normal seasons this will not be necessary. It should be emphasised that to Demaree a stock prior to any signs of swarm preparations being started is not necessarily an infallible method of swarm prevention. Swarming can still occur later in the season, therefore continued vigilance is required by the beekeeper. A different and quite separate system of management can be applied should swarm cells be constructed at a later date. There are many modifications to this system of Demareeing which allows the beekeeper to rear queens, produce strong heather colonies, etc. The basic system has been described above and when it is applied to single brood chamber colonies which have not started to make swarming preparations it has been proved statistically to dramatically reduce the number of colonies which attempt to swarm.

Last month I described a version of the Demaree method of swarm prevention. This month I shall describe the Worthwhile System which is named after Mr A. W. Worth whose background I do not recall.

Worthwhile System

This system of swarm prevention is generally applied to colonies which have come through the winter in the strongest possible condition and are headed by queens not more than two years old which are capable of expanding their egg laying to make use of the extra capacity in an additional brood chamber which is provided. It should be applied before colonies make swarming preparations.

1. The colony is allowed to build up until brood and eggs are found in seven

or eight combs. It should then be given a first honey super of drawn comb above a queen excluder.

2. A second brood chamber is prepared containing not less than two frames of foundation, one on either flank, the remainder being filled with good drawn combs and with a space in the middle to receive the queen on a comb of brood and eggs.

3. The colony is examined to find the queen.

4. The queen together with the frame on which she is found, is inserted into the the space left in the centre of the new brood chamber.

5. The frames in the original brood chamber are closed up and empty comb added to make up the full compliment of frames.

6. The new brood chamber with the queen on the central comb is placed on top of the original brood chamber, with an excluder in between to prevent the queen going down or out.

7. A second queen excluder is placed on top of the new brood chamber and the super is placed on top of all. As few drones remain in the brood chamber containing the queen the excluder remains clear

8. No further disturbance of the brood chambers should be made for fourteen to seventeen days but a second super may be required on top if there is a honey flow. Do not delay in providing sufficient super room.

9. At the end of that period most of the workers and drone brood in the bottom brood chamber will have emerged and the combs in this chamber, cleared of bees, can be put into storage or used to repeat the method of swarm control as the season dictates.

It is found in this method, that the bees below do not realise the separation from the queen and do not raise queen cells. The new brood chamber is protected from chilling and the bees have a large vacant space to fill. They cannot swarm while the lower excluder is in place. If you do not have sufficient empty drawn comb as directed at instruction 2, use more foundation, but there must be at least one drawn comb on either side of the one carrying the queen. As all foundation is drawn above the lower brood chamber, it will be drawn out to the bottom bars of the frames, which is a great advantage. If the brood nest is allowed to become congested, queen cells may be started and a different system of management will have to be applied. If you know that the queen is old or you suspect that she is failing, do not use this swarm prevention system as the queen will not be able to cope with the extra demands placed upon her. Do not operate this system in two consecutive years with the same queen. Other provisions must be made for re-queening every second year.

Swarm Control by Ian Craig. MBE

In my two previous articles this year I described two methods used to try to prevent, or at least delay, the onset of swarming when using single brood chambers. In this article, I shall describe how to control swarming, again using a single brood chamber. This method is also suitable for use with a double brood chamber, as I described in the *Scottish Beekeepers* magazine in May 2007 page 129. Most methods of swarm prevention and control involve finding the queen. Towards the end of this article I shall describe the procedure when she cannot be found. Colonies should be allowed to build up, making sure that they have good unclogged brood combs and plenty of supers. The problem with single brood chamber management, in my opinion, is that eventually the queen will run out of room for egg-laying and the brood chamber will become so congested that queen substance ceases to be evenly distributed. When this is allowed to happen, queen cells will be started and, unless the weather intervenes, swarming will be inevitable. If a colony is found to have fed or sealed queen cells, some form of increase will have to be made in order to satisfy the swarming instinct. It is a waste of time cutting out queen cells in the hope of preventing swarming. The bees will beat you in the end by drawing queen cells from larvae two or three days old and swarming as soon as the first queen cell is sealed in three or four days time. Queens which are allowed to develop under these conditions are considered to be inferior. When queen cells are found, proceed as follows: Remove the brood chamber from its site and set it to one side. On the original site put a new brood chamber containing ten empty drawn combs, if you have them, or at least two drawn combs and the remainder foundation, leaving a space in the middle to receive another comb. Find the queen in the original brood chamber and insert her, on the comb on which she was found, plus adhering bees, in the centre of the new brood chamber between the two drawn combs, first removing any queen cells from the queen's comb. Replace the queen excluder plus supers, plus bees in the supers, on the new brood chamber. The field bees will all join the queen on the original site. The original brood chamber is now queenless, but has queen cells, brood at all stages and bees. A good queen cell should be selected and the frame on which it is found marked with a drawing pin so that its position is easily identified. Remove, without shaking the comb, all other queen cells from this comb. Go through the remaining combs, shaking bees if necessary to get an unobstructed view and remove all queen cells. Alternatively, a few good queen cells can be carefully cut out with a sharp scalpel, prior to any shaking, and used for re-queening other colonies. Give the original brood chamber a floorboard, entrance block to restrict the size of the entrance, crown board and roof. After twenty four hours this original brood chamber will have most of the nurse bees, no field bees and a developing queen cell. The food supply will have to be monitored as it will only have what was in the original brood chamber, if any. Three weeks later the young queen should be mated and laying in the original brood chamber. This new colony can be allowed to build up to honey gathering strength in time for late flows from the heather or Himalayan balsam. If increase is not required, the original colony with the developing queen can be moved one week after splitting to the other side of the new brood chamber, thus channelling more field bees into the artificial swarm possessing the honey supers. Once the new queen has been laying long enough for the pattern of the sealed brood to be evaluated, the old queen in the new brood chamber can be killed and the two colonies reunited. The original brood chamber, containing the young queen, is placed above a sheet of newspaper on top of the new, now queenless, brood chamber and a second sheet of newspaper placed on top, followed by the queen excluder and supers. After a further week, if the new queen is still laying satisfactorily, the colony can be reduced to a single brood chamber. Any combs containing surplus brood can be used to bolster other colonies or nuclei. My preferred method, which requires only one piece of extra equipment, is to place the original colony above a 'swarm board' above the supers of the new colony as described in my series of articles last year. If the queen cannot be found at the beginning of the artificial swarm technique, put a comb containing eggs, but no queen cells, between the two empty combs in the new brood chamber. Put the original colony on top of a swarm board above the supers. The front entrance of the swarm board should be opened and all queen cells removed from this top (original) brood chamber. This brood chamber should be fed because it will rapidly loose its field bees and, if this is the queenless portion, will have to rear a new queen. After one week the front entrance should be closed and a back entrance opened. Check the hive after one week. If the queen is in the top brood chamber, eggs and young grubs will be seen there and queen cells will have been drawn in the bottom brood chamber where their numbers should be reduced to one. If she is in the bottom brood chamber the situation will be as was originally intended, except that new queen cells will have been drawn in the top brood chamber where their numbers should be reduced to one. If the queen had already swarmed when the artificial swarm technique was first carried out, there will be queen cells in both the new and original brood chambers. If the lost queen was clipped, she would drop from the swarm and be unable to return, but the colony will still have all its bees because the swarm will have returned. If she was not clipped, the colony will have swarmed, lost most of its field bees and have received a massive set-back with regard to honey production. That is why I would always recommend that queens should be clipped.

Queen Rearing by Ian Craig. MBE

In the March, April and May 1995 editions of *The Scottish Beekeeper*, I described my Double Brood Chamber method of Beekeeping. Allied to that article, I shall now describe how I select and obtain my young queens. One of the most important and often most neglected aspects of beekeeping is regular queen replacement with well nurtured, properly selected queens. Beekeepers should play an active part in their re-queening programme and not leave the selection to chance. Many of the undesirable characteristics of some honeybee colonies such as stinging, following, and running on the comb can be bred out of a strain fairly quickly. Even a programme of regular culling of queens in colonies which exhibit undesirable traits and replacing them with queens of a better strain will go a long way to alleviating these problems.

In the dim and distant past I have bought queens from other parts of Scotland and England only to be disappointed. I am now convinced that beekeepers should either breed from the best of their own bees or from other bees which are acclimatised to the district. Even moving bees a few dozen miles can create a new environment to which it can take them years to completely adapt.

Hive Records

An essential prerequisite to successful queen rearing is the keeping of hive records. I keep my records in a small loose leaf folder carried in my manipulation box. One page in the folder contains the records of one hive for a season. For beekeepers, who like me, keep their bees in out-apiaries, records should be accessible at home so that manipulations can be planned and equipment made available. This is not possible if records are pinned under hive roofs. An example of a typical hive record is illustrated below.

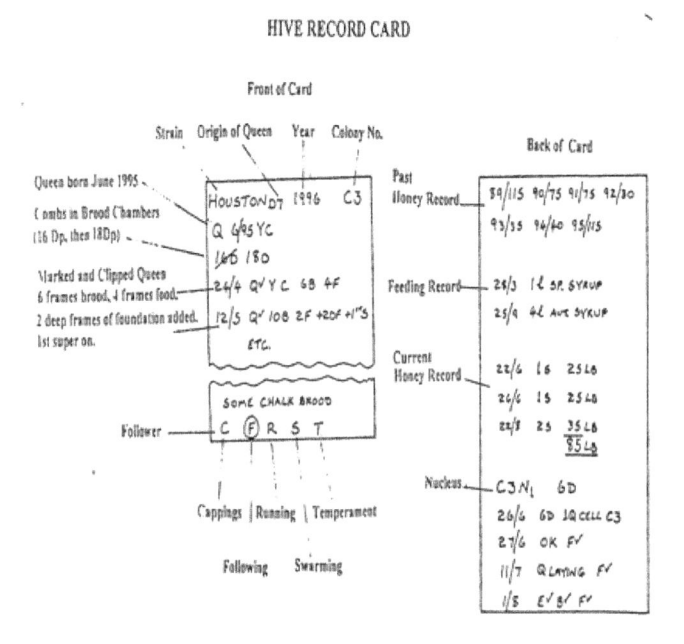

HIVE RECORD CARD

Front of Card

Strain   Origin of Queen   Year   Colony No.

Queen born June 1995
Combs in Brood Chambers
(16 Dp, then 18Dp)

Marked and Clipped Queen
6 frames brood, 4 frames food.
2 deep frames of foundation added.
1st super on.

HOUSTON DT 1996 C3
Q 4/25 YC
1/6 180
24/4 QYYC 6B 4F
12/5 QY 10B 2F +2DF +1"S
ETC.

SOME CHALK BROOD
Follower — C F R S T

Cappings | Running | Temperament

Following   Swarming

Back of Card

Past Honey Record — 89/115 90/75 91/75 92/30
93/35 94/60 95/115

Feeding Record — 25/3 1L SP. SYRUP
25/9 4L AUT SYRUP

Current Honey Record — 22/6 1S 25LB
26/6 1S 25LB
22/7 2S 35LB
  85LB

Nucleus — C3 N₁ 6D
26/6 6D 1Q CELL C3
27/6 OK FV
11/7 Q LAYING FV
1/8 EV BV FV

By studying the hive records at the end of a season, I can then place the queen heading each of my colonies into one of three catagories.

(i) breeder queens from which I will actively rear queens the following season.

(ii) queens which are satisfactory and if their colonies should produce queen cells I will be happy to utilise them; and

(iii) queens which must be eliminated because of undesirable colony behaviour or other factors.

From fifty colonies, I would expect to have four breeder queens, 25 to 30 which were satisfactory and the remainder due for "the chop." About 60% of my colonies are re-queened every year. I have no worries about the other 40%. Often a one year old queen which shows no inclination to swarm won't do so in her second year either. Many of my best yields of honey have been from colonies headed by a two year old queen. Beekeepers who re-queen all their colonies every year miss out on this opportunity. Whatever happens, all queens are replaced after two years, i.e. a queen born in June 1995 would be replaced by August 1997, at the latest. I resist the temptation to retain queens beyond this age and have only lost a total of four colonies during the past twenty winters. None of these were due to failed queens. I do not believe in selecting my so called "best colony" as a breeder, rearing all my queens from it relies on multiple mating with drones from other colonies to prevent inbreeding. Replacement queens, all emanating from one breeder queen would lead to a huge loss of genetic material from the culled queens many of which displayed highly desirable characteristics. This will further manifest itself in the future when Varroosis has wiped out most of the feral colonies which have sustained themselves for many years without any assistance from man and whose drones must play a significant part in the successful mating of our virgin queens.

Selection of Breeder Queens

These queens are selected for their good characteristics, which will hopefully be passed onto their daughters, although there is no guarantee of this because most of us have no control over the selection of drones which will mate with the young queens. Remember that drones are said to travel as much as ten miles from their hive. I do not create special "drone rearing" colonies or practice artificial insemination.

Temperament

This is the most important. It should be possible to handle bees under most normal conditions without getting constantly stung. Beekeeping should be a pleasure. Bad tempered bees are also a menace to the public, a fact which should be remembered in these days when litigation is common.

Productivity

Most of us keep bees in order to obtain pleasure and get honey. The honey record of a strain should be assessed over the past few years.

Swarming

Swarming is the bees' method of reproducing, therefore we cannot hope to eliminate it completely. Breeder queens should be selected from a strain which does not try to swarm every year, nor make more than eight queen cells when it does.

Disease

Avoid propagating strains which have had a past history of disease, even Nosema or Chalk Brood.

Following

Bees which meet you at the gate or follow you more than a few metres from their hive are a menace.

Running

Nervous bees, running over the face of the comb, running around to the other side or dropping off the corner of the frame, make it very difficult to find the queen which may drop off in a "stalactite" of bees back into the brood chamber, or worse still, on to the ground.

Capping

This is important if you are a comb honey producer. Cappings should be complete, white and even, not greasy and wet in appearance.

Productivity and any disease will be noted on the hive record and if any of the other characteristics referred to above are in evidence their presence will be noted by circling the appropriate letter on the bottom of the record.

Natural Queen Rearing

I do not waste time and sacrifice honey producing colonies in order to make special queen making nuclei. I explained in *Scottish Beekeeper* April, 1995 p 92 that I operate all my colonies, except some of those at the oilseed rape, on double brood chambers with 18 or 20 BS combs by mid-June. See figure 1.

Fig. 1

In practice this target cannot be completely met because some colonies start to build queen cells before that date and have to be artificially swarmed earlier. Around 20th June, I reduce all colonies still on double brood chambers to single brood chambers prior to the summer honey flow in order to divert the bees from brood rearing to honey production and also to reduce late swarming. By the end of the manipulation the surplus eggs, young brood and nurse bees are in a top brood chamber which is really a big nucleus, where a ripe queen cell (from a breeder colony) can be introduced. I therefore do not commence ACTIVE queen rearing until 17th June. By that time the colonies are strong, the air temperature is, hopefully, high and there are plenty of mature drones on the wing. If prior to 17th June, colonies in categories (i) or (ii) commence swarming preparations by drawing and feeding queen cells, I artificially swarm the colony by rearranging the brood chamber as follows: Find the marked queen, cage her in a pipe cover queen cage (or similar) in the centre of the bottom brood chamber between two frames of emerging brood with ANY QUEEN CELLS REMOVED, fill up this brood chamber with empty drawn combs if they are available, or frames with foundation and release the queen; queen excluder; at least two, preferably three, honey supers; crown board with a piece of queen excluder over the feed hole; top brood chamber with the remainder of the eggs, brood, queen cells, pollen comb and bees; swarm board top. See figure 2.

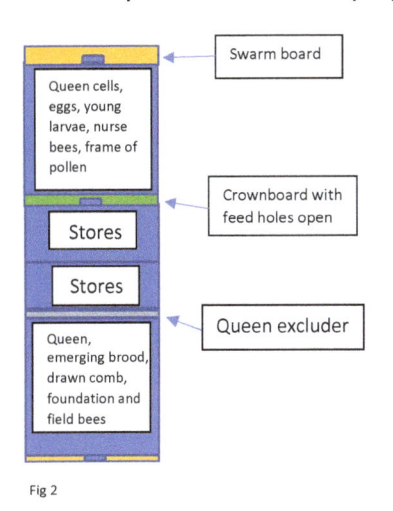

Fig 2

The piece of excluder over the open feed hole is to keep the drones out of the supers. It is unnecessary to shake any bees because of the following day the field bees will have moved into the bottom brood chamber beside the queen and most of the nurse bees will have moved to the top brood chamber to take care of the brood. At this stage the swarm board, with its entrance open to the front, is inserted under the top brood chamber thus cutting off the bees in it from the queen below. If the queen cells in the top brood chamber are unsealed, a feeder of syrup comprising 1kg sugar to 1 litre of water is given above the crown board. (See Figure 3). Also *Scottish Beekeeper* May 1995 p 114.

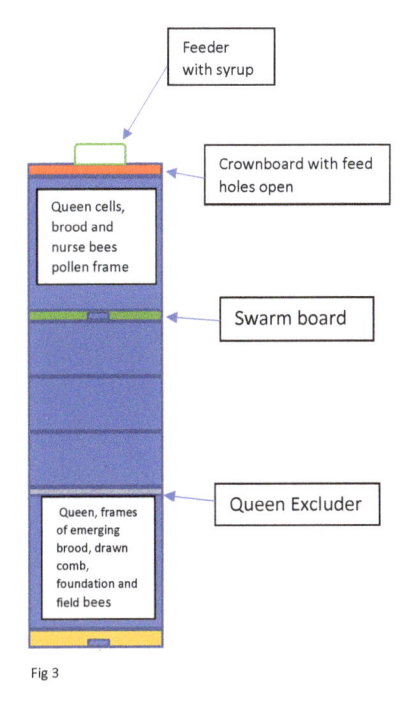

Fig 3

I wait until the queen cells in the top brood chamber are sealed and utilize the FIRST BATCH. These will be the very best queen cells, made in the bees' own

93

time, from properly fertilised eggs and copiously fed. If you break down these cells, the second batch are made in more of a hurry and could even be drawn from old larvae thus producing "scrub" queens. Unless an emergency has arisen in the original hive the first batch of queens will have been produced from EGGS, the larvae from which will have been fed Royal Jelly from the instant they hatched. J. Woke showed that queens produced from eggs were 10% heavier than queens from 24 hour old larvae, also they had 8% more capacity in their spermatheca and 3% more ovarioles. Two days before the queen cells in the top brood chamber are due to emerge, they are reduced to ONE. It is sometimes necessary to shake or brush bees from the combs in order to obtain an unimpeded view when checking for the presence of queen cells. Cells which are destined to become future queens should NOT BE SHAKEN as the embryo queen can easily be dislodged. The front entrance to the top brood chamber is now closed and a rear entrance opened. This filters the "new field bees" down to join the queen, thus strengthening the artificial swarm. The food supplies in the top brood chamber have to be watched because it has lost its field bees. The lower brood chamber can be checked two weeks after the artificial swarm was made in order to ensure that swarming preparations have ceased. If the queen was failing, the bees will still be trying to replace her. The queen cell selected to remain in the top brood chamber will* hatch and the virgin queen will, hopefully, mate and head what is now a large nucleus. Other satisfactory queen cells can be distributed to the top of the brood chamber nuclei of other hives, mainly in category (iii), which have been making swarming preparations and have been artificially swarmed at least one week before.

When selecting sealed queen cells for use in nuclei, the following criteria should be employed:

The cell should be 25 – 29mm in length (longer cells often contain dead or diseased larva); have a ridged, non-smooth surface; should not be surrounded by drone comb. It is also advisable to touch the cell tip lightly with the hive tool to check if a virgin queen has already hatched and the flap swung shut again. I have, on more than one occasion found a dead worker bee entombed in a queen cell. Presumably she was in the act of cleaning out and some mischievous bee sealed her in.

(*Emerging from cell)

Induced Queen Rearing

Breeder colonies in category (i) which have not commenced drawing queen cells by 17 June are now induced to do so. These colonies should now be provisioned and bursting with bees. They are rearranged as follows:

Find the marked queen and cage her between two frames of hatching brood in the centre of the bottom brood chamber, make sure there are no queen cells in this chamber; queen excluder; two or three supers; crownboard with a piece of excluder over the feed hole; second brood chamber with a frame of pollen in the centre, flanked by frames of eggs, then unsealed larvae, then sealed brood and honey on the flanks; swarm board used as a crown board. Any surplus frames of sealed brood should be added to the bottom brood chamber which should then be filled up adding empty drawn comb or frames of foundation and the queen released. The reason for placing only sealed brood in the bottom brood chamber is that it will emerge quickly, thus creating laying space for the queen. The unsealed brood in the top brood chamber will attract most of the nurse bees which will be required for rearing queens.

By the following day the bees will have sorted themselves out, as I have explained earlier. The swarm board with its front entrance open can then be inserted beneath the top brood chamber. (See figure 3.)

The nurse bees which were attracted to the top brood chamber by the presence of eggs and unsealed brood will find themselves cut off from the influence of queen substance and will draw a small number of queen cells. They should be continuously fed with syrup consisting of 1kg of sugar and 1 litre of water. The effects of this syrup and the stored pollen simulates a nectar flow which causes the nurse bees to secrete copious quantities of Royal Jelly to nurture the Queen Cells. After three days check that the bees are drawing the queen cells on the frames which had contained eggs. These frames should have been marked with a drawing pin when they were placed in the top brood chamber. The resulting queen cells should not yet be sealed. Check all frames in this brood chamber and destroy all SEALED queen cells as they will have been built around larvae instead of eggs and will produce inferior queens. A week later, on 27th June, ten days after the colony was split, the ripe queen cells in the top brood chamber are reduced to one the top front entrance is closed and the top back entrance is opened. The surplus queen cells are carefully removed from the comb by cutting well clear of the cells using a sharp knife and distributing them to the top brood chamber nuclei of mainly category (iii) and some category (ii) colonies which are scheduled for re-queening. A section of comb in the centre of the receiving nucleus is cut out and the ripe queen cell carefully inserted. These recipient colonies, which have not shown any inclination to draw queen cells of their own, will have been split one week earlier, i.e. on 20th June. Before giving a ripe queen cell to a nucleus, any queen cells it may have started need to be removed. The bees may have to be shaken from the combs in order to ensure no cells are missed. By following the above method, six to ten excellent queens should be produced. These will, hopefully, mate and start laying in their respective nuclei (top brood chamber colonies) where their sealed brood pattern and subsequently the temperament of their workers can be monitored prior to their being used for re-queening at the time of removal of the summer honey crop in late August. See *Scottish Beekeeper* May 1995 p115.

In this article I have attempted to describe how I re-queen my colonies using methods which can more or less guarantee success. Using the more skilful and time consuming grafting methods employed by commercial queen breeders and some large scale beekeepers often leads to disappointment in the hands of the small scale beekeeper with less opportunity to practice. They also involve the depletion of potentially honey producing colonies which have to be specially prepared for cell building and others which have to be divided for the formation of mating nuclei.

The two methods of obtaining replacement queens just described fit in with my double brood chamber method of beekeeping, require no extra equipment and there is no reduction in the honey crop. Most beekeepers who practice active queen rearing start at least 1 month earlier than I do. However, in practice, my queen rearing starts with the first category (i) or (ii) colony which makes swarming preparations. This has the effect of spreading the workload over six to eight weeks, making the programme less dependent on a bad spell of weather and is, in my opinion, more in tune with the instinct of the bees.

A Simple Method of Queen Production Using a Standard Size Nucleus Hive by Ian Craig. MBE

Beekeepers with a small number of colonies, especially in isolated locations, may well find that their sealed brood has a "pepper-pot" appearance. This could be caused by disease or varroa but is more likely to be the result of in-breeding. Such colonies will never reach their full honey gathering potential. It would be a good plan to prepare a nucleus with a queen cell or virgin queen and take it for mating to another apiary outwith the flying range of its own drones. The bees will tell you when conditions are optimum for queen rearing by starting to produce queen cells but this might be too late for your purposes. Queens should be reared in strong, well-provisioned colonies full of young bees. These conditions should be evident by about the end of the third week in May. Any earlier than this the drones may not have reached their full mating potential (Many textbooks advise "allow 14 days from the date of hatching") and queens should be reared using young bees not old ones.

This article deals with using 'natural' queen cells because grafting is normally used where large numbers of cells are required. In any case, Woyka (1971) stated that "Queens produced from eggs were 10% heavier than queens whose larvae had been grafted at 24 hours old, (a reduction of 8% in spermatheca volume and 3% reduction in the number of ovarioles also occurred).

I shall now proceed to discuss some of the issues involved in producing a nucleus, on full sized brood combs, for removal to a mating site. Mini-nucs are not required for producing one or a small number of queens. In any case, if the newly mated queen remains in the nucleus until it has sealed brood an assessment of her quality can be made. The nucleus could then be built up into a full sized colony for the winter or, since it is on full sized frames, It could be united to your original colony (remember to remove the old queen) using the newspaper method.

Finding the Queen

If the queen has been marked earlier in spring she should be more easily found, using just enough smoke to keep the colony under control.

If you cannot find the queen, move the hive to one side, place an empty brood chamber on a flat board on the original site and shake all the bees in the brood chamber into it taking care not to shake the comb with a queen cell which you are going to use. By using a flat board instead of a floorboard reduces the number of places where the queen can hide. As the combs are cleared of bees they should be placed, temporarily, in their original order in a cardboard box or another brood chamber. The original brood chamber should be carefully brushed out in case the queen is hiding in some recess, or other. Next, put a queen excluder over the brood chamber containing the bees and place the original brood chamber now containing the cleared combs above the excluder. Replace the crownboard and leave the colony for half an hour or more during which time the bees will move up through the excluder to cover the brood, leaving the queen and drones below the excluder where the queen should be more easily picked out.

Obtaining a Queen Cell

The selected colony should be built up as strong as possible by the end of the third week in May. There are a number of methods of obtaining a queen cell or cells. Here are three:

(1) The colony could be confined to a single brood chamber and fed with the intention of forcing it to produce queen cells due to congestion. Not a very good method if it is your only colony and you are hoping to produce honey from it. You will also have put the colony into "swarming mode" so you will have to give it weekly checks during the swarming period.

(2) The queen could be removed to a three comb nucleus for a week thus inducing the colony to produce queen cells. It must be understood that when the queen is returned the colony will continue to produce queen cells as you will have put it into "swarming mode."

(3) A better method would be:

1. Build up the colony on a double brood chamber with one or probably two supers by, say, 20th May.

2. Find the queen and put her in a matchbox with a few attendants.

3. Sort through the brood chambers, placing all the unsealed brood in one and fill up this box by placing a pollen frame and frames of eggs in the centre and sealed brood on the flanks.

4. The remainder of the brood combs are put in the other brood chamber on the original site and the queen released on to them.

5. The hive should be rebuilt with the brood chamber, then two supers, a second queen excluder, followed by the brood chamber containing the eggs, crownboard and roof. Because the bees in the top brood chamber are isolated from the queen with two excluders and two supers means that the transfer of queen substance is reduced and the bees will usually make a small number of queen cells.

6. It would be good to give this top brood chamber a light feed of thin syrup to encourage the start of the queen cells. Do not overdo the feeding in case the food is stored in the supers.

7. After about four days (the bees may well build emergency cells on three-day-old eggs), select a well provisioned queen cell in a suitable position and not yet sealed. Mark the position of the selected cell by sticking a drawing pin into the top of the frame directly above it.

Preparing Nuclei

1. Prepare a five or six frame nucleus hive, or a brood chamber with 5/6 frames and a dummy. The dummy can be made by nailing a piece of hardboard to a separate deep frame. Make sure that the colony from which the frames are taken is disease free.

2. Find the queen and put her in a matchbox with a few attendants or temporarily cage her on the frame on which she is found to avoid the risk of putting her into the nucleus.

3. Select two combs of mainly sealed brood, plus stores, plus bees.

4. Select one comb of mainly unsealed brood, plus stores, plus bees.

5. Select one comb of food (pollen and nectar/honey), plus bees.

6. With the aid of an uncapping fork and a matchstick, kill all sealed and unsealed drone brood and eggs in drone cells from these four combs.

7. Give one empty drawn comb, if available. If you have a sixth frame it can be of foundation.

8. Lightly shake two other combs over the parent colony to dislodge most of the older field bees, then shake vigorously over the nucleus to dislodge the younger nurse bees which are the ones you want for tending to the selected queen cell if it has not been sealed. Try to avoid shaking too many/or any of your own drones into the nucleus.

9. If the nucleus is to remain in your apiary for a day or two, shake in another comb of nurse bees. Remember that the queen cell will be sealed on the ninth day and should not be in danger of starting to hatch until the fifteenth day. You will still have a bit of leeway if you have already, as suggested, reduced the queen cells to one as the virgin is unlikely to fly until it has been hatched for a day or so.

10. If your selected queen cell is unsealed on a frame, place it in the centre of the nucleus flanked by the pollen frame, then the unsealed brood, then the sealed brood. Dig out a hollow in the comb opposite the queen cell to avoid getting it damaged when the combs are pushed together. If you are cutting out the queen cell from another colony, cut well clear of the cell,

make a hole in the comb where it Is to go and carefully insert it into the hole. Avoid shaking or damaging the selected queen cell. Only one queen cell should remain in the nucleus.

11. Take care to provide a small entrance to deter robbing. Do not feed the nucleus for two or three days until all the original field bees, and hopefully drones, have returned to their original hive otherwise it may be robbed out.

12. Use drawn comb, if it is available, to complete the number of frames in the original hive and do not forget to release the queen into it.

13. Remove the nucleus to the mating site before the virgin queen is due to fly.

4. Queen Rearing using mating mininucs
Taylor Hood

Almost at every beekeeping association meeting I attend, you hear of the importance of keeping local bees that have evolved or are adapted to the local area and weather conditions and that good queens are the basis of good honey production and therefore going forward we need to breed our own queens. We are told, it is easy to do and we should have a go.

I had a go last year.

I split the process of rearing queens into producing queen cells and then using mininucs to get the queens mated before introducing them individually into a queenless colony.

Producing queen cells.

I am told I should breed bees with the traits I want them to have (See Holm's Queen breeding) - supersedure; honey producers; docility; resistance to disease; longevity; and from a recent talk purity of strain may be a consideration. To do this you need to observe the bees over time and rate and record your findings over the year and decide from this information which colony or colonies you want to breed from. The young larvae from this colony are the ones you are going to graft and produce queens from, this is called the *breeder colony*.

To produce good queens you want plenty of young bees to feed and look after the larvae in the queen cups. The time you start the process is therefore important, too early there may not be enough bees to look after the queen cells properly, too late there may not be any suitable drones available to mate with the queen. The beginning of May is a good time to start so the emerging queens are being mated around the last week in May and first week in June.

Day 1

Once you have chosen your breeder colony you need to select the colony that is going to produce the queen cells. You should select a strong healthy colony – seven or eight frames of brood with bees almost to the point of congestion. On Day 1 find the queen and place her with a comb of mostly unsealed brood in the middle of the lower brood box and put a comb of stored honey and pollen on either side of this and fill the box with empty comb. Place a queen excluder on top of this and then the upper or a second brood box on top of this with the remaining brood bees and stores of honey and pollen and then feed with 2 litres of sugar syrup – I feed 50% syrup in a rapid feeder but some books recommend a higher concentration of syrup in a rapid feeder. Any supers should be put between the lower and upper brood boxes.

Leave the colony for 8 days only opening up to check and top up the feeder.

Day 8

On the 8th day open the queen cell producing colony at around 10 am weather permitting, check each frame in the top box for queen cells and remove any – remember this is not your breeder colony. (Any royal jelly from queen cells can be used if you are grafting wet.) Remove a frame from the outside of the brood box and make space in the centre and put in your grafting frame with plastic or wax grafting cells. Try to make sure the frame next to this has pollen. The bees will prepare the cells and any gaps in the wax cells will be repaired ready for eggs. Separate the brood boxes with a divisional board, I use a modified solid floor or a Snelgrove board for this. I make sure the entrance is pointing in a different direction from the lower brood box so I can leave a queen cell in this brood box, knowing that the new mated queen will not enter the bottom brood box by mistake and be killed.

Day 8 – at least 4 hours later.

On day 8 around 2 pm I remove the grafting frame and shake/brush off the bees on the cell cups – this is an indication they have been accepted. If they have not been accepted at this stage the chances of the grafted cells being accepted and extended are significantly reduced. I take out a frame of eggs and young brood from the breeder colony and brush off the bees – do not shake. In order that I can see the larvae and can graft ones of the correct age, they need to be less than 36 hours old i.e. about the size of a coma in a book around 2 mm, I like to take my veil off for this and take the 2 frames into my greenhouse, it can however be done anywhere shed, kitchen, car and even outside as long as you have somewhere you can put the frames down and be able to do the grafting. Grafting is the difficult bit – to be successful you need to take care in slipping the spoon like tip of the grafting tool under the larvae and lifting it out of the cell and into the artificial/ wax cell without rolling it against the side of the cell. If you roll the larvae you will damage it and the bees will remove the damaged larva. If you don't get it right try with another larva. There are a number of grafting tools available and you can even make them yourself out of needles, toothpicks, matches etc. My advice is to try as many tools as you can and pick the one you are most comfortable with. The one I use has a bend in the stalk (see photo).

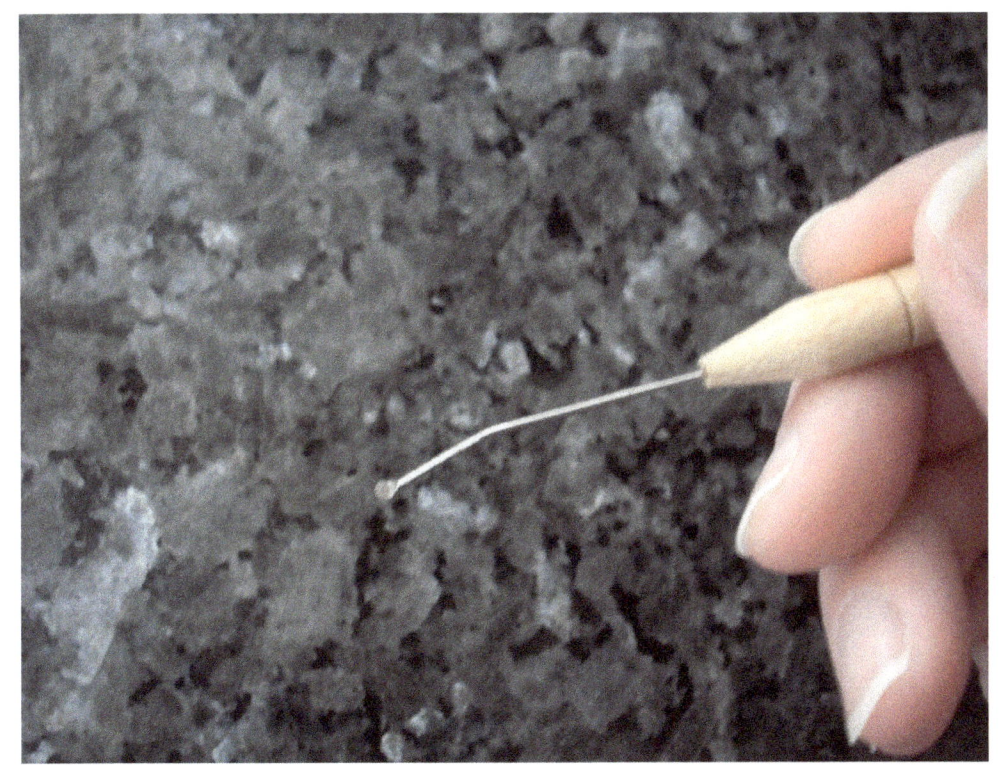

Grafting tool

this makes it easy to see what I am doing when I am picking up the larva and placing it into the cell. I try to use a frame with relatively new drawn comb as this also helps me see the larva. You can also expose the larvae more by cutting down the comb to the midrib using a sharp knife or chisel making it easier to see, pick up and get the larva out of the cell. I try to pick up the smallest larva I can see as this will be the youngest. I am short sighted so can see the larva easily I know other people use a magnifying glass. Some people who sew have magnifying glasses that hang round their necks and these are ideal. You can graft wet or dry wet is where a drop of 50:50 ratio royal jelly to warm water is placed in the bottom of the cell. Or dry when you graft directly into the cup. If you graft dry cover the grafted cells with a wet dish towel and mist the air with a water spray to stop the larvae drying out. I have only grafted dry.

Once the cups have been grafted I place the frame back into the space it was removed from in the cell builder and replace the frame from the breeder colony back into its colony.

Day 9

I check that the grafts have taken and the bees are feeding the larvae and extending the cells. My success rate first time was around 20% of cells being accepted and being extended.

Sandy Cran looking at sealed queen cell on grafting bar

If this has not been succesful you can repeat the grafting process.

Some people may want to try using a Cupkit or Jenter system to produce their queen cells.

Continue to feed the builder colony. It takes 16 days for queens to emerge from queen cells and since the larva are 1 to 2 days old when grafted, the queens should be emerging 11 to 12 days after the day of grafting. Sealed cells therefore need to be distributed on the 10th day after grafting to queenless colonies, nuclei or mini nucs to be mated. This is day 18, and the queen will not be ready for mating until day 22.

I use Apidea mini nucs in preference to the Kieler type mini nucs for queen mating.

Apidea with frame with starter foundation strip

Setting up the mini nucs.

When you get your mini nuc from the supplier, Vaseline all the moving parts so that they slide easily. Add the plastic ventilation grill and entrance/queen excluder to the front of the box and put the feeder queen excluder onto the back.

Make up the plastic frames. The wax channels should be to the inside and make strips of foundation using a Stanley knife to cut the BS foundation to the width of the frames and 1 inch deep. The strips are held on the frames by dropping molten wax on the wax channels using wax and a soldering iron.

With regards to the feeder – use fondant and not syrup and to stop burr comb being produced in the feeder use an 8oz cut comb box with fondant. The boxes slip into the feeder easily. The cut comb boxes makes it easy to replace food when necessary. Syrup, I have found, drowns a lot of bees. If you do not have 8oz cut comb boxes you can use 180g Philadelphia soft cheese tubs with some modification.

Stocking the mini nucs.

I take a mini nuc with a full box of fondant and turn it upside down and open the bottom.

I make sure the entrance is closed with the ventilation mesh open. To stock the nuc you want young bees – they are going to draw comb and feed and look after the larvae. So use bees from supers or shake off older bees from frames before spraying the bees on the frame with water and shaking the bees into a bucket. Spraying with water stops the bees from flying and when you dunk the bucket they fall into a big clump at the bottom. Scoop up 300ml of bees in a mug. (Take an old mug and fill it with water, then measure the volume of mug using a pyrex measuring jug. Many mugs are around the 300ml mark.) 300ml of bees seems to be the right amount for the apidea mini nuc, a heaped mug is required for the kieler nuc ie 350ml of bees. Put the bees into the mini nuc, slide the bottom closed making sure the front is showing the ventilation grill and the entrance is closed. (A Kieler nuc has its ventilation grill on the bottom floor.) Turn the mini nuc over and then, when ready – this may be at another location where your queen cells have been produced - remove the roof and expose the plastic crown board. Lift the flap and place the queen cell through the flap between the frames. If you think the queen cup is going to fall through you can use a plastic queen cup protector with spiky shoulders to keep it at the top.

Apidea, bees and queen cell

For a Kieler nuc box you need to make a crown board and flap. (I used the see through plastic top of the box I bought Christmas cards in.)

I then put the mini nuc into a cool dark place (my garage) for 4 days. Do not do this for less if you are near your home apiary as some bees may return to their original colony leaving the mini nuc short of bees in which case the bees which are left will probably be unable to sustain a new queen and the brood she produces. I spray water 4 times a day through the ventilation mesh. I do this at around 8 am, 1 pm, 6 pm and 10 pm and spray for about 60 seconds. For the Kieler you need to spray the mesh in the floor or you can put a small sponge saturated with water into the nuc. During this time the queen emerges and if you are lucky you will hear her piping.

After the four days, in the evening place the nuc on a hive stand and put a brick on top. Open up the nuc entrance by sliding up the front, this closes the ventilation mesh and the entrance is opened. Make sure the queen excluder is not in operation, you want the queen to be able to get out and mate. It is important that you do not site your nuc in direct sunlight as the nuc will over heat and the bees will abscond. Also I found my mini nucs were terrorised by wasps from early August onwards causing them to become demoralised and in one case to abscond. Do not disturb the bees for at least 1 week then you can check to see your queen, see if she has mated and has started to lay eggs.

Queen and egg on apidea frame

Wait until the cells have been capped to check the queen is not a drone layer then introduce her into a colony in the normal way. I found that by transferring the queen, bees and frames into a mini nuc eke I could unite this to a queenless nucleus or colony using the paper method.

At the end of the season unite your nucs into a normal colony.

The nuc frames can be put into plastic boxes and put into the freezer for a couple of days and then kept in the boxes until the following year. The boxes can be cleaned and then put through a dish washer on a 55 C wash.

I hope you find this article interesting and it helps other newbees have a go. For the more experienced bee keepers please send us your hints and tips on how to improve this process. It would be much appreciated.

Further reading

Queen Breeding and Genetics Eigel Holm

Queen Rearing Simplified Vince Cook

Managing Mininucs Ron Brown

5. A talk and visit with Charlie Irwin on Temporary and Permanent Observation Hives
Taylor Hood

Earlier this year I attended a talk on observation hives by Charlie Irwin and then visited Kelvingrove Museum with him to visit the observation hive.

Charlie is an expert in both temporary and permanent observation hives, he has provided an observation hive at many public events in the Glasgow and surrounding area for decades e.g. the Orchid Fair at the Botanic Gardens, Park Open Days at Glasgow Green and Pollock Park. He has helped groups, e.g. the Bothwell Bee Group at the Bothwell Scarecrow festival, to set up a temporary observation hive for the day. The Kelvingrove Museum permanent observation hive is one of the most visited attractions in the museum.

Charlie is able to speak about observation hives from their origin, from being little more than straw skeps with glass windows, glass bell jars, to the more sophisticated. The Huber leafed framed observation hive – which was very disruptive to the bees. The Nutt Collateral, which was a lovely piece of furniture. The Stewarton hive with slide shutters and windows. The Robb hive which is very difficult to manage and maintain and takes time and patience to set up. Apparently you have to slice down to the mid rib of the honey comb and wax this in sections onto the glass. This allows you to observe the development of eggs, larvae and pupae into the adult bee. Heat lamps at the base of the hive are used to stabilise the temperature, making it easier for the bees to achieve and maintain the temperature for brood rearing within the hive. Charlie tells a funny story of a Robb Hive at the RHS where it was not the development of honey bee brood that was of interest but that of Wax Moth larvae – with one keeking out periodically to see what was going on!

The Beekeepers' Club, Glasgow commissioned Mr Robb to make a hive for them in 1958. However, the hive was seldom used and therefore was given to Mr Willie Robson of Chainbridge. I am not sure if he has it on display, but I know I will be keeping an eye out for it the next time I visit.

At present there are two main types of observation hives used –

The ***Ulster type*** which in effect is a 5 framed nucleus with a display section on top which takes a single frame. The section has access to the lower box but is separated with a queen excluder and thus the queen if put into the display section is always on display. The only drawback with this hive is when the colony in the nuc. box is strong the display frame can become congested with bees, making it very difficult to see what is happening. On the plus side this hive is much kinder to the bees as they are only disrupted at the time of display.

The ***uni comb hive***, this is the type Charlie uses, which lets you see both sides of the comb / frame. This can be a single frame hive, but it is much better if it has 2 or 3 frames, although you can go up to as many as 12 frames if on permanent display.

The observation hive Charlie uses for events is based on the one he used initially, this hive was built by his friend and mentor Gordon Stewart. When Gordon decided to put his observation hive on permanent display in his workshop (Gordon was a cabinet maker), Charlie made one of his own. This is still the one he uses today and is the template for many of the observation hives that have been made in the West of Scotland. The hive is 2 deep and 1 shallow frame in height – it can stand upright in the back of his estate car and is light enough to be easily lifted and transported to events. It has a wide plinth at the bottom which allows it to be clamped to a workbench making it firm and secure when on display.

Sandy Cran standing with an observation hive, promoting beekeeping at a local event

Other considerations when constructing an observation hive are:-

Making sure that all joints are accurate to ensure bees cannot escape.

Clearance between the uprights and frame ends need to be a bee space to stop the bees from gluing them up with propolis or burr comb.

The distance between the front and back glass needs to be $1^1/_2$ inches to accommodate frames that have been drawn on either Hoffman or metal/ plastic ends. If the distance is too small the frames will not fit into the hive and if the distance is too large the bees will produce burr comb on the glass. This is a favourite place for bees to build burr comb and hide queen cells in the observation hive (even when the distances are right) along with the possibility of swarming.

Sufficient ventilation – on hot days the temperature within the hive can build up very quickly, so lots of ventilation holes are extremely important. These holes should be mainly on the top and side of the hive and should be covered both on the inside and the outside with perforated zinc/ plastic. Why inside and outside? This is to stop the bees from putting their stinger through the mesh and stinging someone as happened at a recent Highland Show.

Ventilation holes can be used to produce an entrance / exit hole for the bees. This is generally at the bottom – making it easy for the bees to take out detritus/ dead bees.

Ventilation holes on the top can be used to feed bees syrup from jar contact feeders.

These ventilation holes should only have the perforated zinc/plastic on the outside. Bees don't like drafts and have a tendency to propolise up the ventilation holes, however they need to be kept clear. When bees eat and process honey to produce brood and to maintain the temperature in the hive, a lot of moisture can be produced which can cause damp and the combs to become mouldy. The holes need to be kept open to allow air to circulate and stop this from happening.

6mm Safety glass or polycarbonate needs to be used for safety reasons. Charlie gave 2 examples why this is necessary. At a Park Event at Glasgow Green a young boy looking at the observation hive decided he was going to punch it to see what would happen, luckily to no ill effect to the boy or bees. At another event a young woman decided she was going to throw a brick at the hive – on this occasion the brick bounced off the hive. If ordinary glass had been used a different outcome, detrimental to both the bees and those present would have occurred.

Temporary observation hives to be used at events should be set up the night before or on the morning of the event, the queen should have been already marked to make sure she is on one of the frames taken for display. Charlie uses a colony which has been set up at the beginning of the year for this purpose. Using a hive for display purposes is very disruptive to the development of the colony and therefore you are sacrificing the chance of getting honey from this colony. The bees are out of the hive generally only for a few days and so the observation hive is in effect a sealed unit – with no bees entering or leaving. However Charlie has the option to keep the bees in an observation hive longer by adding a pipe to a ventilation hole to allow the bees to fly. He keeps the observation hive in one of his out buildings. He has found $1^{1}/_{4}$ inch waste pipe works well. The pipe has to be long enough and high enough to ensure bees exiting are not a nuisance to people. A piece of string is put through the tube so the bees can run up and down it as they slide on the plastic.

It is also important that the bees have sufficient food, so having a ventilation hole that a contact feeder can be attached to is important. Charlie also gives bees water during events. Water is dripped through the feeder / ventilation hole onto the top bar, this water is quickly taken up by the bees. If the bees show signs of overheating, i.e. bees running about on the glass, fanning a lot, the hive is put into the shade and water is given to the bees.

Permanent Observation Hives.

Charlie has looked after the observation hive at Kelvingrove Museum since 1985 and was instrumental, after the refurbishment in 2006 in the construction of the current observation hive by the museum carpenter from plans by Bryan Hateley.

Like so many others from the West of Scotland I have searched out the bees at Kelvingrove Museum on visits as a child and then as an adult. So when Charlie gave me the chance to tag along with him on one of his visits I agreed very quickly.

The bees are kept in the museum all year round, so there are a number of considerations that Charlie monitors.

Considerations:-

Food levels

Quantity of bees

Suitability and state of queen

Health of colony

Ventilation

Weather

Manipulations

Swarming preparation

Food levels are very important for the survival of the colony. The bees are continually fed syrup throughout the year as there are not sufficient bees to produce sufficient stores for the colony to be self sufficient. Comb frames of honey are added when the levels of stores drop below a critical level and colony survival is at risk. (This was the reason for our visit.)

Quantity of bees is important as too many can lead to swarm preparation and too few impedes colony development and survival. Charlie monitors the number of bees in the exit/entrance tunnel and if too many, he takes out a frame of bees and adds an empty frame. If by doing so there are not enough bees in the hive he can brush some from the frame being removed back into the hive. When adding bees he uses a lot of smoke to minimise fuss and fighting.

Suitability/state of queen is important. What is wanted is an older "half knackered" queen. Young queens produce a lot of bees, so you are continually having to remove bees and brood as they build up quickly. Too old and the queen can die, usually in the winter as happened in Kelvingrove a few years ago. During this period bees had to be added to the hive until a satisfactory queen could be found and introduced successfully into the hive.

Health of the colony – if the colony becomes diseased the colony needs to be treated or replaced. The Kelvingrove bees had an issue with chalk brood and so the whole colony was replaced. With regards to varroa, the Kelvingrove bees have never been treated. Charlie believes that the varroa cannot thrive or survive the 2 to 3 month break in brood rearing within the hive – with the majority if not all the phoretic mites dying during this period.

Ventilation holes need to be kept open to stop condensation which causes the combs to go mouldy. Ventilation is also important in hot weather to allow the bees to regulate the temperature within the hive.

Weather – At Kelvingrove there are blinds to shade the hive from direct sunlight and a blanket cover with ventilation holes to keep the bees warm when the museum is closed particularly on cold winter nights.

Manipulations are carried out after 4pm when the museum is closed to the public. Most of the manipulations are carried out at the back door, weather permitting. Only once has Charlie carried out a manipulation in the museum – he carried this out in a clear plastic bag to minimise the loss of bees and risk to staff/visitors. This is something he hopes he never needs to repeat.

Swarming – The hive is checked weekly during May, June and July for signs of swarming and appropriate measures taken to prevent/control swarming. Sometimes the bees manage to produce a queen cell undetected. Charlie then gets a phone call from the museum usually telling him that there is a swarm cluster in their favourite tree near the museum.

On the day of our visit, Charlie parked his car at the loading bay. We entered by the back door and collected the special screw driver from museum staff to remove the security screws from the entrance tunnel and plinth of the observation hive.

Photograph of Charlie Irwin removing the Observation Hive in Kelvingrove Museum

A trolley was procured and we proceeded to the observation hive. Screws were removed from the plinth, the tunnel disengaged, and the 2 ends of the tunnel in and out of the hive were covered with cardboard plates so that no bees could escape out the hive or into the room. This sounds easier than it really was and I am glad that Charlie did this so expertly. The hive, which is heavy was lifted onto the trolley and taken down (using one of the biggest lifts in the world) and out onto the loading bay. Outside on this cold afternoon 16 screws from the beading were removed and the glass front moved slightly forward allowing a deep frame to be removed, again not as easy as it sounds and a deep frame full of honey placed into the hive.

Photo of Charlie Irwin opening Observation Hive out in the loading bay

The bees that were on the removed frame were quickly shaken and brushed back into the hive, the glass edges of the display front were covered in vaseline to ensure easy access in the future and then closed up. Very few bees had taken to the wing and those that had hopefully made it back into the hive via the normal entrance. The hive was re-instated – a reverse of taking it out and then checked to ensure it was safe and secure. The whole manipulation took just over an hour.

Charlie checks the bees/ observation hive every few weeks, weekly in the swarming season and carries out manipulations as and when necessary. He tries to minimise disruption to the colony and its development, the hive is only opened 3 or 4 times in a season.

From this visit and his talk I now know much more about observation hives, on how to set up and maintain them and I found it very instructive watching Charlie carry out his visit and the hive manipulation at the museum.

Hopefully this article will go some way in recognising the great work Charlie has done over the years both at Kelvingrove Museum and at events around Glasgow. It is a great introduction and education to ordinary people regarding honey bees and the benefits they generate. Few people forget seeing honey bees in an observation hive for the first time and for some it is the start of a lifelong interest. Charlie has helped many of us in this first step.

Further Reading

Building, Stocking and Maintaining of Observation Hives Bryan Hateley

The Observation Hive Karl Showler

6. RAMBLINGS
Eric McArthur

By RAMBLER 1978
The pleasure of Beekeeping

The magical key to the appreciation of nature at work is our involvement with bees. Of what is this attraction made to the dedicated bee man, the bee is a combination of mystique, respect, challenge, admiration, apprehension and total fascination - this fascination once discovered never varies - even after years of keeping bees at all levels. The sheer joy of observing activity at the hive entrance or opening a hive to check for a particular aspect in it is as real to the man with 2000 colonies as it is with the hobbyist who is quite content with his minimum of 2. Ask any bee man to give up his beekeeping! - might as well request that he amputate a limb - beekeeping and bees become part of the beekeeper's fabric of life. Think about it - life without your bees - not a pleasant thought! Our good fortune as beekeepers manifests itself as a gradually increasing awareness of the experiences which the natural phenomena already spoken of have on the lives and activities of our bees - water, too much of it in the form of inclement weather affects the activity at the hive entrance - with observation we learn to recognise how much rain will keep the bees confined - we learn to our astonishment that contrary to 'lay' belief, bees will fly during quite heavy rainfall on particular occasions such as for example during a nectar flow. I have seen bees working the **lime** in July in weather no self-respecting cat would go out in! - **air,** in the shape of wind, gales, cyclones - high speed air is a pest - besides blowing box vans and caravans over on the Tay Bridge - it also effectively grounds our bees - again by observation we find that our bee will, if not happily, work in winds up to 15 m.p.h. - above this velocity they stay at home **- sun,** in balanced amounts, the giver of Life, without the sun life as we know it would cease to exist - everything that lives and breathes, plant or animal, depends on the sun. As beekeepers we find ourselves concerned with the sun as the people of the land are concerned - the sun is much more than merely the difference between a holiday being simply marvellous, and a dreadful washout. The need for the sun is more superficial than real for the 'lay' holidaymaker. The far-reaching effects of too much sun or too little sun are lost to a 'lay' person until he/she is told, don't use a hose to wash your car "or that the price of bread will have to go up" because the cereal crop was bad. We as beekeepers know that for example for our purposes an arid sunny June will result in nectar sources like **lime** failing in July, but by the same token clover will yield heavily due to this sunshine up to the point where the ground moisture dries up. Very hot sunny weather lays a great strain on hives which are in an exposed location. Observe a hive in the heat of the day and it will be noticed that the front and landing board are practically covered with bees all pointing head down to the entrance and fanning their wings as if their very lives depended on it, and it does because they are pumping ventilation air into the hive where other bees are likewise fanning to create a stream of cool air flowing over the face of the combs to keep the brood and eggs at the correct temperature and humidity and to stop the combs from collapsing. Bees were among the first physicists discovering early that evaporating water will greatly reduce heat levels. This principle is used by bees to help lower the hive temperature. So, by observing a natural phenomenon intelligently we can help our bees by supplying them with water and perhaps be instrumental in saving the colony. To what avail keeping bees without at least some basic knowledge of botany. The enquiring beekeeper will observe activity at his hive and ponders from where the pollen so brightly hued and tightly packed in the corbicula of the hind tibia originates. His/her curiosity will lead gently towards an awareness of plants which previously had been looked upon but not seen.

During the course of the active season for bees the observant bee-keeper will notice variations in the level of activity at his/her hives. To the beginner this activity is initially merely an interesting sideshow to keeping bees. This movement around the hive tells so many stories and a knowledge of the plants growing within a half-mile of the apiary will make the understanding of the stories easier. In the early spring where there are snowdrops around, the early flight of the bees in late January will reward them with orange coloured pollen and some nectar, where there are crocuses in the vicinity there will be masses of orange/deep yellow pollen pellets being carried. Later where there are willow or alder trees the pollen will change to a greenish/yellow colour. Where there are gean trees (wild cherry) the pollen will be tan hued. The next major source of nourishment for bees and the beekeeper is the sycamore with its greenish coloured pollen and then the chestnut, virtually co-incident with the sycamore. White horse chestnut pollen is deep orange/brown coloured but the later variety of the red horse chestnut produces a beautiful deep red pollen. Hawthorn pollen is invariably available around early June, this pollen is a creamy white/greenish hue. Clover is never far behind the hawthorn and is brownish grey in colour, except where it is derived from the red clover where it is a darker brown hue. Lime is the next major tree to bloom producing a yellow hued pollen. Then, for those of us who are prepared to put that extra bit of effort in at the end of the year there is the pollen of the heather (ling) which is greyish brown in colour and provides the final major source of protein for the bees to store for the colony build up next spring. This pollen list is in no way intended to be understood as exhaustive — but it does represent the major sources of pollen available to bees in the West of Scotland. The various pollens of wind pollinated trees like elm, oak and beech are also useful early sources of protein, but these trees secrete no nectar. A glaring omission in the catalogue of pollens given is the dandelion with its beautiful orange yellow pollen and for a long time looked upon as an abundant, reliable source of protein for larvae nutrition. Alas research in the past few years has demonstrated that dandelion pollen lacks a particular protein which renders it useless for brood rearing, if it is only available on its own, although as an additive to other pollen it plays an important role. During the early part of the year — even around mid-March when the bees are gathering pollen well — on the occasional 'better' days of that month — it is good beekeeping practice to begin supplying the bees with very light sugar syrup, proportions of 1 lb. sugar to 2 pints water are ideal. By introducing light syrup at this time, the need for the foragers to gather water will be virtually eliminated. Water is very necessary for brood rearing and this syrup feed will be used by the colony as its needs dictate. A word of warning — once sugar syrup feeding has been started it must be continued until the first major nectar flow occurs, this first flow is invariably from the sycamore in early May. Observing pollen carriers at the hive 'reports' on the actual condition of the colony being observed — but to be of real benefit to the beekeeper the observation should be of a comparative nature — the 'one hive' beekeeper sets him/herself at a tremendous disadvantage — since the assumption that the hive is working normally for the particular season must be made. Where at least two hives are present the activity at one can be easily related to the activity at the other provided of course that both hives are in the same apiary. In early spring after the first cleansing flights, a simple observation will demonstrate the need for water in the hive — where pollen is available and being carried — if the greater proportion of the bees returning to the hive are not carrying pollen, these bees will be carrying water and nectar — give them sugar syrup and the proportions will alter — the feeder should be given right on top of the frame tops so that the bees can move up under it without breaking the cluster —as they would be forced to do if the feed is given on top of the crown board. An empty super placed on top of the brood chamber will 'house' the feeder. Another useful simple observation is comparing the number of bees per minute carrying pollen (large pellets!) at each hive. If all hives are working well and the amount of pollen carried is not markedly different comparing one hive to the other and the pellets are large, everything is as it should be. But if any hive appears to be less industrious or if pollen is not being carried in the same abundance as at other hives then something is amiss — and this hive should be examined at the earliest opportunity — and certainly no later than the first week in April. If the hive is still quite strong but queen less — i.e. no brood or eggs present — a frame having emerging brood and eggs can be given and with luck, if the spring is favourable, the bees will rear another queen and she could be laying by the end of the first week in May. Leaving the initial examination of the colony till later on wastes valuable time and will generally result in loss of the colony or at best requiring it to be united to a queen right colony using the "newspaper" method. Copious amounts of pollen being carried is a marvellous indicator that the queen is laying well and brood rearing is proceeding apace. If the spring is favourable and steady build up is assisted by sugar syrup feeding — the hives can be brought to honey gathering strength more rapidly than in 'let alone' beekeeping — and instead of the colony using the first major nectar source for brood rearing, leaving no surplus for the beekeeper — the bees raised with the help of sugar syrup will, weather permitting, provide the 'involved' beekeeper with early honey.

Ramblings
Simple Management

Knowing when the first major early nectar source blooms can be used as a starting point to work backwards in time to the point where sugar syrup feeding does most good. From egg, first laid, to foraging bee, there is a development time of six weeks, approximately. So, if the first major nectar source blooms during the first week in May (give or take a week to 10 days for seasonal variation), then feeding will do most good if started around the first week in March. This action could be looked upon as the first positive step taken by the beekeeper into actively working "in parallel" with his bees, rather than merely annoying them with aimless prodding about inside the brood nest at indeterminate intervals fixed more by the advent of a sunny day than the need for planned regular inspections. Apart from the beekeeper's first season in beekeeping (where confidence in handling bees must be obtained by practice — i.e. working through the hive regularly every three or four days until working a hive becomes almost second nature) — by working the hive means not merely pulling the brood nest apart and looking unseeingly at the combs, but going into the hive with a plan, something simple to begin with, perhaps to estimate the total comb surface showing sealed brood (each square inch has approx. 25 cells, 50 counting both sides of the comb). Then perhaps next time to estimate the area of open cells containing pollen, next estimate how much comb area has sealed honey. After each operation close the hive and be satisfied. These simple tasks will give point to the initial 'timid' attempts at hive inspection. After confidence has been gained by regular inspections, estimations of the proportions of drones to workers in the hive can be made and will help sharpen observation in the beginner. The beginner starting with a nucleus can 'grow' with his bees during the season without the anguish of having to deal with the 'swarming urge' of the bees in his/her first season, a nucleus will not normally swarm in its first season. Nuclei are also quite pleasant to handle relative to a fully developed 'bustling' colony at 'hoarding strength'. If the beginner keeps notes of his/her, early observations with dates of inspections he/she will gradually build up a solid base of good practical experience — for which there is no substitute. Of course, in the first season the beginner must accept that he/she will be called on to sacrifice a honey harvest in the interests of gaining experience — I can assure you though that the sacrifice is well worth making — if in working bees regularly a skill and confidence is built up which will remove the "terrors" of working bees in all weathers during the active season, in the many happy satisfying beekeeping years to come.

In working bees the use of smoke is perhaps the most important aspect of management — bees normally just cannot be managed without smoke — there are of course strains of bees which are very docile, but these same docile bees, in particular weather conditions can be as wild and vicious as the meanest *Apis adansoni* (The "Killer" Bee). Using smoke is an art and different hives need different treatment, only with practice and experience can a beekeeper learn when to use the right smoking technique — generally speaking on bright sunny days in summer a little smoke blown into, the centre of the hive through the entrance will suffice (the smoke must enter the hive — puffing around the entrance is not satisfactory!!). On dull or wet days, it is better to err on the 'heavy' side and give a good deal of smoke — and on days where there is a threat of thunder — even setting the hive on fire won't keep your bees from going mad; bees and thundery weather do not mix at all. The bees do not like low atmospheric pressures. Hive inspection in such weather must be left to the highly experienced and well protected beekeeper who will have a particular and excellent reason for making such an inspection. To recap, we have been feeding light syrup from approx. mid-March — in areas where there is a sycamore flow! The colony will begin to grow - it is good beekeeping to note the number of frames which the bees are covering — i.e. the number of spaces between combs, filled with clustering bees. If this is done at the start of the feeding programme, the progress of development can be followed remarkably easily. It is also most satisfying to watch a colony develop from perhaps only covering four combs to growing right across the brood box. Where two or more hives are worked, a development comparison can be made and if any hive seems to 'lag' in development relative to its neighbours it should be closely observed during the active year (providing of course that it is indeed queenright!). A colony which develops slowly in the spring and early summer should not be discounted out of hand, so long as its development is progressive. In an area where heather honey is the major harvest the slow steady developing colony is ideal since it will probably reach its peak of development to coincide with the late nectar flow. These 'late developers' are very useful sources of genetic material — since drones flying from such a colony will provide a stabilising effect on the highly prolific "brood producing" strains. There is of course a marked difference between a steadily developing "slow developer" and a colony where the queen is laying poorly. If a colony is not growing or is even showing signs of diminishing in size, then the queen is a 'scrub' or past her best — this hive should be re-queened as soon as possible. By good management where the beekeeper has more than one hive, slow developing hives (by the previously mentioned definition) can be brought up to 'hoarding' strength for the early flow quite easily. The timing of the procedure is quite critical and a good knowledge of the natural history of the bee is very important. In the writer's opinion a good grasp of how the honey bee 'ticks' is worth much, much more than the ability to memorise half a dozen different methods of working bees without really comprehending the underlying principles of the various systems — " Know your bees, read your books . . ." (to the tune of Ol' Man River!!). Before any manipulations involving interchange of bees, brood, combs or equipment are made between colonies of bees, the bees must be checked out to confirm they are free of disease. Assume the beekeeper has two hives, one of which is a "fast breeder", the remaining hive being a "slow reactor" let the colonies build up progressively. In a good spring a "fast breeder" can be ready for a second brood box by mid-April (this will be discussed later). Assume each of the 'fast' hives has seven frames of sealed/unsealed brood at a point in time about three weeks before the first major nectar source in your area. Assess the size of the cluster in the 'slow' hive, taking into consideration the ambient temperature at time of inspection — a warm day will give a false impression of cluster size since the colony will "expand" to take in a larger "volume" depending on the surrounding temperature. The safest time to assess cluster "volume" is in the evening when it is cooler — and all the bees are "home in bed." If the colony to be reinforced is covering not less than four frames at this time, it can quite readily take a frame of sealed brood into its brood nest without danger of chilling. The frame of sealed brood should be added to the colony being reinforced on the first bright day when the bees are flying freely. By making this transfer around three weeks before the nectar flow is due, the colony receives a quite massive "blood transfusion" of approximately 4000 - 5000 young bees which will result in a virtual explosion in the brood nest increasing the number of combs the bees can effectively cover in the brood nest at a time highly critical for colony development. These new bees can assume the 'nursing' duties of other perhaps slightly older bees nearer foraging age thus releasing these bees for foraging earlier than otherwise. The greatest limiting factor to rapid spring build up strain characteristic apart, and all other things being equal is cluster size. The queen can only lay as many eggs as the cluster can comfortably 'brood'. Thus, by making bold management decisions at the correct time the bee colony can be assisted in reaching its peak at the time when it will do most good. The aim of good beekeeping should be to work with the colonies so that they are at their peak foraging strength for the nectar flows which occur in their area — and it is up to the beekeeper to get to know his major sources of nectar and when these sources bloom. It is of course very difficult to attain 100% success with beekeeping management, but the secret of success is to have a plan and work toward making it work. Your plan may not work out, it may not even be feasible but by trying you learn and there is nothing so satisfying as working toward a goal like preparing say for the sycamore in the early spring and perhaps the lime or heather at the end of summer and succeeding in getting colony strength right, to coincide with the bloom and then getting good flying weather to crown your efforts. If the all important weather turns out unfavourable your consolation is that at least you did your homework — and if the weather had done its bit — who knows? By the same token if there is no planning for the nectar flows and the weather turns out marvellous (it sometimes does — really!) and you miss getting the best from your hives, you'll feel like kicking yourself for neglecting to prepare. Crime and Punishment!!

For the beekeeper who is in an 'early' area i.e. where gean, sycamore, chestnut and hawthorn occur in significant numbers the best system of management (using British Standard equipment!) is to maintain the colony in a single brood box, adding supers as required. The secret of working a single brood box is "keep the brood nest clear" — the brood nest being that part of the bee cluster where eggs and brood exist on the combs. During the build up period as the brood nest expands an effort should be made at the regular weekly or nine day inspections, beginning mid-April (for the beekeeper in regular full time employment the weekend is the obvious choice). To ensure that there is no break in the continuity of the brood nest any combs having only honey and pollen should be moved to the outer limits of the "brood nest" if the brood nest is maintained in a state of steady increase swarming may be put off until the first nectar flow has been successfully exploited — for the beekeeper!

For the beekeeper who keeps his bees in a 'late' area, i.e. where clover, lime and heather (ling or bell) occur in significant amounts. Again, relative to British Standard equipment, the most effective system of management is to let the colony build up steadily in a single brood box. Then when the bees are covering five combs of brood, remove one frame having brood, eggs and bees, close combs up and insert empty drawn comb in brood chamber to replace frame removed. Place another brood box on top of the parent brood box. Place frame of the brood, bees, and eggs in position above centre of brood nest in lower brood box and fill top box with drawn comb for preference but foundation will suffice if no drawn comb is available. We now have a colony in a double brood box the lower box having four frames of brood and eggs, the top box sharing one frame of bees, eggs and brood above the existing brood nest and empty frames filling the space remaining in the upper box.

The optimum position of the brood nest in early spring is right in the middle of the brood box. The reason being that, as the colony expands, if the brood nest is at the side of the brood box the colony can only expand in one direction due to being restricted by the side wall of the brood box, but if the cluster is occupying the centre frames the cluster can move outward symmetrically, this applies to both single and double brood chamber management. In working with a double brood box the same maxim applies as in single brood box management - keep a clear brood nest. The ideal method of working a double brood box system, using a regular inspection cycle is thus: As the colony expands the queen will tend to prefer laying in the top chamber (unless in particular good years where she will utilise top and bottom boxes quite naturally, providing of course she is a good, healthy fecund lass!). At a time when the top chamber has at least five, but less than seven frames with brood and eggs, it will be noted perhaps that the lower box has quite a lot of frames which are loaded with pollen and a little honey, but clear of brood — remove these combs from the bottom box and put all frames from the top box which have large areas of sealed brood on them into the bottom box ensuring of course that they are within the bounds of the brood nest leaving frame with eggs and larvae in the early stages in the top box. By working in this way, over a period of weeks the bottom chamber is turned into a 'power house' of emerging new bees and the top box is maintained free and clear for the queen to lay her heart out. The satisfying aspect of this system is that the beekeeper is involved and, we hope in control. As the colony builds up to its peak any deep combs in the hive having only pollen and sealed honey may be removed, undrawn foundation being inserted in its place in the top chamber. At a time when nectar is coming into the hive if the beekeeper replaces removed filled combs with drawn comb the bees tend to deposit nectar in the cells and thus inhibit brood rearing. By using foundation the beekeeper accrues a double advantage since where nectar is coming in the bees will draw the foundation given to produce straight, true, beautifully drawn comb in which the queen will lay and the nectar coming in will tend to go up into the supers which for preference should contain drawn comb, which will lure the bees up that little bit sooner than if they have to work foundation in the super before being able to store nectar. For the beekeeper who is fortunate enough to keep his bees in an area which is both 'early' and 'late' i.e. having a good proportion of Lime trees as well as Sycamore etc. If he is strong willed, sensible, involved, dedicated and possesses all the other positive attributes of the perfect beekeeper — i.e. he wants to get as much honey and pleasure out of his bees as possible — but not necessarily in that order. The "strong willed" aspect enters where the beekeeper is asked to sacrifice the development of some hives for the benefit of other selected hives. Working on the proven maxim that one strong hive will produce around four times as much honey as a weak hive we proceed as follows:—

Observe colony build up in the spring, from around 1st week in March — while feeding light syrup. Where a number of hives are worked, locate the colonies which are showing the most rapid development — the idea being that half of the colonies will be assisted in their development from a particular instant in time (related to the E.T.F.F. — Expected Time of the First Flow). We know that six weeks are required from egg to foraging bee, we know that about $1\ ^1/_2$ weeks are required from sealed larva to 'nursing' bee and we know that the bloom period of any particular nectar source is around three weeks. So, at the very outside we have a period of perhaps 8-9 weeks in which to prepare for the best return from any particular hive or nectar source. By feeding light sugar syrup when pollen is being carried by the bees, colony build up is encouraged.

Using the fact that by introducing brood which will very quickly emerge to perform 'nursing' duties — the slightly older bees will be encouraged to forage sooner. Then for every frame of brood added to a hive being reinforced, after 1 $^1/_2$ weeks have elapsed there are 4000 -5000 extra nurse bees available to the colony for further brood rearing — if reinforcing is done on a weekly basis starting mid April — ensuring that at no time the hives supplying the brood have less than 1 frame of sealed brood and bees covering at least three frames. By the time the early flow is finished (including the Hawthorn), and assuming that the final addition is made at the end of May, the hives being reinforced will have been given at least 32,000 extra bees. The mathematicians among us will now be calculating how the hives being sacrificed can stand this loss, the answer is quite simple. Reasoning that any eggs laid by any queen after a particular date i.e. five weeks before the end of the 'early' nectar flow will produce bees of foraging age too late to be effective as foragers on the early flow, we can utilise these too late eggs in the 'donor' hives - thus, at a point in time perhaps five weeks before the end of the first flow — say 1st week in May — for each frame of sealed brood removed to the box being reinforced — take a frame of eggs and open brood from the hive being reinforced and give it to the hive losing its sealed brood, so long as the bees in the 'donor' hive are covering the three frames already recommended — by the middle of May, the danger of 'chilling' is minimal — we hope the summer has begun by then!! By switching the frames in this manner, it is possible to get bees at the right age, in the right place, at the right time. The continuous 'stream' of brood emerging in the 'reinforced' hive gives the queen enough room to lay. The 'donor' hive which is of course also being fed sugar syrup steadily during the exchange will be able to rear the extra brood 'donated' if the 3 frame rule is not ignored. The foregoing is the optimum system — for the perfect weather pattern (perfect for beekeeping that is!!) even in years where the weather is not optimal — the normal years, the system will give good results to the beekeeper who is prepared to put in that extra bit of effort. A word of warning — the hives being 'sacrificed' should have the entrances reduced to approx. 3 in. wide where the brood nest is located in the brood box, to minimise the chance of robbing. The recommendation that the exchange of brood frames is stopped around the end of May is not without point. Since the Lime blooms around the second week in July we must now prepare for the Lime flow — all hives being worked for the Lime should be on double brood boxes from at least the beginning of June — any early honey should be removed from the hives in areas where the 'June Gap' occurs and the bees should be fed until nectar is again available to them. If the early honey harvest is left in the hives the bees will use this 'surplus' for brood rearing, it is far cheaper to feed sugar where necessary — there being of course no supers on the hive during this emergency feeding. The advantages of working with double brood boxes are many, but there is also a big disadvantage in that the time required to examine a 'double' especially where swarming preparations are underway is considerably greater than in 'single' management. Despite all that has been urged in previous pages the beekeeper himself / herself will soon find out which method suits himself / herself. Anyway, to get back to the advantages of 'doubles' if worked properly — with a good strain of bee, swarming can be hindered greatly due to the vastly greater comb area for brood rearing. This large brood area of course results ultimately in a massive population which can be quite daunting for a new beginner to handle — (use plenty of smoke — wait at least two minutes before even touching any part of the hive — then away we go!!). In working 'double' brood boxes, as already mentioned in these pages, it is sometimes necessary to remove deep frames which have sealed honey and pollen — but no brood — to ease congestion in the brood nest. There is a strong school of thought in beekeeping circles (we are all entitled to our opinions!) which advocate that honey should never be extracted from brood combs for sale (or ever!). The reasons given being that honey coming into contact with cells used for brood rearing will be tainted or discoloured due to contamination from the cocoon residue in these brood cells. With all due respect to that theory — it has been pointed out by scientific observers that virtually all the nectar coming into a hive during a nectar flow goes into empty cells in the brood chamber first then later in the day, when foraging has ended or has ceased to be perhaps quite as hectic, the nectar is moved by the bees to where it will undergo the 'ripening' process before being ultimately sealed for future use. Taking honey from the brood chamber must not of course be done lightly, the beekeeper must be aware of his bees' needs — but honey removed to ease congestion is, so to speak 'fair game' — extract it, eat it yourself (you deserve it!), sell it, give it away — it is perfectly good stuff — and I have lots of satisfied customers who can differentiate between Sycamore, Clover, Lime and Heather honeys effortlessly but they can't tell me which honey came from the brood chamber — and neither can I for that matter!! Brood combs removed to ease congestion can very readily be given to other colonies (bearing in mind the disease factor already mentioned) which perhaps are in a different area which is not doing so well, or they can be given to a nucleus stock to boost its morale — but really sugar will do just as well. Where the 'surplus' brood combs really come into their own is if they are stored in a sealed dry box or tin over the winter period — they can mean the difference between a colony dying of starvation at the end of the winter period or that colony pulling through. To have a few such combs available for such emergencies is very reassuring. More about this later! So we now have our bees working away steadily in a double brood box hive, preparing for the Lime flow, the colony strength is reaching a peak the hive is packed with bees, eggs and brood in all stages. The Lime is due to bloom in about seven days, the supers are in place and the bees have even begun to draw the foundation in the centre frames and nectar is being stored perhaps from the clover or even the dreaded Privet (some people actually quite like Privet honey where it has a dash of Raspberry or Clover nectar blended [by the bees of course !!] into it). The weekly inspection takes place then PANIC! The brood seems cold, DISASTER! the end of the world — Help me . . .!! The first swarm cells, have been built! The colony is getting ready to give birth! (Many beekeepers get the same feeling around this time!) The beekeeper working the double brood box system now comes into his/her own—no panic, no fears of loss of honey harvest, in fact the educated 'double man' has been hoping for just this event to complete his management cycle to prepare for the Heather flow in August.

SWARM CONTROL WITHOUT TEARS

The experienced beekeeper knows that any hive which has overwintered from the previous season will normally make preparations to swarm at some time during the following summer. Depending on various factors, the event may occur sooner (in May!) or later (in August!). The experienced beekeeper also knows that not all hives which show swarm preparations will actually swarm — but he also knows that at the present level of beekeeping knowledge there is no way he can tell for sure if any particular hive means business or is merely bluffing — so every hive which shows the swarm urge (i.e. builds more than 2 or 3 queen cells— more about this later!) should be assumed to be about to decamp. There are more methods of swarm control in existence than there are words on these pages — every beekeeper has his own variation — but if the underlying principles of the swarm urge are understood the beekeeper is better equipped to deal sensibly with this major problem (for the beekeeper!!). Swarming is the honey-bee's method of reproducing its own kind, a colony will continue to issue a number of secondary swarms or casts after the prime swarm with the original queen has left the hive. These casts are headed by virgin queens and depending on the strength of the swarming colony, the colony may ' throw' a number of casts — before the "Spirit of the Hive" (i.e. the collective instinct of the remaining bees!) decides to select the next emerging queen as the new 'Mother' of the colony — killing any remaining un-emerged queens in their cells.

So, our last inspection has told us that swarm preparations are underway (i.e. large numbers of queen cells are present!). "What to do and why?" The most effective way to check a prime swarm from leaving the hive is to deprive it of its queen. A swarm may issue from a hive without the queen actually being present in the swarm. That swarm, like any other normal swarm will settle quite close to the hive perhaps on a tree branch, fence, old upturned bucket — in fact on anything it can effectively cling to and regroup on. But when the bees realise that the mother queen is not present they will return to the parent hive, of their own volition relatively quickly. A method of effecting this condition is of course to render the queen incapable of flying by clipping her wing(s). The disadvantage of leaving the hive until the swarm is in the air is that the clipped queen will invariably be lost since she will merely fall to the ground when she attempts to fly and will wander off into grass or weeds. Going back to the statement — "to deprive the hive of its queen." We should do just that — and at the same time as the queen is removed all the sealed queen cells in the hive should be cut down. At the next inspection the following week all the combs in the hive must be checked thoroughly and all queen cells cut down leaving only one queen cell preferably sealed — the best cells are the ones on the face of the comb near the centre of the brood nest. Don't be tempted to retain the cells built at the bottom of the frame, or the ones built just under the top bar of the frames — or queen cells built in 'the proximity of drone cells — experience is the only real criterion in queen cell selection! When the queen is removed from the colony she should be placed in a nucleus box, or spare brood box, with the frame she is on (provided this frame is not a newly drawn comb containing only eggs!). Ideally the queen should be housed in her new home on 3 frames (virtually a 3 frame nucleus!), one frame having brood in all stages with a good covering of bees, and two other frames having pollen and honey and clear of bees (at least 2 lb of stores/comb) — shake another frame of bees into the new nucleus and close it up — the frames being secured to facilitate removal to another site, at least 2 miles away from the apiary in question. The nucleus thus treated will soon begin to develop, as the queen returns to her laying condition, and with good management will reach quite significant proportions. More about this later!.

Back to the de-queened hive, which had all its sealed queen cells cut at time of removal of queen. Assuming the colony so treated has been housed on a double brood box, the formation of the nucleus described will not really affect the performance of the now queenless colony (with a colony housed in a single box the loss of bees could be critical for the colony performance due to the relatively smaller population in the single box!) The double brood box colony should be treated as follows when swarm cells are discovered: Go through the hive and find the queen, put her in a nuc box as already discussed. Now, starting again with the bottom brood chamber, examine all combs thoroughly, shaking bees from the combs if necessary, to ensure no sealed queen cells are present and cutting down any found. Any frames in the bottom box found to have only honey and pollen should be removed altogether, and placed temporarily in some bee tight container. The only combs to be retained in the bottom box should be more than 50% filled with brood in all stages or eggs. Now deal with the top box which has meantime been laid aside and covered. Examine the frames for queen cells, cutting down all sealed cells. When a frame has been satisfactorily examined, if it contains brood and eggs, place it in the bottom box. Treat each frame in the top box in the same way. At the end of the manipulation the bottom brood chamber should now be full of the best of the frames containing brood and eggs from the original hive, all the remaining combs should have the bees shaken from them into the bottom brood box. The now ' bursting' with bees bottom box should be smoked gently along the frame tops to drive the bees down between the frames — the queen excluder should now be put in place then the supers placed on top and the crown board and roof replaced. We now have a 'single brood' box hive, bursting with bees, loaded with frames of brood and eggs in all stages and no queen. At the next weekly inspection, the brood chamber frames must be thoroughly checked again this time as previously mentioned all queen cells except one which is judged to be most suitable, are cut down and the hive closed up. There is now no need to inspect this hive again for about 3-4 weeks, because it will require this passage of time before the new queen has started to lay after mating. In the meantime, if the manipulation just described is carried out just before the nectar flow, and the weather turns out favourable — stand by with plenty of spare supers! Since the bees having no great amount of brood to rear will store virtually everything coming in, in the supers — the emerging brood (from 11 brood combs) provides a steady replenishing of the natural wastage of foragers, compensating for the 3-4 weeks period when the hive is awaiting the first eggs of the new queen.

Going back to the spare frames left over from the hive after reducing to a single box. These frames can be given to weaker stock or nuclei to bolster them, or stored for use in " Spring feeding" in the following year — as already discussed. From the foregoing, the main advantage of the double brood box system can be easily seen. It will be asked — What about the heather flow? — surely the hive so treated will be in a sorry state from the natural wastage by the time the heather is blooming. The problem does not arise and the hive can go to the heather also in tip top condition.

THE FICKLE TILIA - "*Apis Fanatica*"

The Lime tree is a wonderful "back end" source of nectar — not however as reliable as one would wish. I commented on experiences of this source last year in "A Tale of Two Limes" — submitting; that in years where the bloom period was preceded by 2-3 weeks hot dry weather the tree would not yield nectar. This year has proved to be such a year. The wonderful weather we here in the West Strathclyde region experienced in the weeks from mid-June up to July 8, when the Lime blossom opened, put paid to any hopes of honey from this source. The trees are absolutely yellow with an abundance of blossom but there is not a trace of that elusive limpid green menthol/peppermint flavoured "ambrosia" in any of the hives, despite the fact that the bees have been flying well during the bloom period. The only saving grace is that the bees have done tremendously well on the other late sources like clover and willow herb which seems to have excelled itself this year. At time of writing the second phase Lime in the areas above (North of) the Glasgow suburban boundary have not yet bloomed. I would stick my neck out and predict that if the bees get the chance to fly, these trees this year will yield quite well if their bloom period begins around July 29, since the weather has been extremely wet since about July 15. The only snag about this late Lime bloom is that it precedes the heather bloom period normally by around a week so there is the dilemma — what to do — to remain at the Lime or move to the heather. Heather honey is par excellent stuff — but Lime honey is easier to work and does not granulate readily so is ideal for cut comb or chunk honey — where heather in the comb although it rarely granulates solid, will normally produce "pearls" of granulation in each cell after a few months of storage. The "pearl" is due to other contaminating nectar, since pure heather honey will not granulate in the comb. If bottled however it "clouds" very rapidly.

PREPARATION FOR THE HEATHER FLOW

By the time the lime flow has finished a month will have elapsed and the new queen should have mated and started to lay. The hive being considered will by now have begun to feel the effects of the hiatus on brood rearing, due to the old queen having been removed as the swarm control "device". Many of the older bees will have gone due to natural foraging wastage, but the hive should have a good population of young bees either at foraging age or close to it. The "effective bloom" of the heather (ling) occurs sometime between 7th and 14th August depending on seasonal conditions. (By "effective bloom" is meant the time at which the heather is blooming sufficiently to sustain honey bees.) At no time would the writer recommend moving bees to ling until at least 10 -15% blossom incidence is evident — especially where the bees are being moved to heather (ling) from an area where there is lime blooming. There is just no logical reason to move bees from an area where nectar is available in abundance to an area where the nectar flow has not yet begun.

Some might well ask, "Why move from the lime at all? "Well, if you like honey, after tasting your first heather honey you will know! So, we have gone to our hive to be prepared for the heather. The first thing we have to establish is if there is indeed a new queen present. Before smoking the hive, step back and observe the foraging pattern — if the day is suitable for flying! If as previously discussed the foragers are working well and large pollen pellets are being carried it is possible to say with almost 100% certainty that there is a queen present, either laying or about to lay. Now subdue the hive. Since all the brood will have emerged, the bulk of the combs in the brood chamber will contain pollen and honey, some combs will probably be empty — we hope a few will have been laid in! Now to recap, at the initial swarm control manipulation when the old queen was taken out, we made up a 3-frame nuc with a frame of brood at all stages and eggs, the old queen, and two other frames having at least 2 lbs of stores each, and a good covering of bees — if the nuc has been well managed in the interim period it will have "grown" to be quite an effective force, having had additional frames added as required. This nucleus should now be brought back to the original apiary and placed about 2 feet from the original hive parallel to it, both entrances facing the same way — at least 2-3 days before the planned examination of the original hive. The whole object of the manoeuvre is to unite the nucleus back to the original hive. When the condition of the original hive has been established we can proceed thus — if the original hive is queenless, i.e., no eggs present, and a frame of eggs having been given as a "test" at least five days previously has been used to produce emergency queen cells — then the reinforcing nuc will be given to the original hive "lock, stock and barrel." If, however as is hoped, the original hive has eggs or eggs and brood present indicating the presence of a laying queen, we must remove the old queen from the nuc before uniting. What to do with the old queen is a vexing question and will be dealt with later. To unite the two colonies at this time of year is relatively easy, without any recourse to drastic time and labour-consuming manipulations. It is recommended that the young queen be found and placed in a match box — this match box will be set 'slightly open' - but not enough to let the queen squeeze out! Cut a corner out of one of the older combs and put the match box parallel with the face of the comb — still slightly open, into this corner. It is not however necessary to cage the queen, but it is reassuring to the beekeeper if this is done. Having subdued both colonies, open both hives leaving both brood boxes exposed to daylight for a few minutes, check the hives and assess how many frames with stores are suitable to transfer from the nucleus, along with all the frames of brood. Set the frame with the caged queen in the middle of the original hive, place a frame of bees and brood from the nuc on either side of this frame — blow smoke over comb faces before closing frames up, next place a frame of undrawn foundation on either side of these brood combs, now fill the original hive out symmetrically with the remaining frames of brood from the nuc, ensuring that there is at least 10-15 lb of stores in pollen and honey in the hive to give the bees a start at the heather. Remove the nucleus box, any flying/foraging bees from the nuc will now fly to the original hive and be accepted without question. To recap, we now have a hive which has a young caged queen (the bees will reduce the match box to pulp eventually but release the queen inside a matter of hours and repair the damaged corner of the comb in a few days), a frame of emerging brood on either side of her, next an undrawn frame outside these brood combs and again frames of brood and eggs outside these undrawn combs and any still vacant space filled with frames having a "crown" of stores and empty cells below it, for laying space. The hive population will be foragers from the original hive, foragers from the nuc, young bees near foraging age from the original hive and young bees near foraging age from the nuc and brood in all stages and eggs. The two frames of foundation will be drawn at the heather and the queen will lay in them providing a useful and sure core of young bees to ensure the successful overwintering of the colony. The queen excluder should be laid in place in the now well reinforced hive and for preference a shallow crate of good white drawn comb, with perhaps a frame having a little unsealed honey in it to bring the bees more quickly into the super. Of course, undrawn foundation in the super will suffice, but drawn comb at the heather gives a tremendous advantage — to the beekeeper!!

AT THE HEATHER

The bright orange hued pick-up truck noses gingerly up the rough twisting crater-filled track, bouncing like a cork and growling good naturedly in second gear — up, up toward the girdle of early morning mist around the low peaks ahead. The grasses and ferns fringing the truck are festooned with jewels of dew glistening in the as yet feeble random shafts of sunlight. Nudging its way into the slowly thinning shrouds of mist to the east — the truck seems to sense the promise of yet another brilliant day, and surges, jolting at every wheel turn, effortlessly up the crude track. The Beeman at the wheel, too, senses the reassurance in the settled weather, noting with satisfaction the increasing number of shafts of yellow leaking through the islands of brilliant blue in the grey inverted sea. The still morning air carries the occasional bleat of a startled sheep to the driver — as he dismounts to open and close the field gates encountered on the journey up to the great sprawling dappled purple, green and brown tapestry of heather near full bloom — under the humped mist-garlanded hills. Rounding the final bluff before the moor itself, the driver raises his eyes from the dusty strip of track to the near distance to take in the neat rows of single walled Smith Hives and their discord of multi-coloured roofs ranging from green, yellow, blue, white, violet, with the deliberate variation in the combination of these colours on the individual roof. The hives themselves perched behind their protecting fence of 4 inch mesh 'pig wire' like so many tiny ' frontier' forts on the fringe of a great gently upward-sloping purple void.

With mounting anticipation the Beeman nudges the truck nearer to the orderly miniature city — the thronging population is already abroad — in the seemingly chaotic daily lust for life — the air filled with small dark shapes darting hither and thither on myriad, glistening, flashing wings, emitting a steady soothing characteristic blissful hum of unison — all but to one purpose, the provisioning of the 'city' for the coming annual struggle for survival through the cold damp dangerous months of winter. The beekeeper makes his way on foot among the 'forts' carefully, smoothly, moving easily lest he disturb the idyllic pattern of concerted industry needlessly making mental notes from his visual observations of the activity around him. He notes the movements at the hive entrances with satisfaction — all the colonies are working well and in true rhythm; a black and silver stream, breaking to a seeming disharmony of agitated movement about twenty feet from the row of hives — as returning and departing bees weave past each other in controlled haste — he notes the urgency shown by the returning bees as they land on the hive front or landing board, pushing through the loiterers at the entrances to gain access to the dark sweet-scented interior to depart their loads of brown grey protein, or fill the nearest cell with the rich amber-hued nectar characteristic of the ling heather. The Beeman observes the signs with a warm sense of satisfaction — this is his reward for effort and hard planning. Strong healthy colonies, headed by sturdy young fertile queens laying their hearts out judging by the massive amounts of pollen being carried at each hive. The hives themselves — good solid weathertight constructions, well capable of transportation without risk to man or bees. Colonies which by virtue of their virility and strength harbour no disease and which by virtue of the vigour of the youthful queen will ensure the continued survival of the hive by nurturing the generation of autumn brood which will develop into the adults who will bring the colony safe and healthy through the rigours of our Scottish winter, to repeat the annual cycle in the chain of life in the following year. If the beekeeper is of the right stuff — he will get as much satisfaction from seeing his management pay off as he will from harvesting his honey. To many dedicated bee men the honey aspect of beekeeping, however important, takes second place to his/her pleasure in seeing his colony thrive and prosper during the active year — to come through the winter and be in good heart the following spring. I sometimes ask the question of myself: "Which gives me the most satisfaction — working in harmony with the bees or the end product — honey?" "To put the whole activity into perspective I always finish up asking myself why I don't keep ants instead of bees — and thereby lies the answer. But the pleasure factor is undoubted — it must be there to nourish the beekeeper through the long winter wait till the first days of the next spring — till the colonies begin to fly freely, gather pollen and begin to build. There is no greater pleasure for the beekeeper than being witness to the first timid early spring activity at the entrances of overwintered colonies.

AFTER THE HEATHER

There are so many things to be done to ensure that the bee colony comes through the winter in good condition. The first job is to take time out to do maintenance work on the apiaries where the bees will over-winter after returning from the heather. It is wrong to whisk the hives off the moors as soon as the heather nectar flow is over — if the hives are removed too quickly the bees do not have time to ripen and seal the last of the nectar, and the beekeeper in his enthusiasm to get his well-earned heather honey crop will be tempted to remove too soon, this unripe honey, which will soon ferment. Ideally, the supers should be left on the hives at the heather for at least a week after the 'flow' has ceased, this gives the time necessary for finishing the ripening process. If the beekeeper is working 'single handed' it is sensible to remove the supers of honey a day or so before bringing the colonies 'home' — since if the bees have had a reasonable year a brood box and super or two make an almost impossible lift for one man however 'strapping' he may be. The method of removing honey which causes the least problem to the beekeeper and least upset to the bees is the clearing board method. The clearing board is put on the hive any time after the week, spoken of previously, has elapsed. The disadvantage of using the clearing board is that it entails an extra journey to the heather, but this disadvantage is well worth incurring, since having tried the 'shake and brush' method myself a few times, and been 'chastised' by the bees for using it, I am now all for the easy life! In theory, the clearing board can be left in place on the hive indefinitely, if it is well constructed and put on a well-constructed hive, i.e. it must be 'bee tight' from the outside! The writer is a pessimist, anticipating wind activity, sheep activity — you name it activity! — which could breach the integrity of the clearing board and result in an unholy robbing situation, the beekeeper being the one being robbed! So, I always go back the following evening at the latest. There are usually still a few bees in the super, but these bees are usually so disorientated by their isolation from the brood nest that they stay in the super. I usually keep a nucleus stock in the garden, even when the other colonies are at the heather, just to have bees around me. The bees brought back from the moor in the 'bee tight' supers can be shaken in front of such a colony if done in the late evening they will be accepted by the nuc quite readily. So, the winter apiary has been brought up to scratch, the honey supers and queen excluders have been removed from the hives, and the hive reduced to a single brood chamber. Shut the bees in, in the cool of the evening and ensure that not a bee can get out. Load the hives for transport to the home apiary. On arrival put each hive down on its selected site until all hives are in place. When all hives are in place, replace all the hive roofs before releasing the bees. It is important anytime when releasing bees after transportation to have everything packed away in its place ready for the move off — including the hive roofs being in position. When the bees in a strong colony are released after confinement they are seething, seething mad and they seethe all over the surface of the hive when released late in the evening, but if released during the day when it is still possible for them to fly, have a care! Thousands of angry, frustrated flying bees are a force not to be discounted! After the hives have settled, let them have at least a full day to settle themselves after being released before starting feeding. When the hives are being moved from the heather a note can be made of any colonies which feel lighter than they should be, i.e. 50 -60 lb. total weight, hive (without roof), frames, bees, brood and honey. All hives should be fed thick syrup - 1 pint water to 1 kg sugar, to bring it up to the winter minimum. It is good practice to add Thymol to the syrup fed for the winter to inhibit fermentation in case the bees don't get the chance to invert the sugar syrup given before the onset of the cold weather. An empty shallow crate will house the feeder, which should be given direct on to the frame tops of the brood chamber. When feeding has been attended to, and long before the first frost, the hives should all be made 'mouse proof' by the fitting of one of the many devious devices which beekeepers have devised (see August, September, October, November 1978, issues of *The Scottish Beekeeper* Magazine). After feeding and 'mouse proofing' has been done and assuming the hive has been located in a position where it will be exposed to the minimum direct wind and rain/snow and the maximum winter sunlight and we have checked the hive is true and steady on its stand (the floor board must, under no circumstances, be allowed to rest on the ground, there must always be an air space under the hive to allow air to circulate under the floor board, otherwise the hive will become very damp and the floor board will rot very quickly), the hive can be left to endure against the elements for the next three months (i.e. October, November, December) apart for the weekly visit to out apiaries, to check the hives are intact and roofs are still in position. It is good management to visit out apiaries immediately after stormy weather or high winds, rather than assume, wrongly, all will be well. The honey bee colony is at its most vulnerable during the winter period, but by the same token is also surprisingly tough but, if a hive is de-roofed or overturned, it is as well to make good the damage sooner, rather than later. I have had many a hive blown over by gales and felt heart rending despair at the sight of a colony upside down and askew in the brood box. But in almost every case without exception, the colony after being set to right, has come through to the spring in good heart. So long as the queen is not damaged, the 'un-happy' colony has an excellent chance of survival, if assisted in good time.

WINTER AND EARLY SPRING

As previously stated, between the end September and the start of January nothing really needs to be done to a hive which has gone into the winter period in a well-prepared condition. Around the beginning of January, it is worthwhile quietly removing the roof of each hive in turn (leaving the crown board in place, on the empty shallow super, on top of the brood box) and lifting the hive bodily a few inches off the stand to estimate the hive weight. Some hives will be found to be considerably lighter than expected, considering their original weight at the start of the winter. This weight discrepancy can have many causes, and we can only hope that in the cases in question it will be caused by loss of stores due to late brood rearing at the end of the previous year. Feeding sugar syrup at this time of year is not recommended, but if the beekeeper is of the 'progressive' school of thought he/she can assist the 'light' colony without any great disturbance to either the bees or the beekeeper. Simply place a dampened 1 kg bag of sugar directly on the frame tops in way of the cluster. The sugar will absorb the moisture from the paper of the bag and form a thin shell of crystal sugar — the bees will chew through the paper bag and then start on the sugar which will eventually form a solid crystal lump. The bees will eat their way slowly into the sugar and without breaking the continuity of the winter cluster will gradually eat the contents of the bag, leaving the bag to all intents and purposes intact to the outside observer. When the bees have shown their need for feeding in this way, bags of sugar treated in a similar manner may be given to the colony at suitable intervals. When the bees begin to forage for pollen and water in the spring the feeding of solid sugar can be discontinued and light sugar syrup given — again directly on top of the frames in way of the cluster! Another word about feeding bags of sugar direct to a colony — the sugar in the process of crystallising will absorb quite an amount of moisture, this will have a beneficial effect on the humidity existing in the hive — dampness is one of the major problems encountered in overwintering bees in many parts of Scotland. By overwintering the colony with an empty super above the brood nest and only the crown board under the roof, the hive will come through the winter much drier than otherwise. There is no question about the optimum system of overwintering a bee colony — and that is on a double brood chamber hive, having of course, no queen excluder. If the top chamber is full of good well ripened and sealed stores the colony is virtually assured of always being in contact with an adequate food supply, with the resulting advantage that no additional or emergency feeding should be needed during the 'dormant' period — all other things of course being equal, i.e., good queen, young bees carried over from autumn breeding at the heather and good health. Such a colony, however can still be assisted greatly in spring by feeding light sugar syrup with the onset of pollen gathering by the bees. As the spring progresses it is good management using good judgment, to move combs having honey and pollen, from the sides of the hive in close to the outer limits of the brood nest. It will be found in the spring that the bees will be clustering on perhaps four or five frames — perhaps even fewer — depending on winter mortality, and that the frames in the immediate vicinity of the cluster will either be empty or have only pellets of stale dry pollen in the cells. By displacing these combs in favour of the perhaps untouched combs remote from the cluster, spring build up will be encouraged to proceed more rapidly than otherwise and may indeed be instrumental in saving the colony from 'isolation' starvation. As previously stated in these pages, although bees are at their most vulnerable during the 'dormant' period, they are still nonetheless pretty resilient, and can withstand a surprising amount of disturbance and exposure. Popular teaching has it that under no circumstances should a colony be disturbed during the winter or very early spring period. While this maxim is entirely valid for a colony which is wintering well and suffering no crisis — in other situations there arises the choice — leave the colony alone and lose it, or make an emergency manipulation and give it a fighting chance of survival. In cases where it is obvious that the colony will perish if left to its own devices — the beekeeper has nothing to lose and has also, the chance to grow in stature and experience by making an emergency manipulation. I have seen cases where colony populations have dwindled to virtually nothing, perhaps to covering a mere two frames and these frames being practically devoid of stores — and this at the beginning of January! Providing the colony is basically healthy it can benefit from emergency action. If a mouse gains entry to a hive in winter, not only will the bee population be depleted due to being eaten by the mouse (the mouse chews through the comb in succession until it reaches the fringes of the cluster, then it proceeds to eat the colder, helpless bees it finds there, gradually working its way through the cluster until the colony is killed off and its combs chewed to pieces. Once a beekeeper has had a hive "moused" he normally never lets it happen again!), also the restlessness caused by the mouse disturbance lays great stress on the bees resulting in a high mortality rate. Another reason for a high winter mortality rate may be the hive having gone into the winter with a large population of old bees due to the queen having curtailed her egg laying early at the end of the summer. These bees dying during the winter leave the colony in a depleted state.

EARLY SPRING

Returning to the case of the depleted colony on scant stores — if the hive is opened gently, using no smoke, the outer frames can be re-moved from the hive. It may be discovered that these outer frames have considerable intact stores of honey and pollen — in the case of a hive having been depleted due to the older bees dying off. Remove the frames as quickly and quietly as possible until the fringe of the clustering colony is exposed — gently move the frames having clustering bees on them apart and insert a deep comb of pollen and stores into the "heart" of the cluster, close the cluster up gently using no smoke. Place a comb of stores on the outer fringes of the cluster and at least one more comb on either side of these "store combs," even an empty comb will suffice but ideally also a "store" comb. Close the hive up. This procedure can be done even when there is frost on the ground — and even if the "store" combs given are "stone cold." This situation has arisen on quite a few occasions in my experience and the bees have not only survived but with care and good management, have done as well as any hive which overwintered without problem.

It may be that a particular hive is in danger of "isolation starvation" due to its location in the apiary. The first year where we experienced frost to any degree of severity in the West of Scotland after a succession of mild winters, i.e. 1976-77 cost the writer the bulk of his overwintered nucleus stocks due to isolation starvation. The site they occupied was beautifully sheltered, sheltered from the sun as well as the wind. During the long spells of frost experienced in that first colder year, even on days when the sun came up and broke the "icy grip" of winter temporarily in other apiaries with an open southern exposure, this sheltered apiary remained ice-bound — even the roofs and crown boards could not be lifted from the hives, being frozen solid in place. Rather than upset the colonies by forcing the roofs off, the hives were left until the frost lifted about 2 weeks later — by this time most of the smaller populations were dead — but weaker nucleus stocks in other more sunlit apiaries survived the cold spell. These hives had absorbed enough heat from radiation to warm the hive interior enough to allow the bees to not quite change position between frames but allowed them to move up the frames as a cluster — to stores higher up on the frame above where the cluster had been. The following winter was every bit as cold from a beekeeper's viewpoint but the winter losses were negligible compared to the previous winter. All the weaker colonies and nucleus stocks were overwintered in sites where they would get maximum exposure to any winter sunshine — and I would unreservedly recommend that when hives are placed for overwintering that heed is paid to this most important factor in successful wintering. I would even tempt fate and enter the "What Hive?" arena as a result of this winter experience and suggest that the popular theory abroad in Scotland that it is impossible to successfully overwinter weak stocks or small 4 frame sized nucs., stems from the use of that very popular hive of yesteryear — the W.B.C. This hive is a man-made "sun screen", analogous with the very sheltered apiary mentioned earlier — being in winter a veritable refrigerator for bees since the heat of any winter sunshine has to penetrate a layer of wood before it can radiate across the wide air space to the edge of the bee cluster and by the time any heat has penetrated to the bees — the sun has gone! The W.B.C. really only comes into its own during the "dog days" of high summer, and we don't get them any more!!

There is also a "taboo" which is in normal circumstances quite valid but which in certain circumstances must be ignored. This "taboo" is of course never move hives during the "dormant" period. Never is a long, long time. I had an experience in the last winter, i.e. 1977-78, which caused me a great deal of emotional suffering finally resulting in me taking what I felt at that time was the ultimate in drastic action. At an apiary some hundreds of feet above sea level I had 13 hives overwintering. In previous years this site had proved to be excellent. Last winter (1977-78) we had quite a lot of late snow in this particular area in mid-February — the weather was none too favourable in the apiaries situated on lower lying sites either. In the weekly visits from the middle of January to mid-February large numbers of dead bees were noted on the snow in front of the hives. Each week the number got larger and on examining the colonies by lifting the roofs off, a considerable depletion on the colonies was noted. The top bars of all of these hives were also badly stained with excrement — one of the signs of the onset of starvation! The choice was obvious — leave the colonies here and lose the lot or shift every hive to a more suitable site lower down and give them a 50-50 chance of survival. The weather in February '78 was in no way reassuring but here was a chance to gain experience and find out if the conventional limits as I understood them could be pushed back. So, one dull frosty Saturday afternoon I took the "bull by the horns" and loaded all the hives as they were on to the truck and transported them to the garden apiary where I could keep them under surveillance. Each hive was in the same stage of isolation starvation, the bees were dying but not yet dead — each hive also had ample stores at the sides of the box. The feeble clusters were split and deep combs of stores placed into the centre of each cluster — the hives were closed up and re-examined a few days later. The weather was not flying weather but not freezing. In every hive which I checked, by lifting the roof and crown board, the bees came up to meet me, lively and actively, like a soup pot boiling over, pushing out of the spaces between the frames. These colonies were fed on heavy sugar syrup from late March onward and by the start of May were brim full of bees and "rarin' to go" on the sycamore. The obvious lesson to learn from my unpleasant experience is overwinter on double broods with ample full combs of stores in the top box. Full combs in the top box are very necessary since, if there are empty cells between the cluster at the top of the bottom frames and the available stores on the top frames, if the weather is too cold to permit the cluster to break and reform on the stores on the top comb they will be lost as in single brood box management.

DECEMBER

"December brings the snow, to make our feet and fingers glow." This winter month is also the turning point of the dormant period for our bees. The end of December sees the first tentative start to egg-laying by the queen, after some 2 months of being nursed and nurtured by the colony. The beekeeper should now be starting to think of the steady development of the colonies from this time on until the first spring nectar flow from the gean and sycamore. Hives should be gently hefted, i.e. either raised bodily a few inches off the hive stand (removing the roof but leaving the crown board in place makes this operation easier where single walled hives like Smith or National are worked) or tilted slightly from alternate sides, to estimate the state of stores remaining in the hive. It is also recommended that the crown board be raised slightly to check the position of the clustering bees. This inspection is strongly recommended where single brood chamber management is used. It is best to wait for a reasonably mild day to do this operation — if done in frost it will probably be found that the crown board is frozen in place. Don't try to force it off! If the bees appear to be clustering right up to the frame tops then the colony is in danger of isolation starvation — a cake of candy or one kilo bag of sugar should be given immediately — right on to the frame tops where the cluster is located. It is useless to feed bees in winter if the food is not right where the bees can get to it without breaking the winter cluster. At the first quick tentative inspection the size of the bee cluster should also be noted, i.e. written down and dated! This gives a datum on which to base estimations of development. If the cluster looks settled and the bees are not right at the frame tops all is well. But a weekly check on the lines suggested will keep the beekeeper right up to date with the colony situation. Hives which have been located in shaded apiaries are greatly at risk during the "dormant period" — these hives do not get their quota of winter sunshine and tend to be more prone to isolation starvation. One of the most significant indicators for pending isolation starvation is badly stained top bars and restless, dull-looking bees. In cases of chronic isolation starvation feeding candy is useless as is feeding bags of sugar. The most effective method of saving a colony suffering from isolation starvation is to supply it with combs of honey and pollen — placed right into the heart of the cluster. In an overwintering bee colony the bee cluster, by mid-end December does not cover all the frames in the brood chamber — there are always store combs which are out with the cluster — these combs are the secret for survival for a colony dying of isolation starvation. Simply take some of these combs out. separate the frames which the bees are clustering on — DO NOT USE SMOKE — and insert the new store combs between the frames of clustering bees — close up and then put candy or sugar in bags on top of the cluster. The bees will revive on the honey given and then augment their needs with the candy or sugar given — allowing the natural stores to last longer. The operation just described may sound drastic, it is drastic but also necessary, and if done correctly and in good time the bees will survive, so long as they have honey on which to cluster. Make no mistake, your bees are hardier than you think but only if they are basically healthy, have a good population of young bees and a good young queen. It is worthwhile checking all hives at this time to ensure either the mouse-guards are still intact or that there is no evidence of mouse activity in the hives. Heavy, rough looking debris at the hive entrance is a sure sign of mice at work. Disregard fine powder like debris, this is quite normal as the bees uncap cells to meet their wintering needs in the cluster. If mice have gained access to a hive the most effective way to remove them is to lift the hive body off its floorboard, make sure the mouse is not among the unoccupied frames and set the hive down on a clean new floorboard which will give the bees an entrance height of not more than - 5/16 inch (8 mm) — this operation being done during the hours of daylight. Set a trap for the mouse under the hive floorboard. If the mousetrap is set in the open near the Hive you will catch blue tits, sparrows and robins — not to be recommended!

(The dormant period for the bees can be well utilised by the beekeeper. The cold weather provides ideal conditions for scraping and cleaning queen excluders. The smoker should also be cleaned of all carbon and the bellows treated with a waterproof polish such as "dubbin". The stacks of spare supers and brood boxes should be treated for/against wax moth. In the case of stored brood boxes, the combs can be treated against Nosema spores by exposure to pads soaked in Acetic Acid (80%) inserted in the stack, one pad per 3 brood boxes. One of the most rewarding winter occupations in the evening when the gales are rattling the chimneys and the hives have been secured against being blown over or having roofs blown off, is reading beekeeping lore, beekeeping management and all the other marvellous aspects of the written word relative to our craft — the keeping of bees for pleasure and profit! Readers of *The Scottish Beekeeper* magazine, as members of the Scottish Beekeepers' Association, are entitled to the use of the vast library of beekeeping literature in the MOIR

LIBRARY in Edinburgh. By merely writing for a particular book: the only cost to be met by the borrower is the return postage. Some recommended books for all levels of beekeeping are:— "The Hive and the Honey Bee" edited by R. Grout, R. Couston's - "Practical Beekeeping". Brother Adam's "Beekeeping at Buck-fast Abbey", Dr Bailey's "Infectious Diseases of the Honeybee", Dorothy Hodges' masterpiece "Pollen Loads of the Honeybee" and Dade's "Anatomy and Dissection of the Honeybee" — these books will give the reader a marvellous insight into the broadest aspects of our chosen craft and by diligent and constant reference to such material the so-called "Mystery of the Hive" will be as an open — "dare I say it?" — Bee book.

January – *Apis Fanatica*
Winter Activities (FOR THE BEEKEEPER!)

Where the beekeeper has out apiaries, during the ' dormant' period, i.e. from the end of October until the middle of February, the hives in these apiaries are at greater risk than the hives in the 'home' apiary by virtue of their very isolation. Since in inclement weather conditions, such as gales accompanied by heavy rain, snow or extreme frost, any hives either overturned or bereft of roof and crown board may die of exposure unless the beekeeper makes good the damage in time. With out-apiaries the only sure method of safeguarding the well-being of the colonies is to make regular weekly visits during the normal winter conditions and additional visits after high winds and gales to ensure that all hives are still intact. The condition of the home apiary is a good indicator for the probable condition of the out-apiary; if for instance the wind has been such that hives in the garden have been disturbed, then it is imperative that the out-apiaries be visited even if it costs a couple of gallons of petrol to do so. What is the cost of a couple of gallons of petrol compared with the cost of replacing even one lost colony, to say nothing of the potential loss of a honey harvest from that lost colony (40 - 60 lb)? As the late winter merges into early spring, so then can the beekeeper gradually become more involved with his bees, and instead of merely visually -checking -the hives for soundness against the elements, he should begin to think about the condition of the bees inside the hive and the level of their food supplies. By either direct hefting or merely tilting the hive by lifting from the floorboard, a reasonable estimate of the amount of stores available can be made, especially if a number of hives are being checked at the same time, light ones can be easily identified. Any suspect hives should be carefully watched and fed with either dry sugar in 1 kg. bags or with candy until it is possible to feed heavy sugar syrup (1kg to 1 pt water), (the heavier hives will also benefit from light sugar syrup (1kg to 3 pts water), with the advent of the first observation of pollen being carried in the apiary. At the time of 'bedding' the bees down for the winter it is worth while setting the hive so that the entrance is at a lower level than the back of the floorboard. By disposing the hive in this manner any rain or excess moisture will drain away, giving the hive more chance of remaining dry. In the cold damp climate which we here in Scotland accept as normal, the greatest enemy of the bees is not really the cold but the dampness. This dampness causes any store combs in the hive not in contact with the bee cluster to become mouldy and it is good practice to remove the outermost comb at either side of the hive, thus enhancing the ventilation of the interior of the hive. The removal of the outer combs can be affected after the bees have formed the winter cluster and from my own experience is easiest done around mid-November. If done properly, not a bee will stir — of course it is totally unnecessary to use smoke for this operation. If any attempt is made to remove the afore mentioned combs at the end of the active season around the end of August or mid-September, the hives will be found most awkward to work as the colonies are at their most defensive at this time of the year and a 'deep' inspection will do more harm than good. If the beekeeper has done his year's work well, there will be no need to annoy the bees more than they will be annoyed at the removal of the heather honey in the supers, but if the removal of the heather honey is done using a 'clearing' board and the correct amount of smoke in the cool of an early September evening, again hardly a bee will stir.

January by Rambler

The beekeeping year, having begun with the advent of the heather flow, is now in its fifth month. January signals the start of the most vulnerable period of the honey bees' life in our particular area, Scotland. January sees an increase in brood rearing resulting in an accelerated rate of food consumption. This consumption of stores may be viewed as an indicator for a number of conditions which may exist in the hive. One condition could be that the colony is so depleted in population due to high winter loss that stores are being consumed more rapidly to maintain the heat in the cluster. Another condition — already mentioned — increase in breeding rate, will see a lightening of the hive. In both cases winter feeding should be resorted to as previously mentioned in these pages, sugar or candy for supplementary feeding or combs of honey and pollen for emergency feeding. Any hive where the population has dwindled markedly from its level going into the winter should be noted at the time — the loss could be due to a number of causes, such as the overwintered queen being old, mediocre or a late supersedure queen which did not mate, the overwintering bees will thus be primarily older bees, which by virtue of their age do not winter well. Disease will also cause abnormal winter loss and where disease conditioned loss is suspected, a sample of some 50 bees taken live and placed in a matchbox, without food should be sent for analysis.* Nosema and Acarine are two of the most dangerous diseases of adult bees — and not only during the winter months. Nosema is caused by a spore forming micro-organism called **Nosema apis Zander** which attacks the lining of the honey bee ventriculus or mid gut. This disease reaches its highest level of activity in the spring. Research has shown that a less pronounced peak of activity occurs during the autumn period in infected colonies — this autumn peak plays a significant role in the level of infection the following spring. Acarine disease is caused by a microscopic mite **Acarapis woodi Rennie**. The mite invades the first thoracic spiracles of the honey bee respiratory system and then infests the trachea where it multiplies rapidly, affecting the bee's ability to breath, and sucks its blood. These diseases can be treated. There are a multitude of diseases to which honey bees are susceptible, none, however, are transmittable to humans. One of the many diseases worth mentioning (again!!) is the relatively new disease **Varroa jacobsonii Oudemans**. First discovered in Java in 1904, the progress of this disease, caused by an external parasite (small spider), has been more or less jet assisted across Europe due to the increased traffic in honey bees, commercially. At current time of writing no reports of Varroa disease have been received in Britain. *(This article was written in 1980, several years before Varroa was found in Britain)*

There is no cause for complacency with regard to the disease. The mite is so difficult to detect that a period of some four years may elapse from initial infestation until the effects of the disease show themselves in abnormal population dwindling, deformed adult bees and ultimately dead colonies. There is as yet no cure for this disease. The best cure is prevention!! Although our enlightened agricultural authority has banned the import of colonies and nucleus stocks of bees, queen bees are still able to be imported from countries not listed as affected by Varroa. With the best will in the world, where this situation is allowed to exist we beekeepers in Britain are being needlessly put at risk while this import loop-hole remains unplugged. I will repeat — as yet there is no cure for Varroa — what are we doing sitting around waiting on some queen bee bringing the disease in? Apathy is a terrible thing!! Food for thought is this; perhaps Varroa has already gained access to this country,

---

* At the point of writing the sample would have been sent to the beekeepers' nearest College of Agriculture for analysis. Now this analysis would need to be done by the beekeeper.

even experienced beekeepers of many years standing will have difficulty in detecting the presence of "live" Varroa infestation, many methods of diagnosis are clumsy and cumbersome. One simple method of diagnosis being widely used very successfully in Germany, and recommended by leading authorities on the "Varroa situation", is to place sheets of paper or plastic on the floor board under the colony cluster — during the dormant period, i.e. between end of October and beginning of March. During this period the Varroa parasite also has a winter wastage rate and the victims fall to the hive floor board where they may be detected by close inspection of the debris on the inserted sheet — the sheet should ideally be renewed every month. It is of course not necessary to have the sheet a tight fit across the floorboard, merely wide enough to embrace the cluster width is sufficient — this makes its insertion much easier than full width inserts. It is ideal to treat all hives as recommended, but systematically checking say, one hive in three where many hives are kept, will also give a good indication of the state of the apiary. We haven't got Varroa! But wouldn't it be reassuring if we could prove our freedom from the disease on an annual basis from now on. The mite looks like a spider and measures approximately 1 mm x 1.5 mm a X 2 magnifying glass or greater magnification would help the investigators whose eyes are not what they were. Despite the doom warnings, we in Scotland are relatively lucky so far as the incidence of disease in our colonies is concerned. One of the main factors till the present time, since the period after the "Isle of Wight" disaster, in our favour is the relatively small number of beekeepers in Scotland. Our tough climate must also have some bearing on the level of disease using the analogy that microorganisms don't thrive in refrigerators. To be a successful beekeeper in Scotland is indeed to be a successful beekeeper!

February

In favoured areas many early spring blossoms will have started to show. The snowdrop is normally the first nectar-bearing plant of any importance to bloom, followed by the crocus around the 1st week in February (in West Central Scotland). If the winter is not too prolonged the bees should be out on the milder days — making their first cleansing flights and starting to do their housework after the long enforced confinement. It is around this time that a change in Winter/Spring feeding should be made, if weather conditions are right. When the bees are seen to be gathering early pollen this is an indication to the observant beekeeper that the tempo in the hive is indeed increasing. As breeding increases so the need for water in the hive increases, bees cannot raise brood without water. A switch should now be made from feeding candy or crystal sugar to feeding sugar syrup. It is a matter of judgment for the beekeeper as to what strength of sugar syrup solution he/she feeds. But a good rule of thumb is: if the hive to be fed is quite well provisioned with stores (estimated by hefting, as described previously) then the main function of the feed will be to supply a palatable fluid for use in brood rearing, a strength of 1 kg. sugar to 3 pints of water is recommended. If, however, the hive - to be fed is light in stores, a highly concentrated feed of 1 kg sugar to 1 pint of water is recommended, until the bees have taken down perhaps the equivalent of 2 kg sugar in the sugar syrup fed. The less concentrated feed mentioned can then be resorted to.

By feeding judiciously in this manner two aims are fulfilled: supplying food to either assist, promote or sustain brood rearing and also supplying liquid without which breeding is impossible. An additional bonus for the bees is that they no longer have their problem of carrying water to the hive. I have listened to many discussions around the question of spring feeding and have learned through them that many beekeepers disagree with the concept of what is popularly termed "stimulation feeding." The pleasure aspect apart, the attraction of bee keeping is the procurement of honey, to obtain honey one must have sufficient bees in the hive at the right time, to coincide with the major nectar flows in the beekeeper's area. If the first major nectar flow in any area is the sycamore, which blooms around 11th May in moderate years, to have sufficient bees available for the period one must assist the hives by feeding to reach a population density which will be able to exploit the sycamore. There is an abundance of pollen available in the weeks before the sycamore bloom, but not much nectar — sugar syrup is the bees' substitute for this nectar, thus the rate of pollen collection can be met by the rate of sugar syrup feeding which in turn will regulate to a degree the rate of brood rearing. Feeding also augments the natural stores of honey which would normally be used for spring brood rearing, thus eliminating any possibility of isolation starvation should the weather suddenly turn colder for a spell in the early spring — as is its wont. Now, on the other hand, if spring feeding is not resorted to, even a hive which at the start of February, when hefted felt quite substantial, when brood rearing gets well under way, especially with a good fecund queen, the rate of consumption of stores can be quite staggering, thus leaving the bees in a dangerous plight and threatened with the dreaded "isolation starvation". What really happens to a hive of bees left to its own devices until the first major nectar flow could read as follows: December—plenty of stores, January—plenty of stores, February—plenty of stores, March sufficient stores, April (weather turns absolutely vile) plenty of brood, no stores—isolation starvation! This happened to many hives in Winter/Spring '78/'79. Or if the hive survives through till the first major flow without feeding, it will do beautifully if the weather lets the bees out! At the end of that first flow the hive will be bursting with bees and brood reared on the first flow, but there won't be any honey in the supers for the beekeeper. Take good advice and feed as early as you like, the bees won't get fat! In a late area predominated by lime/heather, the feeding should be aimed at maintaining colony strength so that the bees are in the right condition to be assisted to reach a population peak in July/August — the first rule of beekeeping "Don't do things as a matter of fact or by rote", work with the weather, work - with the flora in your area and most importantly work with your bees. One undisputable fact of spring feeding is that colonies as treated will make preparation to swarm much sooner than hives which were left to their own devices and were fortunate enough to survive. But swarming is a problem which beekeepers have to deal with in the course of the active season anyway, better having them sooner than later. Thus, getting the hives settled in time to score from the Hawthorn, Clover, Lime and Heather. To quote an old saying:

A swarm in May is worth a load of hay;

A swarm in June is worth a silver spoon;

A swarm in July is not worth a fly.

So long as feeding is attended to and the bees observed on the occasional bright day are seen to be taking large pellets of pollen into the hive the beekeeper can rest easy. External observation can tell the beekeeper so much and can give an early warning of problems in the hives either caused by events during the winter or the sudden loss of a queen in a previously queen-right hive in early spring. Don't under estimate the importance of spending that few extra minutes at a hive waiting to see if some of the few returning foragers will have that all important bulky pollen load. Any hive where the foragers, relative to its neighbours are not so actively engaged in pollen gathering, should be noted and observed again on two or three subsequent occasions, to ensure that the bees were not just having an "off" day, before any action is taken.

NATURE AWAKENS   March

This month sees the beginning of the bloom period of the vast powerhouse of plants which provide the major early pollen -sources — the catkin bearing plants — poplars, birches, hazel, etc., the elm too! The great pity of this particular time is that in recent years, where we have had long hard winters and cold "dreich" springs, the bees don't get much opportunity to fly during this period, but when they do get out during the catkin bloom, on a bright day in March, the fronts of the hives are a "marvellous sight" to behold, the returning bees seem to have literally been taking dust baths in the pollen and take on the appearance of mayflies (those marvellous, "yellow duster" looking cadis flies which hatch in their thousands in June/July). The bees seem to sense the great need to harvest as much of this "substance of life" as they possibly can. These vast amounts of pollen entering the hive are a great stimulus for the queen to "hit her stride" and this period really marks the beginning of the massive build up which the colony must undertake to compensate for the pending loss of overwintered bees and in order to be strong enough to make good use of the nectar flows to come, so that the continued existence of the colony and the species is assured by build up to swarming proportions. March and early April are particularly hazardous for bees since if breeding has been proceeding at a rapid rate, and the weather turns inclement the brood must still be fed and at this time precious food stores can virtually disappear "overnight" — this is where spring feeding really shows itself to be worthwhile — if the beekeeper has been feeding assiduously he/she can rest easy during these nasty spells, sure in the knowledge that neither adult bees nor unsealed brood will be at risk from starvation. During the late part of March remedial measures may be taken in the apiary where any hive has given cause for doubt about its condition. During the "active" days of March it is relatively easy to make comparisons in hive behaviour. The importance of running at least two hives in an apiary cannot be overstressed — having more than one hive is good insurance against total loss of any one hive. A hive which seemed more idle than its "fellows" can now with impunity be opened up for examination — if no eggs or brood can be seen and if the queen cannot be seen where the hive is still quite strong it is worthwhile inserting a frame of young brood and eggs into the colony — an examination around 5 days later will determine if the colony is indeed queenless or harbouring a "scrub" queen — if the bees draw queen cells, the hive is queenless and should be united to its neighbour using the "newspaper" method. If the bees do not draw queen cells there is a queen of sorts present and if she still cannot be found, move the hive to a new location within the same apiary on a bright day the bulk of the foragers will leave the hive and return to the nearest neighbouring hive — with the colony thus depleted it will be quite an easy task to locate the "duff" queen and destroy her. The remaining bees should be shaken off the combs on to the hive floorboard and left to fly to other hives after the other parts of the hive have been removed. This operation being done of course on a good flying day. By doing this operation early it will save the beekeeper needless waste of sugar syrup in feeding a "dead loss" and put the otherwise doomed bees back to useful work.

There are many rules of thumb for when to make the first internal inspections of the hive in the spring. While it is not advocated that the beekeeper should embark on a "willy nilly" inspection of hives during the early spring, the examination of suspect hives should be made as early as possible, so that timely remedial action can be taken. A hive can be examined in spring with impunity (it must of course be subdued in the normal manner with smoke — but not a lot!) on any good flying day after pollen has been noted being carried in quantity by the bees in that particular apiary. But don't dally during this inspection and take care not to expose frames having eggs or open brood to the cool atmosphere for too long, otherwise chilled brood will result.

Any work to be done in the early spring such as feeding or colony inspection should be done as rapidly as possible to minimise "robbing". Where "robbing" is noted to have started, work in the apiary should be stopped immediately and entrances of hives under attack reduced to not more than 3 inches wide. Any feeding of hives should only be done in the evening to minimise the risk of starting robbing, any spilled or exposed sugar syrup should be covered with earth, to the same end.

At this time of year, it is important to ensure that the floorboards are free of debris and dead bees. Sometimes, where a narrow "mouse proof" type entrance is used this can become totally blocked with winter rubbish. The bees must be assisted in a case like this or the colony could perish due to being trapped inside the hive. Where colonies have been overwintered on double brood boxes it will be found that the bees and brood will generally be located in the top chamber. It is good practice around the end of March to place this top box on the cleaned floorboard and place the lower box above — giving the queen extra space above her into which to move as the colony expands. By having this "laying space" above the established brood nest the bees are encouraged to expand upwards, the queen will follow and lay in the top box — the bees will expand more readily upwards than in a sideways direction in the spring. By expanding upwards, they rise into an area already warmed by the rising heat of the colony, thus lessening the possibility of chilled brood if the colony in expanding sideways has to retract because of a sudden cold spell in the spring weather — not unusual in this Scottish climate of ours. It is really worthwhile at this end of March period to take notes of the size of the cluster in the hive as a basis to assess whether the colony is expanding at a rate compatible with the apiary average. Poor performing or damaged queens can be quickly pinpointed by this simple management device and the hive can thus be timeously re-queened.

May

We are now approaching the beginning of that period of beekeeping where the joys and sorrows of our craft can be experienced almost simultaneously. The joy of seeing colonies thrive and expand. The sorrow of being caught out and missing the signs of swarming in a colony and then losing the prime swarm. The best and most effective method of swarm control is based on regular inspection of the colonies during this and coming months. Swarming occurs sooner or later in the season depending on the weather conditions during nectar flows. If the bees have done well in the early part of the year on perhaps gean and sycamore then swarming will probably occur around the middle part of the sycamore nectar flow. If the late spring and early summer nectar flows don't materialise then little or no swarming will occur in May or June. The problem of late summer swarming then arises, if the clover and lime flows are good then the bees will invariably prepare to leave around the end of the first week of the lime flow i.e. mid July (these dates relate to West Central Scotland). The problem of very late swarming, i.e. at the heather, is perhaps the most unfavourable time for both bees and beekeeper. The event of an August swarm is usually an act of sheer desperation on the part of the bees perhaps having been held in check by previous inclement weather or non-consistent swarm control measures by the beekeeper or both! For a colony swarming in August or later, unless the beekeeper is handy and witnesses the act, the fate of both the swarm and the parent hive is "doom" — the swarm will not normally be able to build up enough to either produce the necessary brood to supply the young bees required for good wintering, or provision itself enough to overwinter successfully. The parent hive will have a virgin queen with little or no hope of mating — thus the hive population will gradually waste away until the hive dies either in the winter period or early spring.

Swarming is a phenomenon which the beekeeper must learn to live with and to work within its limitations. Every colony which has overwintered from the previous year will normally make preparations to swarm at some time during the active season. These preparations do not always result in actual swarming — but until we know a bit more about bees we must treat every colony which shows swarm preparations, i.e. builds more than 2 or 3 queen cells at any time during the active season, as a potential 'risk'. Although swarming should in no way be encouraged, the bees need to swarm or reproduce must be respected in successful beekeeping. Every year we beekeepers engage in the annual contest with nature — we strive to maintain our colonies intact by manipulation while the bees strive to deplete their strength by swarming.

Since time immemorial beekeepers have asked themselves the same question generation after generation, "Why do my bees swarm ?" We modern beekeepers are a bit better educated in this respect, thanks largely to the continuous efforts of our "apicultural boffins," who incidentally should not be written off as being on some remote or abstract beekeeping plane. These people are the same as we are — enthusiastic beekeepers — but with a more specialised interest in bees than we "grass roots" beekeepers. In ancient times beekeepers looked forward to the swarming period with anticipation — this was their only way of increasing colony numbers. Due to our better understanding of honey bee genetics and natural history we do not require to work within the restraint of the "natural swarm" today. The bees have made it easy for us to anticipate their intentions and deflect them from their aims, when it comes to swarming. The building of queen cells signals the imminent departure. Normally, the swarm will not leave until there are sealed queen cells in the hive gives us another advantage over our charges. Since, on noting eggs in the always present "queen cups" in the hive, the beekeeper knows that his bees will swarm any time after the 9th day from laying of the eggs. Thus, he has time to think about what action to take. And action he/she must take! If the queen, has been clipped, then the time scale margin for the beekeeper will be extended to the 14th or 15th day after discovering eggs in the queen cups. Since even if the bees do swarm, they will soon return to the hive, the queen not being able to fly will be lost on the ground. Bees will not leave the hive in a swarm without a queen being present! The beekeeper, however, must do something positive inside his 14 odd days grace. If he does not, when the first virgin queen emerges, the bees will swarm, and what a swarm! The original prime swarm bees will be reinforced by the bees which would have constituted the first "cast" i.e. — swarm with virgin queen(s)! — making a mass of bees which is quite enormous — the loss of which will render that particular hive a productive write-off. There are a million and one methods of swarm control/prevention, but any system which does not deprive the "swarming" bees of a queen able to fly is doomed to failure, one way or another! Each of us eventually works out his/her own particular modification of the various systems of swarm hindering — not all are successful, but those which utilise the limitations of the bee's natural history at least have some chance of success. A very successful swarm control method is to cage the queen in a cage, made from queen excluder material, in the brood chamber, cutting down any queen cells at the time of caging and then again cutting the next wave of cells at the next inspection — inside a 9-day period. The queen can be released on the next visit — again around 9 days later. This system is not 100% fool proof, sometimes the bees kill the queen when she is released, sometimes they merely begin building swarm cells again. A disadvantage of this method is that the queen is out of action for around 18 days with the resulting loss of egg production — 2000 x 18 = 36,000 lost bees! For the hobbyist beekeeper with not too many hives the most efficient control system is to find and remove the queen to a nucleus hive, shaking about 2 frames of bees into the nuc with her. Remove this nuc to another apiary, the queen will soon begin to lay her full quota of eggs again. The sealed queen cells in the parent hive should be cut down, leaving all queen cells with eggs or unsealed brood. Nine days later return to the parent hive and cut down every queen cell save one (preferably one on the "face" of a comb! Cells at the top bars or at the bottom bars are not recommended!!). Leave the hive for 28 days then check the hive for eggs — it is not necessary to find the queen at this inspection — if eggs are present all is well! It is recommended that all hives have queens marked in the first spring (around mid-May) after their first winter. Marked queens make good beekeeping sense!! They are easier to find during hive inspections. Cutting queen cells down any sooner sometimes results in the bees using an unsealed larva to produce a queen cell — the bees then swarm with the good queen from the cell left after cutting cells at 2nd visit. The late larva cannot of course produce a queen and unless timely action is taken the colony will decline.

June - SWARMS

The beekeeper is not always as efficient as he/she may be when it comes to "swarming time" mid-May to late July normally. If a swarm occurs then, depending on circumstances, the method of apprehending it must be varied accordingly. Normally a swarm will issue in the late morning, around 10.30 a.m. - around noon. The bees seem to sense they have to give themselves time to re-group and reorganise — but a word of tempering — bees do not do things invariably! There are numerous reasons for bees swarming and the reason will govern the behaviour of the swarm. The main types of swarm are:

1. Prime Swarm.

2. Cast.

3. Mating Swarm.

4. Hunger Swarm.

The Prime Swarm is the most important swarm for the beekeeper, in so far as if the bees have not been frustrated in their efforts to swarm by the beekeeper, it will be the biggest of the naturally occurring swarms mentioned. The prime swarm consists of the parent queen (old queen) of the hive, and the number of bees in the swarm may be 20,000 to 30,000, the percentage of the population leaving the hive may vary between 50% to even 90%. The age of bees varies between 4 and 23 days old mainly, but bees of all ages will be found in prime swarms. Drones will of course also be present. A prime swarm will not normally leave the hive until at least one sealed queen cell is present (nine days after the egg was laid in the queen cell!). The initial instinct of a swarm of bees on leaving the hive is to cluster and regroup. The bees can be encouraged to cluster at particular points by the beekeeper placing either old dark frames on posts or by suspending combs from tree branches in or adjacent to the apiary. Or by covering an old brush or mop with sacking smelling liberally of old comb, then sticking the mop or brush handle into the ground in front of the hive(s), perhaps 10 or 12 feet (3 or 4 metres) from them. The prime swarm also usually occurs after the first early nectar source is finished — if the weather is favourable,

The Cast is the swarm which emerges, usually 7 or 8 days after the issue of a naturally occurring prime swarm, the cast will usually be smaller than a prime swarm and may contain one or more virgin queens. A hive may "throw" several casts, depending on how strong the parent hive is. Sometimes a cast will emerge and form a number of small clusters of bees each having its own virgin queen or queens. Occasionally some of these small clusters will unite with another after some time has elapsed. This is most probably due to the cluster, having split from the initial main cast containing several virgin queens, finding itself without a queen and remedying its error. The mention made of "frustrating" the prime swarm, previously can have significant consequences. If the beekeeper is of the progressive school he/she will have a clipped, marked queen in each hive. If this beekeeper, however, is not consistent in checking for swarming on a nine day basis of inspection he/she may miss the swarming signs in the hives preparing to swarm. The bees will swarm, cluster and then realise they have no queen with them. The queen, by virtue of her clipped wings, not being able to fly will have fallen onto the ground and become lost. The bees will soon break the cluster and fly back to the parent hive — they will indeed be feeling frustrated and are liable to be in stinging mood at this time. The climax to this "prime swarm" attempt comes 7 or 8 days later — weather permitting — in the shape of an enormous swarm, comprising the bees of the original prime swarm and those bees which would have formed the early casts.

To see such a swarm emerge is an awe inspiring sight, to lose such a swarm leaves a great empty void where the beekeeper's stomach used to be — the voice of bitter experience speaking here!

The Mating Swarm may occur at any time from the beginning of May until October. It is the writer's belief that "mating swarms'" do not occur in nature, or only very rarely. In progressive beekeeping however, the mating swarm is quite a frequent occurrence. Where the beekeeper practices queen rearing and where use is made of either the "mini" nuc. or small "colonies" of broodless bees to accommodate a virgin till she is mated, the incidence of mating swarms is very high. When the virgin flies to mate the small colony often goes with her. These mating swarms are usually quite small affairs, but nonetheless disheartening since the beekeeper loses about a month's work when he loses a mating swarm. The late or very late mating swarm may also be quite small, but occasionally it will be as large, perhaps, as the size expected of a cast. The reason for the late cast is loss of the parent queen late on in the season for whatever reason. In itinerant beekeeping the loss of a queen can easily occur either going to the heather or returning from the heather, and usually a three week period elapses from the loss of the queen until the emergence of this last desperate attempt at survival by the unfortunate colony. The bees case is hopeless since the queen will be an unmated virgin, and it is not normally worth the effort to "take" this swarm. It will be quite difficult to locate the hive having swarmed so late in the season. The following spring is the earliest the now queenless hive can be pinpointed for we mental (thinking) beekeepers. There are two possible methods of establishing the hive in question though. Given luck and reasonable weather conditions; if the swarm is taken and the queen found and killed, (she will be an unmated virgin!). Then the bees can be dusted with flour and dumped on a clean, dry surface. They will soon realise the queen has gone and begin to return home. Another method which I do not have the courage to follow through is; in the evening after the colonies have settled for the night, give each hive in the apiary a few hefty raps and listen to the disturbed bees (ear of course to the hive side!). Queenright colonies settle down quite quickly after being so disturbed. Queenless ones not so quickly. If, and when the queenless hive is located it should not be united to a queenright hive until the health of both hives has been checked out.

The Hunger Swarm is thankfully a rare occurrence in progressive beekeeping occurring more in nature than beekeeping husbandry.

The hunger swarm only occurs during the active season, in summers where the dearth of nectar is so severe that the bees are forced to cease rearing brood.

In the writer's experience, hunger swarms have only occurred in the summer of 1972 where hives not fed were starving during late May, June and early July. That year many hives had no brood in June. Returning from a fortnight's holiday in June the writer was appalled at the condition of hives in a particularly badly hit apiary. I witnessed three hunger swarms on the one day that June. These swarms formed very straggly clusters and without fail all returned to their original hives. These hives were examined after the swarm had left and apart from a few forlorn looking young bees they were clean as the proverbial whistle — polished empty cells — the absence of dead bees was quite striking. Massive sugar feeding was instituted and although there was little honey that year the bees built up and got through the winter. Those were the days when the winters were not so vicious and long.

TAKING A SWARM

The classic swarm issues at 10.30 a.m. on a Saturday or Sunday, preferably Saturday. Most of us enjoy a "long lie" on a Sunday! It then settles at the end of a low branch (5 ft from ground level), forming a compact tight cluster, in a tree in the garden, or the garden of a very near neighbour with whom we are on "speaking" terms. Such a swarm is the beekeeper's dream, taken in less time than it takes to tell — not by shaking into a box, but by using sharp secateurs and holding the branch with the swarm between the swarm and the secateurs. Thus, the beekeeper is left holding a perhaps 2 ft long twig, loaded with an undisturbed cluster of engorged bees which can be lowered into a swarm box or hive without the loss of a single bee. Veteran "swarm takers" and "hivers" will know that when a swarm is hived its instinct to cluster is very marked. The temptation to leave an empty shallow crate or brood box on top of the brood box with combs given to the swarm on hiving, is very strong because the tidiest way to hive a swarm (especially if the weather turns cold or wet or both, a not uncommon occurrence in Scotland), is to invert the swarm box over the frame of an already prepared empty hive, with floorboard and fitted with frames, using a shallow or deep box as a sort of filler funnel. Then, after the bees are all in, merely putting on the crown board and then the roof. If this procedure is used the beekeeper on examining the hive some days later will find the colony nicely housed in the hive, having built nice new combs, which it has filled with the balance of the engorged honey from the parent stock. The new combs will however be attached to the underside of the crown board — and the beekeeper has a problem which could result in the queen being killed or maimed while the "wild" combs are being cut down. Never leave a wide air gap between frame tops and crown board when hiving any kind of natural swarm. Unless you have reason! The ideal time to hive a swarm after it has been taken is evening when all flying has ceased. The swarm, having been taken in the afternoon or morning, being left in the swarm box until then, in a shaded ventilated position — if this shaded ventilated place chosen is in the garden or anywhere in the open — please remember Copernicus — he was the man who twigged that the earth went round the sun. What may be a shaded spot in the garden at 1.00 p.m. or whatever, may be exposed to bright sunlight a few hours later. A swarm so treated is a very, very sorry sight indeed — my one and only experience made me feel so incompetent that I almost gave up keeping bees. I was ill for days! Once the bees are in the new hive and the cluster is looking like a slightly disturbed "clooty dumpling" lying on the frame tops, by using a little smoke judiciously the swarm can be encouraged to "melt" into the gaps between the frames — the swarm "twig" or branch can now be removed, shaken over the hive and the crown board placed on directly over the frames, then the roof. There is a "trick" which can be used which eliminates the need to use smoke to get the bees down into the new hive and which allows the beekeeper to leave the "filler funnel" super or deep in place. If the new hive is made up with a frame of open brood in its centre and the swarm is "dumped" in way of this, they will very quickly settle around the brood. As a matter of pure fact, a swarm queen will actually begin to lay sooner than usual if the swarm is so housed. In normal hiving procedures it will be found that the swarm queen can take up to 15 days before she is inclined to start laying. This is a natural safety device to allow a swarm to establish itself before more or less taking on the responsibility of raising a family! Swarming bees are peculiar animals, to homo sapiens any way. It is virtually impossible for even the experienced beekeeper to predict how a settled cluster will behave, i.e., how long it will remain in its position, how far it will fly when it lifts and in which direction!! A gem of a story comes to mind which probably quite a number of members already know. At a beekeeping demonstration, in a nameless shire and by a nameless, but well-known member of our great fraternity, a swarm issued from a hive in the demonstration apiary great excitement of course ensued — the day was bright the bees settled quite quickly — the apiary owner was seemingly in no hurry to take the bees. Some time elapsed, the demonstrator finished his work with the hives, and went over to look at the quite splendid, near classic swarm — examined it very closely peering with interest at the activity of the scout bees on the outer shell of the cluster. He stepped back, and said quite evenly to the interested group: "The bees are about to lift, they should go in about the next 5-10 minutes, and... He paused, looking around at the near distance. Then throwing up his right arm like a direction indicator he continued in the same even voice: "They will head for that church tower about 400 yards from here". The disbelief was evident for all to see. Around eight minutes later a swarm of bees was in the air and now yonder church tower has new tenants!! The lesson here is take your swarm right away, otherwise the bees may beat you. A general rule of thumb is: on a bright, hot, windless day a swarm will lift from its clustered position quite quickly, sometimes inside half an hour, but most certainly before two hours have passed. Where the weather changes quite dramatically for the worse after the swarm has issued the bees may sit tight for up to three days. By this time they are no longer a swarm, comb building will have been started and so far as the bees are concerned they are home. Such a colony can survive and prosper during the summer and even get through the winter in the open but, as soon as brood rearing begins in the late winter/very early spring, the bees are doomed to starve. The age of a swarm is a very important factor, and when answering a "swarm call" in the neighbourhood, the beekeeper is well advised to ask of the interested onlookers who will have been there since the swarm arrived — most probably a disconsolate police officer, who has been "detailed off" to protect the bees from the neighbourhood children and the neighbourhood children from the bees and themselves — why some fool invariably wants to throw bricks or poke sticks at honey bee swarms I will never understand! Anyway, if the swarm is less than three hours old — if it is an awkward position, the bees can literally be scooped up — gently does it! — using bare hands, and no veil — newly swarmed bees are not inclined to sting. This can be useful Information if called to a swarm from work or play without your "gear." Once the queen has been "shovelled-in" the rest of the swarm will follow — just like marching ants. A day old swarm is not so pleasant and unless the weather at the time is pretty inclement is liable to "explode" when badly handled, i.e., shaken into a swarm box. Such a swarm should either be gently cut down if this is possible, or gently smoked into the swarm box. A comb of stores in the swarm box placed near a swarm of this age is about the most effective method, but rather time consuming. Before taking your swarm, weigh the situation up, know exactly what you are about, although unfortunately, the only way to get experience, is to get experience! Happy swarm catching!

August - Hive Products

Most beekeepers keep bees for pleasure and honey. Whether the pleasure aspect takes first place or the honey aspect takes first place depends on the beekeeper. I wonder how many beekeepers have ever considered the marvellous beneficial force which they have "under their hands" and how many "gifts" bees offer mankind. Honey is the obvious hive product which comes to mind when bees are discussed, but there are many other hive products with which we beekeepers in the more temperate zones do not care to concern ourselves. Products like royal jelly, propolis, pollen, pollination wax, honey dew, bee venom and bees themselves, or rather bee brood. The uses to which honey can be put are many. As a very palatable food honey is most popular, despite its ever increasing cost. Honey is a marvellous wound dressing, being an antiseptic, due to various chemical properties. It is hygroscopic, in that it is capable of absorbing water. This characteristic makes it virtually impossible for the pathogens, which are specific to man, to survive in honey — all living organisms require water for growth. Honey in absorbing any water in the presence of bacteria and other micro-organisms deprives them of the water they need for development. Despite the apparent sweetness of honey, it is actually very acid. Its pH value is 3.9. In fact, the number and types of acids occurring in honey is quite staggering. The most important acid present in honey is gluconic acid. The list reads like a chemistry lesson: aspric, malic, succinic, formic, acetic, butyric, lactic, pyroglutamic, phosphoric and hydrochloric acids. Honey also contains trace quantities of at least 16 amino acids, these amino acids form the building blocks of proteins which we require for growth and development. There are also many vitamins in honey, albeit in trace amounts, substances such as thiamine, riboflavin, ascorbic acid, pantothenic acid, pyridoxine and nicotinic acid. Most of the minerals which we require for development are present in honey: potassium, chlorine, sulphur, calcium, sodium, phosphorus, magnesium, iron, copper and manganese. Many of these minerals play important roles in the enzyme systems of the human metabolism. Honey is widely used in baking due to its sweetness and hygroscopic quality — where honey is used instead of sugar, baking remains fresher longer — the honey draws moisture to cakes, etc., where these items, when baked with sugar, are more likely to give up their moisture. Pipers know about the hygroscopic quality of honey — it is used inside the bag of the pipes to keep it soft and pliable. The two most important sugars in honey are dextrose and levulose. Dextrose is another name for glucose and is a pre-digested energy source which is assimilated directly into the bloodstream, making it an ideal pick-me up for invalids, babies and sportsmen. Levulose (fructose) is not so readily assimilated as dextrose and due to this characteristic is a backup source of energy which "comes on stream" as the dextrose begins to lose its effect. Honey contains about 32% dextrose and 40% levulose.

Royal Jelly is the secretion which nurse bees produce in their hypopharyngeal glands (head glands). It is fed to queen bees throughout their larval and adult lives. It is the richest palatable source of pantothonic acid, apart from yeast. Pantothonic acid is an important amino acid for humans. Royal jelly is highly prized by "natural food" people and is very expensive. It is a creamy white colour and tastes rather sour, due to its high acid, nitrogenous content. The Japanese actually have a thriving industry based on the harvesting of Royal Jelly.

Propolis enjoyed a brief period of fame a few years back, where it was highly sought after and therefore very expensive. It is used in pharmacy in the treatment of burns and skin complaints. Propolis is bacteriacidal and plays an important role in the hive, not only as a "draught excluder", but also in helping the hive stay healthy. There have been cases reported of dead mice being found in overwintered hives, these mice were completely "cased in" with propolis and showed no evidence of decomposition.

Pollen is at present enjoying a great deal of popularity as a highly nutritional stimulant in Denmark and Germany. Pollen is the male germ plasm of plants, and forms the chief source of protein, fats and minerals in the honey bee diet. Chemical analysis of pollen has shown it to be rich in lipids (fats), and to contain free amino acids, carbohydrates (sugar, starch, cellulose), minerals similar to those found in honey. In fact, it is highly probable that the minerals found in honey are due to the large numbers of contained pollen grains in honey — this pollen gets into the honey by being accidentally shed from the hair of the bodies of foraging bees, while they are in the hive. Pollen also contains numerous vitamins, similar to those in honey. Pollen is also rich in pantothonic acid. Although pollen has been demonstrated to have direct nutritional benefits for man where it is consumed, the most important function of pollen in nature is as the fertilising agent in plant reproduction. The importance of honey bees in this pollination process cannot be too strongly emphasised.

Pollination is becoming increasingly important as a hive "by product." Agriculturalists and horticulturalists the world over are now beginning to appreciate just how critical pollination is in farming. As the numbers of naturally occurring insect pollinators, like bumble bees, solitary bees and solitary wasps diminish due to the destruction of their habitat and the abuse of agro chemicals, so the role of our honey bees in pollination increases in importance. So much so that orchard operators and fruit growers are now being forced to pay premium prices for the privilege of having hives of bees available during the blossoming periods in their acreages. I would go further, and state that the role of bee keeping in the agricultural economy of the world will become increasingly more important and could become one of the most dominant factors in the efficiency of our producing industry. As a well known fruit grower in the south once stated in the writer's presence, and I quote, "I hate beekeepers, but I simply can't do without their bees!"

Beeswax is a true wax and is non-rancid. Research has shown that to produce a pound of wax, bees have to consume, on average, 8.5 lbs. of honey, by a simple calculation the cost of 1 lb. of pure beeswax at 1980 prices for honey works out at £8.50 per lb. Beeswax consists of hydrocarbons, monohydric alcohols, diols, acids, and at least two paraffins (not Esso blue or Alladin pink!) The current production of beeswax in America alone runs at around 5,000,000 lbs. This figure represents only half of the amount of beeswax which is needed in American industry. Beeswax is used widely in the cosmetics industry, where it forms the base for such concoctions as lipstick, cold cream, ointments, pomades, rouges, etc. The church also uses a great deal of beeswax in the form of candles. The beekeeping industry is itself, of course, a massive factor in the demand for beeswax for the purpose of foundation manufacture.

Honeydew is not a significant hive product in Britain, but where it is harvested it should not be either discarded or underestimated — honeydew contains many of the higher sugars such as maltose, erlose, melezitose. These higher sugars are important in medicine. Honeydew is extremely rich in potassium. Honeydew is not recommended as a food for overwintering, but if returned to the hives in the spring with adequate water supplied it can make an excellent contribution to spring feeding. In Europe, honeydew honey, termed "Wald honey" is looked upon as a delicacy and is expensive and highly prized. Beekeepers actually migrate their hives to the spruce and pine forest areas to obtain this much vaunted "by product" of aphids and lachnids.

Bee venom is stored in the poison sac. New born bees have very little venom. The amount of bee venom produced by the bee gradually increases and accumulates to about 0.3 mg. in a 15 day old bee. When a bee has reached the age of 18 days old, no more venom is produced. Thus, when the poison sac is emptied it cannot be replenished in a bee older than 18 days. Bee venom is now widely used in the treatment of certain types of arthritis and in desensitisation of people who are hypersensitive to stings of any kind.

Bee Brood is a very important nutritional factor in countries like tropical Asia, Africa, South America and Australia. Certain cultures habitually consume larval and pupal stage honey bees. Honey bee larvae are rich in fat and protein, vitamin A and vitamin D are present in considerable quantities also. In Canada, where many beekeepers kill their colonies in the autumn because of the greater economic advantage in buying package bees from the South in the spring, there is therefore a vast potential for bee brood if the market were to be developed. Bee brood has been found to be an excellent fish bait, and pet food. Honey bee larvae could be produced on a tonnage basis if a market could be found since the average bee colony can spare up to 1 lb. of brood every six days without detriment, during the active season. The wholesale decimation of the brood of our bees is not being advocated here. The statements made are from a purely interest aspect and do not constitute acquiescence on the part of the writer. The vast storehouse of benefits which bees represent for mankind have only been touched on here. But suffice to say — that the world needs its beekeepers — more than it knows! Perhaps even more than the beekeeper himself knows!!

PREPARING FOR WINTER

The bees have already been at the heather for at least four weeks. The beekeeper's year has begun again! The late success or failure of the colonies now preparing for the cold long haul through the dreaded Scottish winter (has to a greater or lesser degree, been decided. Did they get the chance to work well at the heather? How good or bad was the weather? Did the colony raise the population of young bees which will be a major factor in the colony's ability to overwinter with the heather pollen? Did the colonies have good fecund young queens present while at the heather? So many questions, and many more still to ask and answer before the bees are brought back from the moors. How do I know when to take the bees off the heather? I have been asked this question many times. The heather bloom normally endures four weeks, five at the out most, so any time at the end of the fifth or sixth week from the beginning of the bloom period won't be far out. The important fact about removing hives from the moors is: the bees should be given time to seal the honey in the supers before that honey is removed. By leaving the bees "put" for around the week after the "effective" bloom has ended, the danger of taking off "watery" honey which will soon ferment is greatly obviated. By observing bee behaviour and observing the heather bloom it is quite easy to tell when the heather "nectar flow" has ended. By noting activity at the hives on any bright early September day the availability of nectar in the field will be adequately demonstrated — if the day is mild, dry and sunny and the bees at all hives appear listless — the nectar flow is over — the honey bee by virtue of its nature and industry will not expend energy needlessly. By looking at clumps of heather at varying altitudes in the vicinity of the apiary, the hive observation may be confirmed.

Spent heather blossom begins to tinge brown, so where the major part of the observed heather blossom appears brownish it is time to take the bees home. If all the heather blossom shows the brownish tinge the bees should have been home at least a week ago. Or at least the honey should have been removed a week ago. Where the beekeeper works single handed, as I do, it is good practice to clear the honey from the hives using a "clearing" board. If the clearing boards are placed in the late afternoon or early evening the honey supers can be removed the following morning, or at latest the following evening. In any event no longer than 24 hours. Once the supers are clear of bees the honey in them becomes very vulnerable — especially if the clearing board is not bee tight. Honey bees are, as we all know, inveterate robbers. The home apiaries should have been tidied up and stances renewed where necessary. By stances is not meant some "earthquake proof" structure of 4 in X 4 in, although the beekeeper who takes the time and effort to build such stands will be adequately rewarded by the ease of working his hives at a higher than back breaking height. But a good sturdy stance can be affected by a piece of 2 ft x 2 ft (550 mm x 550 mm) corrugated iron merely placed on the ground such that it is tilted slightly forward for draining purposes. A simpler but just as effective stand can be provided using three half bricks — two at the front of the landing board and one to the rear. In all cases the hive itself should be tilted slightly forward to facilitate drainage. The most important factor which must be considered when stands or stances are being prepared is: at all times the hives should be set down facing south, when they will be well sheltered from the prevailing wind and exposed to the maximum winter sunshine. Bees can survive long periods of confinement during the winter — so long as they have adequate stores, adequate populations and are healthy. An important factor in maintaining health in over-wintering bees is that they have ample opportunity to void their gut at least every six weeks during the winter. No matter how severe any winter may be, there are always the occasional mild days, even during December and January. If hives are placed exposed to the maximum sunshine the bees will be able to detect the rise in temperature outside as the sun heats the front of the hive. Next time it is bright during the winter, place a hand on the front of the hive after the sun has been out for around an hour or so — you'll be surprised. Since I learned the lesson of "south facing" in winter, my winter losses have been drastically reduced! When bringing hives home from the heather it is useful to make a note of any colonies which feel relatively light — these colonies should be fed at least 15 - 20 lbs of sugar. All other colonies should be fed at least 10 lb of sugar as a matter of course. All feeding should be finished by early October, unless of course the weather is very mild as it was during this period in 1979. But where late feeding is undertaken, for whatever reason, if the feed is sugar syrup, thymol should be added, around a teaspoonful dissolved in ether/gallon of syrup, to inhibit fermentation. If the colonies can be wintered in double brood boxes with full deep combs of honey in the top box. Then the beekeeper can feel easy that his bees will come through to the spring in good shape all things being equal and providing the hive is not located in a bog or a frost trap. The hive entrances should all be checked around this time to ensure that the entrance is not more than $5/16$ in (8 min) high. Beekeepers have all-got their various ideas about ventilation in a hive in Winter. Ventilation during the dormant season is of paramount importance in Scotland. Dampness is more of a killer than cold in this part of the world. With due regard to the direction of the prevailing wind, a full width entrance is recommended so long as it is not higher than $5/16$ in If the end combs are removed around mid-November on a cold day — it will save them from becoming mouldy. By having an empty super above the brood box, topped by a crown board under the roof, dampness will be virtually eliminated from hives which are unlucky enough to be overwintering in even the dampest apiary — providing air can circulate round the hive. Under no circumstances should a hive be overwintered with the queen excluder in place where there is honey above the excluder — the cluster will move up gradually into this honey leaving the queen below the excluder!

EFFECTIVE UTILISATION OF MATURE "HEATHER" BEES AFTER HEATHER "FLOW"

When a bee colony is brought back to its "winter" stand, if it has had a favourable August, it will be literally "bursting" with bees. Many of these bees are bees which will have been foraging for the duration of the colony's stay on the moors and will be relatively old and weary from their labours — after battling against wind and weather on the exposed moorlands. There is an old adage in beekeeping which comments on the beekeepers' wont to pack all sorts of insulation on top of his clustering bees to keep them warm in winter, i.e. — old carpets, jackets, blankets, etc., etc. — it runs: "the best winter packing for bees, is bees." I would like to, if I may, elaborate on this most important piece of advice as follows: "the best winter packing for bees, is young bees."

Normally, in a fully developed colony returning from the heather the queen will have virtually ceased to lay, having probably laid around 25,000 eggs at the heather flow. There are thus plenty of bees, young and old in the hive at this time, to see the colony through the winter. I suggest that the older bees are completely superfluous to requirements and that if they are eliminated from the hive the colony will overwinter much more satisfactorily than if they remain in the hive.

To substantiate any seemingly sacrilegious statement I would draw the attention of the observant beekeeper to a phenomenon which he has probably seen but not "rationalised." Recall the autumn of 1979, after the atrocious summer and mediocre heather flow. During mid-late September and October the weather was mild. The hives were active longer than normal, before the winter clusters formed. Now cast the mind to the spring of this year, 1980. How did the winter mortality rate, i.e. dead bees on the floorboard of the hives, compare with seasons where the past heather period was unfavourable for bee flight? In my own hives the floorboards were surprisingly free of dead bees in the spring this year. Due, I would suggest, to the late flight activity last autumn, the bulk of the otherwise "winter" wastage occurred in September and October before the winter cluster formed. Conversely nucleus colonies which had come back from the heather covering 3-4 frames, when examined at the end of October 1979, were covering at least one further frame. All colonies were of course being fed during September and into October. The nucleus stocks had obviously continued rearing brood during September, when pollen was still available and they had utilised part of the late feeding to rear more brood, which seemingly had offset the "late activity" losses of the older bees of the nucleus. Another factor which points to the sense of eliminating the older bees at the end of the active season is the appearance of the brood combs out with the winter clusters position, in the spring of the following year. The overwintered colony will normally be covering not more than five frames by mid-January, probably less, (but if the brood combs on either side of the cluster are examined it will be seen that they are not quite full of stores — even the outside combs —if they were not removed to improve winter ventilation) will either be empty or have only a "crown" of stores under the top bar. The stores in the centre of these outside combs will have been consumed by the older bees which subsequently died off during the winter period. To remove the older bees without risk to life or limb at the end of the heather bloom is easy. But most beekeepers will baulk at the seeming heartlessness of eliminating these older bees and elect to take their chances in the traditional way i.e. leave well alone! If these unwanted bees could be utilised in nucleus stocks to boost the cluster density and also stimulate the queen in the nucleus to lay at the end of August - a double advantage is obtained. The fully developed colonies will have been relieved of large numbers of bees which would otherwise deplete the colony stores and congest the hive entrance with their bodies due to winter wastage. The nucleus colonies will receive a massive injection of labourers who will forage, warm, and convert fed sugar syrup into either brood or honey and will diminish in numbers each day in the "field" as they work to build the size of the nucleus. Such that by the time the bees settle for the winter the older bees will have gone to the "happy hunting ground, where clover blooms for 6 months of the year, and heather blooms for the other 6 months", but the "young bee" population will have grown. To get these older bees into the nucleus, just go to the bees at the heather on a good bright day in the first week in September, while all the hives are active. Block up the hive entrance, using smoke where necessary, and load the hive(s) onto your transport. Place a queen-right nucleus on the site of a removed hive — the flying bees from the original hive will home into the nucleus as will the foragers of adjacent hives which may have been removed at the same time. The nucleus will of course be housed in a full sized brood box with a full complement of combs. It is well to go into the nucleus before doing the "switch" and cage the queen of the nucleus in a matchbox. This will safeguard her during the transition stage after the nucleus has been put down. Bring the nucleus home any evening after that — the sooner the better — and start feeding it hard — 10 to 15 lbs. sugar in 1:2 syrup (one of water; to two of sugar). The combined results will, I think, surprise any beekeeper prepared to have a go! To make a thoroughly good job of the "colony switch" all hives concerned should be checked out for disease.

NOVEMBER

I have always had great difficulty in trying to visualise, without a diagram, the parentage of the drone caste in the honey bee colony. To the lay, non mathematical mind, like mine, the fact that the drone is one generation in time, out of step, so to speak, is rather mind bending. A drone, although produced from an egg laid by the queen of the colony, like the worker, who is also produced from an egg laid by the queen of the colony, is not really the "child" of the queen in the same way that the worker is the "child" of the queen! In the honey bee system of reproduction, a virgin queen leaves the hive to mate either in a cast or as a supersedure queen, or as the last queen to emerge in the hive after the prime swarm and the various casts have gone. The virgin queen mates with between 5-10 drones in flight in rapid succession: how this event is brought about still remains somewhat of a mystery to me. I have read the books and thesis viz. Professor Woyke, Professor Ruttner, Brother Adam and others. I have also seen the signs at the hive, as will the observant beekeeper, namely the mating sign, and the subtle variations in the appearance of different generations of workers in any particular colony over the active period. The appearance of overlap in worker variations makes the observation easier, a colony will appear to have a number of distinctly differing groups of workers in the hive at any one time, at most three variations, but normally two. The groups will have subtle differences such as banding of the abdomen different in colouring or disposition. The abdomen may appear longer and sharper. Even the character of the hive may change with successive generations. If observations are carried out diligently the differences are easily noted — in the worker caste. The drone population, however, changes chronologically but not physiologically, right through the various generations, i.e. a spring drone will look identical to a late summer drone. Size may vary, but not body colouring or marking characteristics (unless it is a "stray"). When the queen of the hive returns from her mating flight she can, after a certain time lapse, some 5-7 days, lay eggs which being fertilised will produce worker bees which, as the beekeeper knows, are really infertile or sterile females. These workers are genetically directly related to their parent queen and the respective drones with which she mated. Not so, however, the drone. The drone is produced from an unfertilised egg, by a phenomenon called Parthenogenesis, and discovered in honey-bees by a Polish clergyman, Dzierzon, in 1845. Drones can be produced by the queen without her having to be mated. These drones are naturally, normal and complete in all respects. And, being produced solely by the queen without any male agency, the drone must be identical in genetic make up to the queen. The queen however owes her genetic make-up to her mother and the drone with which the mother mated. Thus. since the queen and drone in question have the identical genetic structure, the drone is really the son of the queens parents and therefore, genetically speaking, is the queen's (from whose egg he was produced) brother! It still gives me a headache when I think of it!! Very interesting you say, but so what! Well, to take the "drone syndrome" a stage further. The average beekeeper when he wants to increase his stocks in summer will hive any prime swarm he gets from his hives, in another hive and if he knows what he is about he will go into the parent stock (which will of course have no queen, but plenty of queen cells) and do one of two things; he will either cut down all the queen cells present save one (the cell to keep is one on the face of the comb, not near the top bar of the frame, nor too close to the lower edge of the frame, and certainly one which does not have drone cells in its vicinity) if he only wants to increase slowly.

Or he will save the best of the queen cells and make up nucleus stocks for them, if he intends to increase rapidly. If he leaves more than one queen cell in the hive after the prime swarm has gone the hive will probably cast — and that is bad beekeeping! To continue on the "drone syndrome", queens may be produced more or less at will by beekeeper manipulation of hives. Either very early in the spring or late in the autumn. The main problem with early or late queens (virgins that is) is that mating is either difficult or nearly impossible. In the Autumn, the difficulty due to fall off in drone rearing and availability of drones of the correct age a drone is not sexually mature until it is at least 10- 12 days old (i.e. 36 days old from egg laid!). Drones older than 4 weeks are not to be recommended for mating purposes (i.e. 64 days old from egg laid!); in early spring it is nearly impossible because of the natural scarcity of mature drones and favourable mating weather.

In traditional beekeeping we call for a fecund mated queen in our hives. Consider the "unconventional", a fecund unmated queen. She will lay eggs like a normal mated queen, she will be genetically similar to any sister queens which were correctly mated, therefore, the eggs she will lay will be genetically sound eggs — but they will all be drone eggs. A virgin queen which does not mate within 4-5 weeks of emerging (i.e. 49 days after egg laid!) is no longer capable of mating, due to permanent changes in her spermatheca (the semen storage vessel!). Any virgin queen can be physically restrained from flying to mate by the simple device of caging, shortly after emergence from cell. In the autumn, end of July, a good unmated queen could be maintained in a mating apiary by being supplied with ample food and a frame of sealed worker brood every 11- 12 days and where the bulk of the comb available for laying is drone comb. By using an 11-12 day cycle for worker brood introduction, the previous frame given can be withdrawn at the same time, since all brood will have "hatched". The drones thus reared will "flood" the area and give mature drones of correct age for mating right up to the end of September. Similarly, in early spring, if the drone laying queen stock is managed well it will overwinter without trouble, giving the facility that as soon as any brood rearing begins drones will be reared automatically (and from early January), to be available for mating-purposes around the first week in April where weather permits and where the beekeeper has made the manipulations necessary to produce early queens. All that has been suggested I have already either proved or observed. I have seen drones flying right up to the first week of November in mild years. I have had queens mated as early as the end of April in really exceptional springs without using a "drone bank" — all the ingredients are there. As the old proverb goes in the case of our "drone layer", we will have created a virtue out of what in beekeeping was previously considered a sin. Ideally, we should have at least three drone producing queens to limit "in-breeding". A word of warning about virgin queens mated after mid-August — they do not normally lay in the current year. The following spring is the earliest their correct mating can be established.

BEE PROTECTED

It goes without saying, almost!! That the act of working a strong colony of bees should never be lightly undertaken. Surprisingly though there are many beekeepers among them the over-enthusiastic beginner and the very experienced old timer who, for their various reasons scorn adequate protection. I heard of a newcomer to the craft a couple of summers ago who would not wear a veil. He would turn up on a Monday morning for his work with either a black eye or a thick ear or with sundry lumps on his forehead and chin. I have never met the man personally but the tale just related is nonetheless true. His reasons for the foolhardy "idiot"-hardy! behaviour appear to have been deeply seated. After being impressed by reports in some book he had read, of beekeepers working bees without veils or gloves. This, he seems to have convinced himself, is man against nature in the extreme, and in his ignorance he rose to the challenge. The 'hero' of the book of course was obviously one of a number of things: either he was a charlatan who was not quite telling the truth, or he was a very experienced hobbyist beekeeper who knew how to read the signs of the weather and the hive or he was perhaps a foreign beekeeper working the gentle bees of Greece (*A.m. cecropia*). The bees were most certainly not A.m. Mellifera, Ligustica or even Carnica.

Carnica is the now predominant honey bee in Germany, reputed to be gentle, but the reports are not strictly true. I have worked with Carnioleans in Germany, wearing a veil of course I can assure you, despite the reports, that some strains of *A.m. carnica* are not averse to stinging. Carnica incidentally was introduced to Germany from Russia and Eastern Europe many years ago. It was reputedly a more prolific bee than the indigenous *A.m. Nigra* which is still to be found in its pure form in the Luneberg Heath region.

Getting back to our beginner again, I can sympathise with him to a degree because I too in my early "Don Quixote" days in beekeeping scorned gloves, and a proper beesuit — I wore a veil however! I reasoned "I am giving the bees a hard time so I must be fair and give them a chance to get back at me" — and they did. Once I got past the 4 hive stage I decided enough was enough — I could no longer pick my days to work the hives. I became a "Bob Couston" disciple and acknowledged the importance of the "Nine Day" inspection system.

I very quickly learned to protect myself. I bought a "zip" bee suit, a sturdy pair of bee gloves and a good pair of "wellies" (which I learned quickly, very quickly, to seal off at the top: Having the soles of your feet stung to bits is no way to enjoy beekeeping!). I also found out the temptations of the "Sahib's Sun Helmet" supplied for use with the square visor folding veil — I kept walking into low tree branches, because I couldn't see above waist level with this hat. I now use the type of safety helmet used widely in industry and can see for miles!! It is extremely important to ensure that the bee suit is completely bee tight, i.e. all access through the suit (for bees!!) like side slits and any button holes are sewed up and that the veil put on before putting on the boiler suit, and then the zip pulled right up to the veil. It is also a good policy to mark the gap above the zip by wearing a light scarf. So clad, you are ready for anything. The unexpected can always happen when working bees — you can drop a frame loaded with bees, you can stumble carrying a strong hive or otherwise upset your "charges." who will proceed to give you a "leathering" if you are not well protected. When I take my bees to the heather it is normally my policy to return the following week and check the hives for queen rightness — at this time of year pollen carrying merely indicates the bees are still feeding young brood.

The queen could meanwhile have been fatally damaged in transit to the heather. It is also a good practice to remove at this time perhaps two frames from the hive, which have only honey and pollen on them and no brood. These frames can be replaced with two frames of undrawn foundation which the bees will draw out and in which the queen will lay — ensuring that at least 8,000 new bees will be there in the spring when they are most needed.

One particular year, 1977, I was accompanied by two other beekeepers. The date was the 12th August, the previous week the weather had been idyllic. We went to the bees to make my normal management manoeuvre. In the early stages the bees that day were merely "awkward" but soon the whole apiary was in an uproar. The bees were completely berserk, and my two companions soon began to hop about —they were being cruelly stung — a number of hives were open, but the whole apiary was seething, Adansonii had nothing on my bees that day.

My two friends decided they'd had enough — I indicated that I was not being stung, but we were all covered in angry, frustrated, "hell bent on murder," frenzied bees. I remained in the apiary to put the opened hives to rights and to tidy up. My companions hopped, skipped and jumped to where one of the cars had been parked — about ½ mile away and the bees followed, reluctant to let their tormentors go. I worked away steadily—it took about 15 minutes to rebuild the opened hives, lift all the spare gear on to the truck and make my escape — every second was a year of apprehension in case the bees would find a chink in my till now at least "bee tight" armour. I drove the truck away, stopping about halfway between the apiary and my companions and got rid of the bulk of my escorts, stopping again about 100 yards short of the parked car to re-de-bee myself and remind my companions to put their veils back on again. We re-grouped at the parked car systematically eliminated the remaining bees from myself and from the inside of the cab of the truck, and took stock. Their outfits were suits that had been used successfully under normal beekeeping circumstances for many years — but the side slits for access to inside trouser pockets were open — the bees had very speedily found these openings and "Tally Ho"!

The next day it rained fit to burst and the rest of August was a complete wash-out. The bees had known this was going to be the case, probably due to their sensitivity to low pressure as the barometer fell, heralding the imminent wet weather, and had reacted accordingly.

The lesson of course is no matter how expert you are or how experienced you are, drop your guard at the wrong time and the bees will show you who really sets the pace.

I am not of course advocating that beekeeping manipulations should always be carried out in a "bomb proof" rig out. There are operations which dictate that certain precautions must be temporarily dispensed with. In early May it is good beekeeping sense to clip and mark queens — this can in no way be carried out wearing gloves. The gloves however should be left somewhere handy! Also, on particular days when the bees are busy and the sun is high, with nectar flowing freely, it is a real pleasure to discard the gloves and really get close to your bees. Indeed, gloves should be dispensed with at all times when the bees will allow it. When for instance one is taking or hiving a newly emerged swarm there is no need for gloves. With a newly emerged swarm a veil can even be dispensed with but only at the beekeeper's discretion (which should be tempered with experience!). Beginners are advised to go to their bees expecting the worst and then modifying their approach in the light of experience (which can be quite expensive and painful in beekeeping!).

A word of warning — when going to a "strange" swarm, always ask the people concerned how long it has been there. Swarms are normally extremely docile. On leaving the hive to swarm, bees engorge themselves with honey. This honey will supply them with provisions for about 3-4 days depending on weather conditions. Any swarm which is hindered from establishing itself, perhaps due to the onset of bad weather after it emerged, is, after 3 days have elapsed in a potentially dangerous condition and liable to react viciously to attempts at "taking" it.

Better safe than sorry. Protective gear can be "progressively" discarded in working with "strange" swarms after their "temper" has been established!

BEEKEEPING — THE GROWING PAINS

Beekeeping is an enigmatic art, in that it appears to follow the law of diminishing returns. To explain the preceding statement, I will quote not only my own experience but the tempered experiences of not a few fellow pursuants of the craft. In the beginning I possessed the precarious number of one hive of bees, and knew everything about bees, i.e. "You get a hive of bees and you put it in your garden, and at the end of the year you go and take honey from it". This statement will be recognised as the typical lay approach to beekeeping

— as a matter of illustration of the layman's level of the understanding of the craft of beekeeping, I was once informed that the beekeeper was entitled to charge a sensible sum for his honey because he had to wash all those jars.

Many beginner beekeepers find to their surprise that despite sheer ignorance they can get honey —unless the year happens to be a 1979er. This initial success results in a reaction akin to a nuclear fusion process — it sets off a chain effect whereby the novice is motivated to grow — i.e., acquire hives up to his elbows!

When the sensible minimum number of hives is attained, i.e., two, the moderate will draw himself/herself to full height, survey the two hive units and be happy.

Alas in beekeeping the need to reach saturation has resulted in the word moderate being struck from the "glossary of terms" in the manual of beekeeping — with two hives again the following year honey is obtained.

Visions begin to materialise — the Midnight oil is burned — ponderings are pondered, questions are asked in the deep recesses of the mind — Will I try double broods next year? "Will I put the hives side by side and move one away to another site in the apiary, thus re-enforcing the foraging population of the remaining hive with the flying bees of the moved hive? Will I try two broods on the same stand each with a queen, upended by a shallow crate then the two hives could work with the same super and fill it up quicker?

"How can I divide my hives and still get honey?" "How would a queen excluder under the brood box during the swarm period work?"

The management ploys fly thicker than a prime swarm leaving the hive — who needs the midnight oil, the mind can see in the dark!! The appetite thus, whetted by hypothetical apiculture creative management, many of us proceed to expand beyond our limited ability and experience as beginners and then the law of diminishing returns asserts itself — the two-hive successful novice becomes the four hive panic stricken beginner — all the hives thrive beautifully in that spring and early summer, the population build up fine, the honey surplus is growing daily — then suddenly all four hives are found to have queen cells.

Not all hives however are preparing to swarm, some hives are superseding but the newly fledged four hive man has an emergency on his hands because he expanded to the limit of his equipment capacity with the four hives.

Our beginner fumbles from hive to hive cutting here, thrusting there, until at the end of his hive examinations the ground around is littered with amputated queen cells, and perhaps even a crushed dead queen, killed in advertently in the process of corrective "management".

After the initial massacre of unborn queens, our beginner heaves a proverbial sigh of relief, only to find five days later that, despite his nine-day inspection routine, the hives which really were preparing to swarm have indeed swarmed in his absence. Later he finds the hive or hives which did not swarm but were superseding are queenless because either the queen was killed during the hive inspection or her last precious fertile eggs which had been utilised for the queen cells, he cut down, were indeed her last precious fertile eggs.

So, with luck the sheltered beekeeper has one queenright and three other hives in various stages of depletion.

That year he gets less honey than he did before he even started keeping bees — he was buying it then!! But experience gained, prepares the way for the following year where he/she embarks on an eight-hive disaster. The reasons for the eight-hive disaster are the same as the reasons for the four-hive disaster, lack of experience, because the management system for two hives is inadequate for four hives and the system for four hives is not suitable for eight hives — because our experienced novice does not yet have the depth of insight which comes from repeated failure. The word 'repeated' is used with deliberation since if the failures are not 'repeated failures' the beginner beekeeper concerned has probably bought himself a camera or a set of golf clubs!!

This discourse now drawing to a close is not intended to dissuade the 'growth' beekeeper or discourage the beginner — but it does highlight the potential mental suffering about to be incurred by some of us in the coming season.

If the beginner, at any stage, gets his projected aim into perspective, and accepts that in expanding he will inescapably encounter problems, then this will help to control the panic emotion. While the beekeeper is expanding he will continuously be faced with new situations.

Too many new situations in series, or worse in parallel, will result in setbacks. The Chinese have it, "softly, softly, catches monkey"!! From bitter experience the most vulnerable aspect of 'growth' beekeeping is equipment shortage. At least 50% spare equipment is needed at the beginning of each spring for normal 'static' management, where growth is hoped for 100% spare equipment is just less than ideal.

Also, the fact that, next to the weather factor, the queen bee is the most important factor in beekeeping, has in all its ramifications to be fully comprehended and acted upon for success in the craft. Once stability of numbers has been attained, peace of mind is at hand!

GOOD BEEKEEPING OR BAD! -*Apis fantastica.*
JANUARY RE -QUEENING!

There is an inherent weakness in the management of honeybees using single brood chamber overwintering. I have no doubt that many beekeeper exponents of this method found this weakness out for themselves this last winter season, to their cost. The net result is a dead colony, unless the beekeeper is aware of the limitations of single brood chamber management. I have been using single brood box management since the beginning of my beekeeping career — and although I recognise the dangers I find it economically non-viable to work my bees on doubles, except during the spring build up prior to the first nectar flow in May (sycamore!).

Working single handed it is out of the question that I attempt the transportation of massive two storey hives, the sacrifice to the beekeeper of leaving so much honey in the double storey hive I also find unacceptable.

Let it be known that like all sensible beekeepers I feed sugar syrup in copious quantities to all hives on their return from the heather, — the snag here is, if the bees have been retarded from rearing brood at the heather the copious feeding done in August/September can be converted to late brood — with subsequent loss of feed sugar for overwintering purposes and if the feeding is left too late the bees can't "ripen" the syrup before the onset of the winter cold, thus the feeding of sugar syrup in autumn, while of critical importance, does not always give the result desired by the beekeeper, especially in single brood box management.

It is now a well established fact that an overwintering colony does not call a halt to rearing brood. The queen continues to lay eggs, albeit in diminished quantity, right through the dormant season. This circumstance is of paramount importance to the bees and beekeeper where "single" management is practiced.

In a long winter situation, i.e. where the cold, inclement weather persists right up to late March early April, bees are dangerously susceptible to "isolation starvation". Literally the colony starves to death in the midst of plenty — particularly relevant in "single" management, due to the fact that in extreme cold conditions and where the temperature remains below freezing over a prolonged period, the colony cannot change its position ON the combs — consumes the stores in its immediate vicinity and dies of starvation due to circumstances mentioned. Another aspect of isolation starvation in prolonged cold weather is that even where the bees do get the occasional opportunity to change position when the temperature rises — they will be reluctant to relinquish this position due to the fact that they have brood on the combs — which they will not lightly desert — to their extreme danger!

In double brood box management, provided the stores in the top chamber fill all the deep comb above the wintering cluster so that there is no large gap between the top of the frames in the bottom box and the lower fringe of stores in the top box, then as the bees consume the stores in their immediate vicinity in the bottom chamber, they can move gradually upwards into the combs in the upper chamber, which incidentally will be warmed by the heat rising from the cluster below. The bees in a "single" unfortunately do not have this facility.

The warning signal to the beekeeper that trouble is imminent in his single storey colony is when on visually inspecting his hive from above by judicial lifting of the roof and crown board, in **cold** weather, it is observed that the cluster is right up at the frame tops. In a well provisioned hive, i.e. a colony where there are adequate stores above the clustering bees — in cold weather the bees will be well down on the frames and difficult to see!

At this first warning the beekeeper must make some attempt to ease the bees' dangerous situation, by feeding heavy sugar syrup ($1^1/_2$ lb sugar dissolved in water to give a pint of sugar solution! **Not** 1^1/$_2$ lb. to the pint of water!!) or bee candy or a kilo dry solidly crystallised bag of sugar with the bag torn where it makes contact with the bees to give them direct access to the contained sugar crystal.

The best temporary solution to the problem is to lay a comb of honey flat across the frame-tops, directly over the cluster and in contact with it — this will give the bee temporary relief, successive combs given in this way can save the day. However, if the beekeeper is tardy or that most (in my humble opinion!) uncommendable of beekeepers, the let "alone" man, where a winter peep into the hive is definitely not on the curriculum and on inspecting a hive as described, later in the winter or early spring, staining is noticed on the top bars then the bees will be either dead or dying, but most definitely if alive, in a critical condition — and most likely not suffering from Nosema, but on the contrary **"isolation starvation."** (Get the bees health checked anyway).

In a situation of this kind the only sure procedure which gives the bees a chance of survival — which indeed they have (provided they are not at the black, greasy stage!) is to open the hive, remove the outermost combs — which you should find to your surprise and relief are chock full of stores, split the cluster and slip the full combs of stores alternately between the frames of dying bees — bruising these combs slightly to expose their stores will save the hard pressed bees that bit of extra work — which could be the difference between survival and loss, then slide the frames up close and put the hive back together again.

If in the interim the weather has not been continuously extremely cold, after about a week these otherwise doomed bees will be alive, well, bright eyed and bushy tailed and will survive.

As a result of continuous observation and awareness of the dangers to bees in single brood chamber management, I have over many years devised a virtually fail-safe method of winter management for strong colonies in single brood boxes. The management of overwintering nucleus stocks is carried out differently and the survival of nucs is to a degree a calculated risk which I am prepared to take. In years where the winter is moderate, nucs have a remarkably high survival rate — but in winters similar to those of 1981-82 where the severe cold persisted for a very prolonged period, the mortality rate of nucleus colonies can be high.

No matter how meticulously attention is paid to autumn feeding, there are always some colonies which are more prodigal with their winter stores than others. It is the prodigal colonies which have to be husbanded through the early spring months. Where I find a colony looking dangerously close to isolation starvation I opt straight off for drastic surgery, i.e. I move the intact side combs into the heart of the cluster as previously described — anytime from early January onwards — I have to date lost no colony which has been so managed.

A few months ago (January 23) I checked the hive in a particular out apiary and found a few threatening isolation starvation — these I operated on. In the course of the procedure I observed sealed brood in all hives.

One hive in particular gave me cause for pause — the sealed brood was drone brood but reared in worker cells — the typical evidence of a sterile queen — she must have exhausted her supply of spermatozoa at the end of August. The colony was good and strong otherwise I marked the hive and made a mental note to check this hive again the following week to see if the brood pattern had changed to worker brood.

Meanwhile I checked some light nucleus stocks I had in the home apiary and found one nuc which had diminished so much in population that the cluster about the size of a large golf ball had crawled out of the combs to cluster on the inside surface of the empty super above the brood box. (This surface was the one facing south!!.

I realised the bees were doomed: so, I split the cluster with my fingers and found the queen, which I removed with about a dozen attendants into a match box. This match box I placed on the shelf above the living-room heater in the house. I made up some "queen candy" using honey and powdered granulated sugar and eased this into the match box. Three days later I placed a match box with fresh workers from a strong hive face to face with the original old original ones.

On January 80 I went to the out apiary with the suspect "drone layer" opened the hive, checked the brood - more drones! I went through the hive, (smoke was used) I found the queen, removed her and taking the match box with the "rescued" queen, I placed it slightly open - (enough for proboscis and front legs to project — no more!) right on the frame tops of the hive — where the bees chewed the match box to pieces, released the queen and accepted her.

How do I know? Because I have two full frames of worker brood in the hive to date (March 27) and the hive is going well.

What's that one about the early bird?"

SPRING MANAGEMENT — SURVIVAL

February already, the colonies **should** now be beginning to stir, brood rearing will have been underway on a moderate scale since early January. Time now to reaffirm that the hives have sufficient stores to cope with the steady increase in the queen's egg laying rate, and thus the number of mouths to feed.

Candy or a solidly crystallised 1 kg bag of sugar, placed right on the tops of the frames in direct contact with the cluster, is a reassuring measure against isolation starvation at this time – ("isolation starvation" results from the bees eating out the stores immediately enclosed by the winter cluster and then being unable, due to prolonged cold weather, to move over to adjacent frames which are well filled with honey and pollen).

This "hard tack" will give the beekeeper a good indication of the morale and vigour of the colony. If at any time during the late winter/early spring, staining (due to dysentery) appears on the top bars of the frames in way of the cluster, or on the paper bag of supplemental fed sugar — then the colony is in the early stages of starvation. In such an event the bees must be given honey immediately if starvation is well advanced the bees may be unable to move from their clustering position, due to the cold and or weakness. In the harder winters where wintering colonies are on single brood boxes starvation is a real threat and the hives should be visually checked internally from around mid-January onward by lifting the crown board and observing where the winter cluster is positioned on the frames. A quick look, without of course using smoke, will suffice. If the day is cold when the inspection is carried out and the top of the cluster is right up at the frame top, this is an indication that the bees are hungry It is essential to feed quickly either candy or a frame of stores (deep or shallow frame) placed horizontally on the frame tops directly over the cluster — dry crystallised sugar is not recommended in this case. This colony must now be observed regularly until the time when sugar syrup feeding may be resorted to (after pollen has been noted being gathered by that colony). In cases where the colony is obviously clustering on empty combs it is advisable to give them at least two good deep store combs placed into the cluster. This can be done easily by removing two outside combs from either side of the brood chamber — these combs will most likely be quite well filled with stores — by parting the cluster gently (do not use smoke.) and sliding these frames alternatively between the frames of clustering bees and then closing up the frames to the correct spacings. The colony will be given a new lease of life. Even if the given combs are ice cold it does not matter, the bees will soon penetrate the cappings and get to the life-giving honey and pollen underneath.

To return to the combs removed from the side of the box, if these frames are completely filled — right to the bottom bar — with stores then they can be placed in another brood box which can then be placed directly on the hive, positioning the full frames directly above and in contact with the clustering bees — the bees will gradually percolate upward into these upper frames and survive.

A word of Warning — it is not advisable to do this manoeuvre if the combs in question have merely crowns of the sealed stores above the usual arch of cells left empty by the last brood to emerge — the gap between the cluster below and the honey above will be too great and the bees will still be in trouble.

At this time of year, it is worthwhile checking the previously inserted Varroa " inserts," i.e. the sheets of stiff white paper or polyethene, slipped onto the floor board under the cluster. The winter debris if closely examined will give an indication of any external parasite infestation on the clustering bees.

Varroa is a small mite about 1 mm x 1.5 mm Braula, which is also an external parasite of the honey bee is easily recognised due to its size and dark brown to orange colour — but if in doubt, get expert advice.

This time of year is ideal for doing chores like scraping and cleaning queen excluders, the adhering brace comb and propolis being cold and brittle is easily removed by the hive tool. A particularly messy job can also be carried through at this time i.e. the cleaning and preparing of any frames from which heather honey combs were cut at bottling and pressing time. The bothersome wasps which make this job quite hazardous in late September are all long gone and there will be no bees flying, thus there will be no danger of a mass invasion of hungry bees during the operation. All the frame scrapings should be washed and strained and the wax thus reclaimed saved — if the beekeeper has more than ten hives making it worthwhile to purchase a wax mould, his hoard of wax at todays (1981) prices is worth around £8.50/lb compared to buying deep foundation. Any out apiaries should of course be visited at least once a week during this dormant period more often in the aftermath of gales and high winds. Where there is a danger of roofs being blown off or hives being overturned these visits are well worth the extra petrol — the cost of a couple of gallons is still negligible compared with the cost of replacing a colony which has been lost due to prolonged exposure.

A check of the accumulated debris at the hive entrances at this time of year can be very reassuring — the normal debris to be expected at the entrance is the "dust" of wax cappings and small flakes of dry waxy debris. If the debris looks more like rubble than dust or flakes, and the hive entrance is more than ³/₈ in (9 mm) high then the activity indicated is most probably a mouse or shrew — a baited trap left under the hive at night should do the trick in eliminating the marauder. The trap, if not sprung should be removed in the morning and replaced again at night, otherwise the victims will not be the intended mice, but small hungry birds like robin, sparrow or tit. Hive entrances in winter should always be not more than ⁵/₁₆ inches (7mm) and not less than ¹/₄ in., (6 mm) high. This will keep even the small skulled pygmy shrew at bay.

MARCH - SPRING MANAGEMENT

The month of March is potentially the most important month in the beekeeping calendar — for beekeepers who are lucky enough to be domiciled in areas with good stands of Sycamore trees. The Sycamore begins to bloom around the end of the first week in May in normal years and by the end of May it is virtually finished as a nectar source. For a hive of bees to do well on the Sycamore the population must attain its Spring peak around the middle of the second week in May. The eggs laid in March will produce the young nurse bees which will help maintain the population level, which tends to diminish as the older overwintered bees die off as the demands of early foraging in adverse weather conditions begin to take their toll. These young March bees will also determine the rate of growth of the colony in early April, since no matter how energetically the queen wants to lay eggs, she can only sensibly lay the number of eggs which the population can effectively 'brood' — i.e. cover. The smaller the colony population coming out of the winter the longer the bees will take to reach their population peak. The feeding of sugar syrup is highly recommended from as early a date as possible — as stated in earlier articles — as soon as pollen has been seen being carried in reasonable quantity in early Spring. A colony which is heavy in stores, i.e. stores that are accessible to the cluster, does not require to be fed a heavy sugar syrup, merely give such a colony a palatable drink, proportions 1 kg sugar to 2 litres water. A colony which is light in stores should be fed a 1 kg sugar to 1 litre water or stronger solution of syrup. That bees need not only honey or sugar syrup and pollen for brood rearing, but also water is easily established by observation. During the active season when there is a shortage of nectar, (nectar contains a large percentage of water) where brood rearing is underway, great numbers of bees may be observed at water sources around the apiary. But as soon as a nectar flow occurs -these water foragers quickly disappear - because the water which they require in the hive, can be obtained from the incoming nectar. In fact, the observant beginner who is still not completely familiar with the flowering cycle of nectar-bearing plants in his/her area can use the discontinued water foraging as an indication that there is a nectar flow in progress. Not to feed bees sugar syrup in early Spring, before nectar starts coming in, is analogous with the man who decides to live off his bank balance — eventually he becomes bankrupt. Bees not fed in the Spring will eat out their "bank balance" — they won't go bankrupt though. They'll die!

As previously advocated, March is very important and in this month the quality of being a good beekeeper will pay dividends. Good beekeeping in my opinion begins with acute observation, by observing the behaviour of the returning foraging bees at the hive entrance, the state of the hive can be determined quite accurately in Spring. Any hive which over a period of a week seems less active than its neighbours is worth noting either mentally or in a "log book". It might be a slow starter, but if seen to be idle when other colonies are actively engaged in pollen collecting, then there is a problem in the hive either the bees are dead, dying, very weak, queenless or have a scrub queen. It could also be infected with disease. Such a hive should be opened and examined on the first mild day even if it is cloudy (using smoke of course) after pollen has been observed being carried at other hives for a minimum of three days. This examination should determine if brood is present in the hive — do not bother looking for the queen — brood is the important factor. If no brood or eggs are present, the colony should be given a frame with some eggs or very young (not more than 2 days old) larvae. By checking the hive the following week all will be revealed — if queen cells are drawn the colony has no queen — if the brood given is all sealed or open worker brood, a queen is present — if by the 3rd week in March, where bees have been able to forage and pollen has previously been seen being carried by other colonies, the suspect colony still has no worker eggs or open worker brood then the queen present is a "scrub". A drone laying scrub queen's brood has an irregular "bumpy" appearance on the face of the comb — as has that of laying workers. The first priority now is to establish if the eggs have been laid by a "scrub" queen or laying workers . This is quite easily ascertained so long as the beekeeper is able to see eggs. Eggs laid by a queen are attached to the base of the cell pointing vertically upwards, when newly laid. Eggs laid by laying workers are found adhering to the sides of the cell some distance from the bottom of the cell — due to the shorter abdomen inhibiting the laying worker from reaching the bottom of the cell. Usually laying workers place more than one egg in the cells in which they are laying.

With a queenless colony, if it is still strong in bees, and the beekeeper is a "taker of calculated risks" — not a gambler! — it can be worthwhile to allow the bees to raise another queen. A queen raised from eggs in late March (3rd to 4th week) will be well capable of mating up to the 1st week in May. If eggs are not seen in the hive by 12th - 14th May, the queen either did not mate or was lost on her mating flight The calculated risk factor involves the anticipation that the Spring weather will be favourable to early drone rearing and give some warm days in late April/early May where the queen can get out and drones are flying. Such was the Spring of 1980. If the queen "misses" then another attempt should be made immediately and with luck the hive will have a mated laying queen by early June. To maintain colony strength during this period of "luck pushing" a frame of brood given weekly or even fortnightly will suffice. A hive with a scrub queen or laying workers requires a different approach.

SPRING MANAGEMENT — (CONTINUED) SURVIVAL

Where it has been established, by the process of elimination recommended in the March issue, that a hive in Spring has a Scrub queen or laying workers then, dealing with the laying workers first of all; if laying workers are present it is an indication that the hive has been queenless for quite a long time, the queen may have been killed in the return journey from the heather, or she may have been more than two seasons old and died of old age during the winter (treat with quiet forbearance any stories beekeepers may tell you of queens achieving 6 or 8 seasons — unless the queen in question is marked, clipped and has a date in her mouth!). It is normally futile for the beginner beekeeper to try to re-queen such a colony. These manoeuvres are best left to the man of experience. But re-queening can be carried through quite successfully if the beekeeper knows what he is about. In the event of deciding not to re-queen, due to the hive being very weak in spring, if there is a queenright hive adjacent to the hive in question, by using the tried and true "newspaper" method, placing the queen right box above the one with the laying workers, the whole operation will generally result in 100% success. It is virtually impossible to introduce a queen on her own into a hive which has had laying workers for some time — she will be killed forthwith — but if this queen is introduced on a frame of open brood with a good covering of bees — the frame being itself carefully enclosed in a sealed newspaper wrap and placed into the "heart" of the queenless colony the introduction will usually produce the same high success rate as the other more standard "newspaper" method. After about a week go to the hive and remove the remainder of the newspaper wrap — which will be pretty tattered by this time — a quick check for eggs at the same time will reassure the beekeeper of the success of the operation.

For the beginner the safest way of effectively dealing with a laying worker problem, where there are other hives in the apiary, is the "dispersal method".

On a bright day when the hives are all active and the bees are flying freely, go to the queenless colony, subdue it with smoke in the usual way and then move the whole hive across the apiary away from its original stand. If there are only a few bees left in the colony at this time the bees can be shaken onto the ground providing it is dry or shaken onto the roof of an adjacent hive — they will fly to the old stand initially. Feeling completely disorientated (and this disorientation is important) — they will then fly to the entrances of the other hives in the apiary and by behaving in the subdued manner which is characteristic of disorientated bees they will be accepted without fight. Where the hive is still quite populous merely move the colony as suggested and leave the bees to fly to forage. They will return to the original stand, then drift again, disorientated, into other hives in the apiary.

When the colony has been sufficiently depleted in this manner the hive can be taken away and the bees shaken as previously stated — but, remember, the bright day! This gives the bees a chance to gather their wits, if it is dull or wet they will probably cluster at the old site and die, or if they try to enter the other hives, the day being dull, the hives will have most of their bees at home and be in more defensive mood and less amenable to accepting lost bees.

Before thinking about uniting or dispersing bees always check a sample of newly dead bees for disease. This extra effort will pay manifold dividends, since if a union is made between healthy bees and sick bees, the beekeeper is the ultimate loser. This willingness to put that little extra bit of effort into your beekeeping will assure you of success where another beekeeper will have a failure on his/her hands.

To deal now with the hive having a scrub or drone laying queen, which has to be removed. A scrub queen is not easy to find in a hive which still has a good population of bees, albeit old bees. In a case where the hive is quite populous, by merely moving the hive to another stand in the same apiary, the population will be reduced as in the case of the laying workers discussed earlier, the scrub will then be easier to find. If she cannot be found even after this manoeuvre merely shake all the remaining bees off the combs, they will behave in the same way as the disorientated laying worker colony. The scrub queen will also of course fly, but she will be immediately killed if she attempts to enter any other hive. It is not recommended that a hive having a scrub queen be united to a queen right stock, the scrub queen should be found first. Many beekeepers unite hives having older queens to others with young queens — leaving Nature to solve the equation. This is not, in my opinion, progressive beekeeping and this again is where the 'extra effort' rule comes into play — best to find the unwanted queen.

There is, however, another method of dealing with the queenless hive or a hive where a drone layer scrub has been found and removed. If this colony still has quite a good population, by giving it a frame of eggs and brood, even early in the season, i.e. March, the population can be maintained at a level where the bees will pull through to the summer. If a full frame of egg/brood is given to this ailing hive on a 14 day cycle it will raise its own queen ultimately. A check should be made the week after each subsequent frame is added to establish if queen cells are drawn. Look for eggs in the hive around 3 weeks after seeing the first sealed queen cell. If no eggs appear by this time give another frame having eggs, if the bees draw more queen cells the hive still is queenless — the queen cell either did not hatch or the queen was lost. Leave the second lot of cells for another three weeks then check again. If you are lucky and did the job well, you will have a laying queen in what would otherwise have been a write off — by mid-May or at least by mid-June.

If after that initial first cell hatches and no eggs appear by the third week after seeing the queen cell, place a frame having eggs into the hive. If the brood is sealed normally, there is a queen of sorts present. Examine the frames for eggs — if none are found give the hive another 2 weeks, and if flying weather occurs during this period — with luck your queen will mate. If, however, she has not started to lay eggs at least 5 weeks after she hatched, you will have to find her, remove her and start again with a frame of eggs. Try it, it is not as complicated as it sounds — but make sure, the queen of the hive from which the eggs are being taken is not on the frame you are removing!!

COLONY EXPANSION

In May the growth in a colony of bees which has a good queen and plenty of laying space provided, can be quite dramatic, especially if the weather during April was primarily mild, dry and windless — and the bees were being fed a suitable sugar syrup during that time - a light solution, 1 part sugar to 3 parts water, for a colony still quite heavy in stores, i.e. a typical well-provisioned colony overwintered on two brood boxes; and a heavy solution, 1 part sugar to 1 part water, for a colony light in stores, i.e. a typical well-provisioned colony overwintered on a single brood box. As the colony expands it is good beekeeping practice to always have a good stores comb next to the outermost comb which the brood nest bees are covering — thus ensuring that there is always honey and pollen within the reach of the nurse bees to feed the rapidly increasing number of mouths. By gradually moving the outermost deep combs, when required, adjacent to the expanding brood nest, isolation starvation of the colony will be safeguarded against, if the weather should turn nasty in May/June, a not unusual occurrence in Scotland!

Care must however be taken that the brood nest is not split by inserting the comb being given too far into the brood nest. Such an action sometimes results in the queen restricting her laying to the frames on the one side of the inserted comb, being reluctant to cross the barrier of honey presented to her by the full comb of stores. The net result of this is restriction of laying space due to congestion of the brood nest which naturally affects the queen's laying rate. Congestion of the brood nest is one of the "triggers" for the swarm impulse!

Spring feeding should normally continue until the first major nectar source blooms, i.e. Sycamore, but the beekeeper should not automatically desist from feeding as soon as the Sycamore bloom is on. If the weather is bad during this bloom period, although there may be an abundance of blossom the bees won't get to it — so feeding should be continued in colonies which are light in stores until the bees are seen to be working the Sycamore well.

All things being equal, if a colony is noted at this time of the year to be "marking time" or even diminishing in size there is definitely something amiss with it. It will probably be found, on examination, to be either queenless, have a failing queen, have a drone-laying queen or have a scrub non-laying queen. A colony in this condition should have been spotted earlier, but the measures to be taken described in the March and April issues still apply in May/June/July.

May is a marvellous time for the beekeeper since in this month he/she has the chance to observe the virtual rebirth of his/her colonies after the long haul through winter and the unreliable early Spring months. With the advent of May, we can hopefully say goodbye to the possibility of frost or snow recurring, and work our bees to bring them to their population peaks for the respective nectar flows in our "domains". Where the Sycamore is the only major source in an area, unless the beekeeper is prepared to "migrate" hives and manipulate early in the Spring he/she will always be at a disadvantage, since colonies will not normally be strong enough to exploit the Sycamore which blooms usually in the final three weeks of May. If the hives are left to their own devices in May then by the end of May the colony will be chock full of bees and brood with barely any surplus honey for the "master". But if, however, particular hives are selected around the end of April and reinforced with even two full frames of brood from other hives — the acceleration of build-up in the reinforced hives will result in a bigger force of field bees at the correct time, giving the beekeeper a far better chance of obtaining early honey than otherwise. The colonies which were "sacrificed" can be brought up to strength for later flows by removing frames of brood from the hives originally reinforced — thus producing a twofold benefit, i.e. the weaker hives are in turn reinforced and swarming in the more advanced colonies will be, hopefully, discouraged, especially where the frames of brood removed are replaced with frames of undrawn foundation. When "juggling" frames between colonies the beekeeper should be projecting himself/herself to anticipating colony conditions at a time three weeks after the manipulation. Moving frames of brood around does not give instant results.

It is quite a good ploy, between nectar flows, to place a second brood chamber on a hive — even where the beekeeper is a committed "single brood box" person. By giving the queen more laying space and even feeding lightly where there is a shortage of nectar, i.e. in the June gap, colonies can be encouraged to expand and perhaps even discouraged from swarming. Just before the start of the next nectar flow, i.e. clover or raspberry, the colony can be reduced to a single brood box which, if the doubling up was done correctly and timeously, will be able to be filled right out with 11 or 12 frames full of eggs and brood in all stages, producing a powerful stock of foraging bees and a steadily emerging army of new young bees to replace the natural wastage of the older bees as foraging takes its toll.

If the queen is removed from the hive at this time and placed in a nucleus with an abundance of stores and enough bees to cover two frames, with or without brood, and removed to some other apiary at least two miles from its original stand, by the time the parent hive has worked to the end of the current nectar flow, accumulating, we hope, supers full of honey due to a massive field force of bees and virtually no brood to feed, the nucleus hive will be going strong with three or four frames of brood in all stages, which can be used to reinforce the now depleted colony in good time for the heather flow.

A word of warning where use is made of this "Utopian" system — the parent stock, deprived of its queen will rear its own new queen and settle down beautifully — or it will raise a number of queens and swarm! So, at the time of reducing back to single brood chamber, check the hive for queen cells if ,any are present, cut all **sealed** cells down. Check hive again seven days later, cut all queen cells down except the best looking cell, i.e. one on the middle of the face of a comb near the centre of the colony, away from drone cells. There must only be **one** queen cell left in the hive. Close the hive up replace supers. At the end of 28 days check colony for eggs. If no eggs are seen, place frame of eggs into hive. Check seven days later, if bees have drawn queen cells, colony is queenless. Unite nucleus to it using newspaper method. If queen in nucleus is old, replace her before taking colony to heather or in good time for overwintering. If bees on the other hand seal the brood given, hive is queen right, you have a new season queen which should give you plenty of young bees to carry the colony through the winter and perhaps even give you a surplus of heather honey too. A colony so treated need not be examined again for swarming signs, but it should be periodically examined, like all colonies, to establish that the colony *is* in good order.

JUNE

Re-queening should in progressive beekeeping, be carried out if not annually, at least bi-annually. Although a virgin queen is reputed to mate with between 5-10 drones on her mating flight(s), thus receiving around 1,500,000 spermatozon, which in theory should be sufficient to carry the queen over a period of around six active fruitful years of lay, assuming that she lays 250,000 fertile eggs in each year.

Considering that in the active season at the height of her laying, a queen honey bee lays more than her own body weight of eggs each day, her metabolic rate is something to marvel at, coupled with the fact that she also has to cover vast distances as she lays her eggs in the traditional concentric pattern, as opposed to queen termites or queen ants which are sedentary, the stresses and strains imposed on her by nature are quite fantastic. Notwithstanding, however, the queen bee is undoubtedly a sturdy animal. Nature though, in her wisdom has decreed that the honey bee colony will, every summer, swarm. In unusual weather situations which influence colony growth at critical times, a honey bee colony may be inhibited from swarming (in nature!!), but this colony will, if it survives the intervening winter, without fail swarm the following summer. The implication for the beekeeper from this natural phenomenon is self evident. In nature if an organism is not vigorous and efficient (it is unable to obtain for its sustenance sufficient nutrition for its daily needs) it dies!

Since bees have been on the planet for around 80,000,000 years it is quite obvious that they know a thing or two about survival. It is good beekeeping sense to heed the natural instincts of our charges. The question may now be posed, why, if bees are such clever little b's, do we beekeepers find, and quite often, supersedure queen cells in our hives? The question has actually just answered itself. The key phrase is "find in our hives"! In nature, or in beekeeping where the "Let Alone" method is practised, without fail, weather permitting, each colony will swarm each year. Rarely will supersedure, or rather emergency supersedure (in the swarming act the parent queen is actually being naturally superseded) occur in nature or in "Let Alone" beekeeping, unless the "Let Alone" beekeeper insists in running the swarm back into the parent hive and the following week to ten days are wet.

Depending on the age of that particular queen — i.e., if the queen is already in her second season — if the colony survives the winter the bees will probably supersede in late April the following year. Queen bees produce pheromones discovered by Butler, which have been shown to be the fatty acid 9-oxy decenoic acid. These pheromones are produced in varying concentrations by queen bees. A fecund, young, healthy queen by producing the correct amounts of 'queen substance' maintains the balance in the hive. As a queen 'ages' her ability to produce 'queen substance' in the correct amount diminishes and sooner or later she will be superseded. In hives where superseding queens have been found working virtually side by side with the old queen — the old queen on being examined has been found to be producing a mere 25% of the 'queen substance' produced by a young healthy mated queen. These queen substances are produced in the mandibular glands of the queen and on other parts of the queen's body. They are taken up, incidentally by the attendant workers and distributed throughout the hive — thus every bee receives a proportion of queen substance which seems to signal that all is well within the colony.

Supersedure may occur in a bee colony where the queen although still quite young becomes damaged. If she is damaged in such a way that her egg laying capacity is reduced, which in turn depresses her ability to produce 'queen substance' the bees will replace her. However, in so saying not all damaged queens seem to be adversely affected in their egg laying capability. I have seen queens with a rear leg dragging or a missing middle leg and even once a queen with a quite noticeable dent on her thorax, which appeared to be able to carry out their duties unimpeded. In cases like this I will normally leave well alone until I begin to prepare my hives for the move to the heather. They are then replaced by young mated undamaged queens lest their injuries cause their demise during the late autumn or winter. In such cases "better safe than sorry" — with so much at stake it is bad beekeeping to risk a colony having a damaged queen, however well she may appear to be performing.

Re-queening a colony of bees is a relatively simple procedure providing certain conditions of the hive are understood and fulfilled. It is pointless trying to re-queen a hive which already has a queen in it, whether drone layer, scrub or old queen — if there are even laying workers in the hive it will be impossible to re-queen unless particular methods are used. Procedure for re-queening hives having drone laying or scrub queens or laying workers were discussed in the spring issue. However, to do a straightforward re-queening of a colony where the queen is old or damaged but still laying, it is only necessary to remove her and introduce a mated laying queen (not merely a mated queen!!!) in some kind of cage to allow the queen to re-orientate to her new surroundings.

Introduced queens are accepted or rejected by the bees of the colony, not only due to her condition, but also on how she behaves in the new colony initially, once a suitable queen has accepted food from the new bees she is virtually assured of survival — the cage helps in so far as it stops the bees from overwhelming the queen in their eagerness to tend her, before she is ready for them. The use of words "suitable," and "laying" regarding introduced queens is not accidental. By suitable is meant a queen which matches the condition of the queen being replaced. That this introduced queen is also in lay is also important — a queen which has just spent the previous 3 - 4 days couped up in a mating cage is completely out of condition and if introduced into a strong colony immediately she is received from the postman even in her 'mailing cage' she will almost certainly be killed. Such a queen should be introduced to a weak queenless colony and left for a few days until she has been released and is laying, only then can she be introduced into a strong stock where the queen however old was still laying.

PREPARATION FOR THE HEATHER - JULY

July is the month when the "heather" man begins to make his plans for the prospective move to the moors in mid. August (in "normal" years). The eggs laid by the queen from the start of July until perhaps the 21st of that month are critical for the hive population during the heather flow. The heather normally blooms for 5-6 weeks, but the middle 3-week period is the important time i.e. weeks 2, 3 and 4 of a 5 week bloom period, these are the weeks when the heather is at its best. The first week of the heather bloom is not usually very productive, and if there is a choice to be made between leaving hives "on site" for the last few days of the lime flow, against the first few days of the heather

— leave the bees on the lime! To prepare a hive of bees for the heather entails much more than merely ensuring that the hive is capable of travelling without bees escaping. It is also good practice to take virtually all stocks to the moors, except perhaps very weak ones covering less than 2 frames, but everything stronger than 2 frames should be taken — better to take a calculated risk that the heather will yield well and thus save the feeding of too much valuable sugar, than leave them "put" fully aware that there is nothing really worthwhile for the bees after the lime is finished.

To get the population of a hive going to the moors at the right level at the right time requires a bit of effort on the part of the beekeeper. The following manipulation around the end of June beginning of July, will repay the effort many fold, if the "heather weather" plays its part, and the hives worked are blessed with the proximity of good stands of Tilea (lime). For a colony on a "double brood box," subdue the bees of course, check through lower brood box and remove all combs which have no brood in them. Examine top brood box and take as many brood combs containing brood as will fill out the bottom box from which the broodless combs were removed. Place top brood box back in position over lower brood box, any frames which still have brood in them in the top box should be moved systematically to either side of top box, i.e. four frames having brood. Put two to one side of box and two to other side of box. Now take 4 frames of foundation and alternate them with frames having some honey and pollen in them, so that between each frame of foundation there is a drawn comb at least. Replace queen excluder and supers. The bees will draw out the foundation and the queen will lay in it. The adjacent frames will be used partly for laying in and partly as storage combs directly adjacent to where masses of brood will be reared. Drawn comb on either side of the given foundation also encourages the bees to draw the wax out straight and evenly. The hive should be examined on the "Couston Cycle" (i.e. every 9 days — for the working beekeeper, every seven days is just as good) for swarm preparation, from now onward. Around the end of July/first week in August, subdue hive again. Examine lower brood box, remove all frames (into a spare brood box) which have either no brood or only small patches of brood. Examine top brood box and remove all frames which are well filled with brood into lower brood box, if the lower box can be filled out with filled brood frames so much the better. But space should be left for, preferably, 2 partly drawn frames of foundation in the lower brood chamber (if undrawn foundation is given the bees tend to leave a gap between the underside of the comb and the bottom bar on drawing the foundation out — in a single brood box!).

The condition of the lower brood box now should be that it is virtually solid with brood, having the equivalent of at least two brood combs full of honey and pollen on these brood combs, and two partly drawn combs one on each side of centre comb in the box. (Undrawn foundation should be used if no partly drawn comb is available. On no account should empty drawn comb be given — the bees will merely fill these up with honey in a good flow and congest the brood next). Now shake all the bees from the top brood box into the bottom brood chamber, smoke the frames clear, replace queen excluder and put on a shallow crate of empty drawn comb — your double brood box hive is now a single, heaving with bees, chock full of brood, having adequate stores and the queen has space to lay initially in the shape of the foundation given. The empty drawn comb makes the hive easier to transport by being lighter, and also gives the colony extra clustering space during the move to the moors. The spare combs having small patches of brood should be given to weak stocks or nuclei — this will give them quite a boost too. For a single brood box the procedure is much the same, but there will be fewer combs with large areas of brood than with the "double box." Remove all combs having no brood and replace them with preferably partly drawn foundation (for reasons given previously), alternated between the combs in the middle of the brood box. Replace queen excluder, place empty shallow of drawn comb on top and close hive. The honey removed in the course of the operations described should of course be extracted or if the honey is in shallow frames and not quite ripe it should be given to the hives going to the heather for finishing. It is good practice to mark these frames to identify them from combs which will be filled at the moors with we hope, pure ling honey!

Any deep frames with good areas of pollen should be given to weak stocks or nucs. With deep frames of honey the choice is the beekeeper's.

Either extract and give wet combs to other hives or retain for use at some later time, or use to bolster the stores in nucleus hives.

Deep frames of sealed stores Stored dry over the winter is a marvellous "peace of mind" device, since these can be given to hungry stocks in the spring — with rewarding results. If the manipulation of placing the empty shallow with drawn comb on the hive is done at the right time, i.e. about 4-7 days before moving to the heather, the bees will have re-propolised the hive parts quite effectively making the hive itself more secure for the migration operation. Securing hives adequately for transportation is critically important for the success of the operation, and the beekeeper must use his judgement, depending on his type of transport. There are beekeepers who move hives without even blocking up the entrances — but this takes skill and good judgement — luck also plays a part in this method: breakdowns, punctures.

Moving very early in the morning is the trick here. The hive can be wedged, pinned, screwed, banded, roped, taped, you name it, but the key word is secured. Happy migrating!

P.S. — For best results the queens going to the heather should be not more than one winter old.

AUGUST

Honey is money, is honey. This statement of the apparently patently obvious is, however, not as true as one might imagine. Honey to the uninitiated is a delectable substance which can be obtained at a slightly higher price than best jam from the shelves of supermarkets. It has a mild sweet taste and can either have the appearance in the jar of peanut butter or be a translucent light amber colour (water white on the shelves of American supermarkets!).

The colour and taste of most honeys obtainable on the supermarket shelves are no mere accidents — these honeys are the result of years of extensive market research into the "tastes" of the great British honey eating public. The honey which most of the commercial packers supply is carefully blended to produce a homogeneous flavour and colour year after year, in fact, the bigger the packer the more standard the product. Many lay people who have a "honey tooth", when presented for the first time with naturally blended honey (blended by the bees!) immediately on tasting it, ask the identical question of the beekeeper: "What is the difference between your honey and supermarket honey?" From purely legal considerations there is no difference — whether honey is blended by a human agency or by an apical agency — if it is being sold as honey — it must be pure, and the sole product elaborated by honey bees with no impurities or additives.

There are, however, certain rules which must be adhered to where honey is **being** bottled and in bulk packing of honey the limiting factor in efficient packing procedures is the viscosity of honey. Viscosity is the measure of stickiness at particular temperatures — depending on its temperature honey will either flow readily or it will indicate the honey viscosity is such that it will just not flow at all. In the application of heat, honey can be irreparably damaged either nutritionally or edibly. (Research has shown that honey heated to a temperature of 135°F (56° Celsius approx.) undergoes irreversible physical and chemical changes if this temperature is maintained longer than a particular time. Heating honey affects the diastase content, diastase is an enzyme found in honey, its function is not fully understood but it is very easy to measure.

Germany places great importance on the diastase level in honey and its level is used as an index to establish the heating history of honey — if the index is too low the honey is downgraded either to commercial use or to be sold merely as invert sugar. In recent years though it has been recognised that the diastase content of honey can be naturally low and also, that the length of storage of an unheated honey can reduce its diastase index.

Another heat sensitive material present in honey is H.M.F. (Hydroxymethyl-furfuraldehyde), when honey is heated to increasingly higher temperatures the H.M.F. content also increases — this is a much better index for tracing the heating history of honey than diastase. Apart from the need to pack honey efficiently, the application of heat may be used to destroy all the dextrose crystals entrained in honey — the elimination of even the most minute of these crystals is very important where honey is being marketed in a form where granulation would ruin the product, such as chunk honey, or honey bottled in fancy containers with narrow necks — these are mostly found on the American markets. If heather honey is heated too much it performs in much the same way as the white of an egg as the egg is boiled. The heather honey caramelises due to a protein in the honey reacting to the applied heat. Caramelised heather honey is virtually unpalatable. In selling his/her honey direct from the hive the hobbyist beekeeper's honey incurs none of the potential damage which heating can cause to honey, provided the beekeeper strains his honey to remove the entrained wax flakes, bees' legs and occasional particles of propolis, and uses no heat in the process — he/she can unashamedly market honey as the pure natural unprocessed product of the hive.

To avoid the hard natural granulation which will occur with many unheated honeys a "seed" of a soft previously granulated honey can be introduced into the honey in the ripener or storage drum, stirred well through the honey over a period of perhaps a week. When the honey begins to look "porridgy" it is time to bottle it, such a honey will granulate like very thick cream and not solidify to "concrete" conditions.

One more very important reason why honey, which is intended for the supermarket, where its "shelf life" is of the utmost importance, requires to be heated to prevent fermentation. Honey which has a water content of around 17% will not normally ferment — as the water content increases, the risk of fermentation increases. Where a honey has granulated the risk of spoilage through fermentation is dramatically increased, due to the percentage rise in the water content of the remaining honey (primarily fructose!) as the dextrose forms crystals and comes out of solution.

Fermentation is caused by yeasts which are entrained in the honey. By heating to around 145°F for very short periods of time these yeasts are totally destroyed — thus eliminating the prime cause of fermentation. We beekeepers operating on a small scale need not worry too much about fermentation, since we either eat or sell or give away our honey to be eaten, quite quickly. Unless of course we extract and bottle raw nectar - which is easily recognised in the comb by virtue of the ease with which it can be shaken from the cells when handling combs during a nectar flow.

The main differences between honey straight from the hive and commercially packed honey are:

(1) Flavour — The beekeeper working his hives for different nectar flows can produce honey having the flavour of the dominant nectar bearing plant in the area at a particular time of year, i.e. sycamore, hawthorn, clover, lime, bell heather, willow herb, ling heather, etc.

(2) Nutrition — Most commercial honey is highly filtered to give it its eye appealing sparkle. This filtration removes the bulk of the entrained pollen grains present in the honey. The beekeeper does not filter so finely — if at all — best to leave the honey to settle out naturally, thus allowing the wax and other debris to float as a scum to the surface of the ripener.

(3) Interest—The customers buying from the beekeeper, get personal contact and a lesson on the natural history of the honey bee and an insight into the mysteries of the hive, if they ask a few pertinent questions during the handing over of the jar!

(4) Quality — The hobbyist beekeeper will by dint of his small number of hives lavish more care into the selection of his honey for sale and will only sell the best honey extracted from combs which are fully sealed or have a minimal number of open cells on the comb.

BEE THINKING

The honey hunter, in his natural state, on these islands is a historical fact. We modern beekeepers are all "honey hunters" in our own right. But not to be confused with the "intrepids" of yore who could smell a bee tree for miles in any direction. Despite the fact that chasing around the forests looking for "bee trees" has gone a little out of fashion, perhaps due to the shortage of suitable forests, there is still adequate scope for the bee hunter of today if he is prepared to keep his ear to the ground and his eye on the bough, so to speak. Many beekeepers know for instance where feral colonies exist, but due to any number of permutated reasons these colonies are merely interesting landmarks for these beekeepers. This is where the ear to the ground bit comes into its own, casual conversations at local beekeeper meetings sometimes turn up the necessary snippet of information regarding the approximate location of a wild colony. The prospective hunter must now search among the "boughs" in the vicinity indicated and with luck he/she will locate the prize. Joking aside though it is quite a challenge getting bees out of an attic or barn or tree bole. The whole process is actually surprisingly easy if the correct approach is made and "Lady Luck" enlists her support. First find your bees. Then in the light of the following recommendations, modify your approach by tempering the whole proceedings with common sense. Safety is the first pre-requisite. Timing is most important, for best results. Equipment is minimal but slightly specialised. Required normally, will be a sturdy ladder of adequate length; woodworking tools e.g. hammer, saw, nails, a length of rope; approx. one square metre of fine wire mesh (less than a bee can penetrate); a helpful friend; the usual beekeeping equipment e.g. smoker, veil, gloves. In addition a putty type of material, e.g. plasticine or playdough (where the beekeeper has young family). Having located the flight hole being used by the "about to be captured bees", the object of the exercise is to render the flight hole such that egress is available but ingress is if not impossible, nearly so. To do this take the wire mesh screen and roll it into the shape of a truncated cone, e.g. no apex, the top end having a $^5/_{16}$ inch (7 mm.) access hole. The base of this open ended cone should be large enough to totally cover the access area being used by the bees for their nest. By putting the cone over the flight hole and sealing the base using putty or whatever, the bees are forced to leave the colony via the top hole in the mesh cone.

Here is where the carpentry comes in, a ledge or shelf now has to be constructed close to the top of the mesh cone. After the shelf has been constructed, the next stage in the procedure, not previously mentioned is that a queen-right nucleus stock is then installed adjacent to the mesh cone. A queenless nucleus with eggs and adequate covering of bees would also suffice. The procedure described is of course out of place: first build your platform, then fit your mesh cone, then install your nucleus, making sure it is secure against being blown over by wind and weather.

Now the situation we have, is a wild colony of bees, wild, because the foragers now can't get home. They can come out through the cone but they can't get back down again, providing the hole is small enough. Over a period of time the foragers from the feral colony will drift with the nucleus stock, until after perhaps four/five weeks most of the "adult bees from the feral colony will be in the nuc, leaving the queen and some young bees in the wild colony.

Depending on the size of the united colony a larger hive may have to be provided, unless the congestion is relieved by removing frames of bees from the nucleus. Anyway, at the end of due time the mesh cone can be removed, leaving the nucleus in place, the bees from the nucleus will soon proceed to root out any honey remaining in the feral colony and ultimately all the bees and the honey from the wild hive will now be safely in the custody of the beekeeper/ bee hunter — who has not only done himself a good turn but also his neighbouring beekeepers since any feral colonies in an area are a potential reservoir for disease. Although there is a theory abroad that feral colonies are free of disease. I would temper this statement by making the point that most bee diseases are endemic, only becoming epidemic when the colony becomes "one degree under" and is thus incapable of resisting the latent threat of endemic infection. A stated timing in the foregoing procedure is important — if done at the end of a nectar flow, the labours of the bee hunter will be rewarded with the capture of bees and honey. If carried out during a nectar flow the bees of the feral colony are disrupted and gather no surplus and the loss of young bees which would otherwise have been produced in the wild hive during the nectar flow will be lost due to non-materialisation. In an early area, i.e. an area with good sycamore forage but having little available in June, the end of May is an ideal time to set the wheels of the hunt in motion.

QUEEN REARING FOR THE HOBBYIST

With the present accent on home reared stock, due to the ban on imported bees, which is a direct result of the Varroa threat, it is now of paramount importance that we beekeepers in Scotland make a concerted effort to propagate and improve from our own stocks. The omission of the word "indigenous" to qualify the phrase "our own stocks", was deliberate. Since the incidence of only native British bees among our present stocks is highly debatable.

The qualities of pure strain British "Blacks" were, according to my father and other veteran beekeepers, legendary; an ideal "Heather" bee, in being a leisurely developer, reaching its "peak" population at around the beginning of August; it is reputed to have had a low incidence of swarming and a high degree of longevity/stamina; it also produced pearl white cappings on its combs — a honey judge's dream-! As we know the "Isle of Wight Disease" sounded the death knell for the bulk of the native bees on these islands. Varroa could do the same for our present stocks; if we are not successful in keeping Varroa out of Britain.

It never ceases to astonish me that beekeepers have so much trouble with the raising of new queen bees. Since if the operation is undertaken at the right times, and proper attention is devoted to the condition of the bees i.e. population size and food supply, the only major problems apart from not having any control over the sires of our queens are weather and predators.

A simple management procedure, beginning with the major requisite that the stock available is worth propagating, would be to commence feeding selected stocks from the advent of the first available pollen — in good springs mid-late February. A moderately concentrated sugar syrup (around 1 kg sugar: 1 litre of water) should be used initially — reducing to a solution of 1 kg to $1^1/_2$ litre water as the spring progresses and the selected hives show signs of being well provisioned with stores — thus imparting a feeling of well-being and security in the bee population — starving bees will not raise brood!

Around the first week in April subdue the stocks in question in the usual way — selecting a good bright day. Having prepared an alternative brood box filled with drawn combs, remove a brood comb containing eggs and emerging brood, make good the loss of the comb by inserting a drawn comb in the space created at the side of the original brood box, when the remaining combs in the hive are closed up. Place the comb of eggs and brood previously removed, into the new brood box in such a position that when the new brood box is placed on top of the original brood box the frame with the brood and eggs is central above the brood nest in the lower chamber. Replace the feeder and close the hive up. The colony is now in a condition where, encouraged by the presence of brood in the top chamber, the bees will move up into the adjacent combs — and into their own generated heat from the lower box and thus be encouraged to expand the brood nest more rapidly than otherwise if kept to a single chamber. Feeding should be maintained and by judicious transfer of drawn comb from edges of brood nest in the top box into the centre of the brood nest the colony will be encouraged to expand even further. By late April, early May the selected colonies should be rather busy — heaving, is probably a better description. At this stage the colonies can now be split. The selected colonies each now having around 22 brood frames all well covered in bees, will provide enough frames to make up between 5 - 7 three frame nucleus colonies, each with at least one frame containing eggs and open brood in the middle. It is not necessary to find the parent queen at this stage! These nucleus stocks now have to be removed to a remote apiary (at least 1 1/2 miles from the original stand) to raise their own queens. Due regard must of course be given to the most important requirement that sufficient drones are available to effect adequate mating of the new virgin queens on the third week after the nucleus stocks have been made up. The nucleus stocks with the embryo queens will all naturally still require to be fed steadily. With a bit of luck and the right weather conditions, a month after the nuclei were made up the first fertile eggs should be present.

To make really effective use of the nucleus stocks another stage of management could be undertaken. Providing it is possible to make up further queenless nucleus stocks; these nucleus stocks can be given queen cells cut from the previously made up nuclei — thus by timely intervention a plurality of queens can be raised depending on the needs of the beekeeper. Of interest is that depending on the strength of each original nucleus stock created, each nucleus will produce anything from 2 - 5 queen cells, depending on time of creating nuclei! The earlier in the season the fewer cells will be built, but invariably a minimum of two cells will be raised.

In a word, queen rearing is not some mysterious magic operation, outwith the scope of the hobbyist, providing the loss of honey production is accepted.

The would-be queen rearer must naturally be prepared to invest that little bit of extra effort into his/her beekeeping than otherwise. Also, the need to feed can be a bit "off – putting" with the current price of sugar, but, considering the cost and limited availability of queens — when they are needed! — a try at rearing your own queens will undoubtedly be worthwhile materially — and the additional experience gained will make the beekeeper a better beekeeper.

OCTOBER

The bees should now all be back in their winter apiaries (after the migration to the heather, where practised) the hives should be sitting on a level support and should also be tilted slightly forward to facilitate drainage. Other preparations for the long haul for bee and beekeeper, through the winter, number; ensuring that the hives are weatherproof and that roofs cannot easily be blown off by high winds. Where the beekeeper uses deep roofs the problem of wind blown roofs usually does not arise but the shallow hive roof is very susceptible to removal in high winds. I myself use very shallow home-made roofs and I find it very reassuring to place a full sized building brick (or even two) on top of the hives. Mouseguards or entrances reduced to no more than $^5/_{16}$ inch (8 mm) high should have already been resorted to. With the entrance to the hive so reduced it is a good safeguard to leave the entrance "fullwidth".

This ensures good ventilation throughout the hive. Due regard must of course be taken of the prevailing wind direction, in this respect. Too much stress cannot be laid on the importance of over-wintering hives being exposed to the south. In apiaries where hives can benefit from the rays of winter sunshine, it is worthwhile troubling to face the entrances south, this procedure can be a decisive factor in a colony's survival chances, it also incidentally takes care of the prevailing wind problem. A simpler "device" can be used to check the benefits of hives exposed to maximum winter sunshine, thus, on a day where ground frost persists all day, but when the sun is shining from one of these classic crisp winter blue skies; around noon, go to any hive which is standing in a position where it has been continuously in the shade, and try to lift the roof up slightly, nine times out of ten, the roof won't budge, because the moisture under the rim will be frozen and the roof literally "cold welded" to the hive. Now go to a hive which has been exposed to that winter sunshine for some time and repeat the previous exercise — the roof will lift up with ease! That the hive is by this time adequately prepared for the winter is extremely, if not critically important. But the most critical factor in overwintering bees is of course the condition of the bees themselves. Very rarely do we beekeepers ensure that our colonies go into winter in ideal condition. Providing the hive and location have been adequately considered with the colony in the correct condition, over-wintering will hold no terrors for the "aware" beekeeper. The first prerequisite is that the colony be healthy.

The next prerequisite is that the colony be headed with a queen which has not laid for longer than two full seasons. That the queen be attended by a populous, young bee, population should be almost axiomatic with having a young queen present. Hives returning from a reasonable season at the heather will normally have this young population, this population having been reared during the hey days at the moors.

The colony should now be equipped with adequate stores, fed in such a way that the bees have time to evaporate the excess water from the syrup, and invert the sugar to honey before, at the latest mid-October — it is sensible if late feeding is resorted to, to add Thymol to the syrup solution to inhibit fermentation — the Thymol dosage/gallon of syrup is given in Bob Couston's admirable book "The Principles of Practical Beekeeping" which in the opinion of the author, no self-respecting Scottish Beekeeper (or world-wide beekeeper for that matter) should be without.

The considered optimum quantity of stores in an over-wintering colony of bees varies from beekeeper to beekeeper — to my mind beekeepers' opinions are purely subjective and while always well meant are sometimes unreliable (about bees that is — I hasten to add!). The hobbyist bee-keeper with only a few hives is in an excellent position, relative to the commercial or semi-commercial beekeeper, to adequately and optionally feed his bees, and merely by letting the bees decide when they consider they have adequate stores. The hobbyist merely requires to maintain his feeding schedule until the bees refuse to take any more syrup down and this they will do, but a word of warning — a good strong colony going into the winter can shift a gallon of thick syrup in a night in many cases! A good rule of thumb for assessing adequate stores is to heft the colony either by raising the hive (roof removed) bodily and "guesstimating" the contents (by comparing the weight with an empty hive full of empty combs a reasonable idea of stores quantity will be achieved — practice makes perfect!).

In my own " beekeeper's opinion" between 30-40 lbs of stores is the minimum amount of stores for successful over-wintering. Providing spring feeding is resorted to as early as possible (during the dormant period a 1 kg bag of sugar directly on top of the brood frames is a marvellous safety factor), the bees will pull through this "minimum". Feeding sugar syrup should be commenced with around mid March but definitely by the start of April at least, and most definitely after pollen has been seen going into the hives over a certain period of time (at least three days). In winter the bees will regulate, their intake to suit their needs, but only while the populations are small. Strong stocks will tend to store any surplus — give more laying space! The water in the sugar solution is an invaluable asset to the bees at this time of year, providing them with the necessary moisture required for brood rearing.

Opinions differ widely about the best method to prepare bees for overwintering, especially in the question of insulation and ventilation. Although both ventilation and insulation are of paramount importance in overwintering, those requirements are not always compatible with each other. Where the beekeeper practices top insulation, the placing of quilts and carpets etc. on the crown board is just not conducive to good ventilation and this can be adequately demonstrated when in spring the sodden insulation is removed from the hive (unless the winter has been particularly dry — not a normal occurrence in West Scotland!). Bad ventilation and dampness result in mouldy combs outside the cluster of overwintering bees. In single walled hives adequate ventilation without dampness can be achieved by merely placing an empty super on the brood box, topping it with a crown board, then the roof, and leaving the entrance wide open but reduced in height, as previously recommended.

NOVEMBER

Winter is upon us, the land is slumbering, nursing its precious store of life, dormant until the first warming rays of spring sunshine. Our bees are, synonymous with the slumbering land, resting, conserving their energies for the spring with all its promise. Looking back to spring this year, and thinking of previous springs, I never cease to be astonished at how different each year, the behaviour of the seasons really is. We had a marvellously mild spell of weather in December and January and hives examined then indicated large patches of brood on the combs. This did not augur well for the wellbeing of the colonies should the weather have turned cold in the early Spring. Just as the crocus began to make itself felt the weather changed and continued cold and frosty for almost three week. This cold spell checked the development of the crocus and snowdrop, but not the development in the colonies. The brood reared in January and December rapidly replaced the normal winter losses and the clusters in most hives actually indicated growth, but the weather being so cold hindered the bees from altering the cluster position — ideal conditions for 'isolation starvation'.

Bags of sugar were already on all of my hives, normally the bees do not make much inroads into this crystal sugar until late January early February, this year, however, by the third week in December the stronger colonies were ready for their second bag, placed of course right on the frame top in way of the cluster.

By mid-February some colonies were well into their fourth bag of sugar, and on making internal examination of these colonies, it was quite horrifying to note just how badly off the bees were for stores with which they had direct contact. Some of the colonies were actually covering up to nine frames but brood rearing had diminished dramatically relative to December and January, due to the cold weather experienced in February. I dread to think how many colonies I would have lost but for the feeding of bagged sugar. Incidentally all colonies had been fed at least ten pounds of sugar, in syrup, going into the winter.

The weather eased again around mid-March and the bees began working the retarded crocus and in many areas snowdrops which had also been held in check. Sugar syrup feeding had been started around the end of February —in desperation in some cases. With the advent of steady pollen carrying, sugar syrup feeding was undertaken in earnest to bring the colonies up to scratch for the sycamore.

During late March and early April the colonies really made fantastic progress and many hives were sporting the second brood box by 4th April. For the first time in 15 years of keeping bees, I noticed catkins, crocus, snowdrop and dandelion blooming simultaneously — not of course all in the same area.

Another first was also noted in that on March 27 1981, fully fledged drones were noted in the brood nests of numerous colonies. By fully fledged, mature is not implied since drones require at least '8-10' days after emergence from cell to be mature enough for mating purposes. But seeing these drones so early in the season gave reassurance that a very early start could be made with queen rearing and the first of the new season's queens emerged on April 12 1981, she commenced laying on April 19, 1981, this first queen was followed in rapid succession at intervals by nineteen others which all commenced laying by May 17 1981. This Spring was one of the very few Springs where the bees were able to take almost full advantage of the crocus. They spent a goodly number of days hammering crocus pollen into the hives and then, timeously, the salix catkins bloomed just as the crocus was showing signs of wear and tear, and then around the middle of April as the catkins began to wilt, dandelions and wild cherry filled the gap. The succession of pollen bearing plants in spring in Scotland makes this land of ours a veritable treasure house for bees — when the weather is right.

A footnote worth adding regarding the production of early queens is that in the early part of the summer i.e. end of April to around the third week in May, the incidence of our summer feathered visitors; the swallows, swifts, and house martins is either relatively low, or non existent. The absence of these birds is most important in successful queen rearing. Since raising queens is only half the job in breeding, the queen must be able to return safely after her mating flight. In my experience the percentage of successful matings in spring is much higher, nearly 100%, compared with the success rate in mating later in the summer.

On a bright day in June/July take note of the aerial activity directly above the apiary — it as no accident that swallows and swifts seem to be present in relatively large numbers in the sky above the hives at this time of year. They are eating our bees, and there is nothing at all we can do about it, except perhaps by breeding queens early, ensure at least that they are not eating queens.

This year the mating of queens in the later part of the spring was adversely affected by the persistent strong winds which made a mockery of the normally quite reliable sycamore, nectar flow.

In my own apiaries in the west of Glasgow it was touch and go with my 'second wave' of queens which were due to mate after May 12, 1981.

The bees themselves, let alone virgins and drones just could not get out due to the gusty conditions. These windy conditions persisted well into June, giving the very rare day where mating could be hoped for. Despite the weather being so adverse the success rate for mating during the period May 12, 1981, to June 20, 1981, Was around 50% — not good, but much better than expected, due to the adverse, weather which prevailed for the bulk of the period mentioned.

I am tempted to be philosophical here and hypothesise that the queens which mated successfully during this period were acting within the law of natural selection, and will produce progeny well suited to the adverse conditions found in Scotland, since many of the colonies which constituted the 50% failure were found to have beautifully developed queens which were laying lovely brood patterns — of drone brood!

These queens did not venture from the hive to mate because the conditions were too adverse. Thus according to the law of natural selection in nature these particular strains would have disappeared to be replaced by a more hardy strain better suited to the ecological conditions in their area. These drone laying queens were of course eliminated by myself and the colonies either re-queened or given a frame of eggs from a suitable queen. The lesson here is that the beekeeper should leave nothing to chance; and should check the colonies regularly to ascertain that the colonies are in good order — especially at critical times like the re-queening period. The failure of the early nectar flows was more than adequately compensated for by the copious amounts of lime, privet and willowherb nectars Which the bees poured into the hives in July.

Proving yet again that by being eternally optimistic and maintaining hive strength by supplementary feeding during a bad spell — when the right conditions occur the bees are ready for them.

In August the bees and the beekeepers were again gifted with superlative weather and the foresighted who made the effort to the heather won hands down. All that is needed now is a short, sharp, hard winter, ending around the beginning of March, followed by a mild windless, moist spring followed by a Mediterranean type summer, followed by an Indian summer in August, followed by . . .

DECEMBER

This being the last article to fall from the Rambler's pen, after a spell of three years contributing to the magazine. It is difficult to find a suitable subject on which to end.

Perhaps the situation could be likened to a colony going into the winter cluster, a much appreciated rest period is about to begin. I wish my successor fewer sleepless nights than I, in dreaming up topics to write about.

The hives should all be settled now for the dormant period and yet is the winter period really so dormant for the colonies? Bees as we know do not hibernate, they remain conscious but with a decelerated metabolic rate, in a state of continuous movement. This continuous movement has a very important function in so far as the necessary warmth for survival is generated in this manner. For many years it was thought that the queen stopped laying eggs altogether during the 'dormant' period, but recent investigations have shown that she continues to lay right through the cold period. This finding explains certain anomalies which have been noticed in the condition of hives coming out of the winter, over the years. Some hives which were well fed and fully provisioned at the end of October in any year, at the end of February, showed themselves to be astonishingly light in stores. Considering that it was (is) believed that bees consume relatively small quantities of honey during the winter period, how could this lightening of the hives be satisfactorily explained? The disappearing stores were quite obviously being converted into young bees by queens which were so fecund that they just could not cease or reduce their egg laying capability to decorous levels commensurate with the wintering needs of their colonies!

An interesting letter appeared in the April 1981 issue of 'Bee Craft' where the beekeeper had systematically noted the weight of colonies over the winter period. The decrease in weight, at monthly intervals was quite staggering and he had to resort to emergency feeding very early in the year. These hives obviously had good fecund queens. For this reason, it is good beekeeping sense to check by lifting the weight of stores in the overwintering colonies. Even a quick look under the crown board is recommended, since if early in the spring or late winter the colony cluster is right up at the frame tops, this is a sure indication that the bees are getting hungry, and steps must be taken if the colony is to survive to the spring.

Despite all this colony activity during the winter, the honey bee Colony is at its most vulnerable at this time. Where out apiaries are maintained, it is mandatory that the bees be visited at least once a week, to ensure no damage has been incurred by the hives, animals can topple a hive quite easily, a dog or perhaps a sheep, or even a human - vandals are not such rare animals these days and are capable of much damage to an apiary, and not merely in winter.

Bees although vulnerable in winter are surprisingly hardy, but if they are left exposed to the elements over an extended period, where a hive has been toppled or a roof has been either removed or blown off, they will surely perish. Thus, it is advisable even if a regular weekly visit is made to the bees, to make an extra trip to the out apiary after periods of high wind or gales. The devastation severe winds can wreck on an apiary has to be seen to be believed, especially if the apiary is situated in a location where it is bordered by high walls but open to the direction of the gale, the turbulence caused by the sudden arrestation of the wind's forward progress can lift not merely hive roofs but also hives themselves, this, I know from personal experience. In such an event, all is not lost, as previously stated, bees are extraordinary resilient, and even where a hive has been toppled and the combs disarranged, nine times out of ten the queen will survive, and so long as she pulls through, after the hive has been reassembled, and fed where necessary the colonies will pull through and greet the spring "bright eyed and bushy tailed".

There appears now to be more official interest in beekeeping in this country since Britain became a member of the E.E.C. The subsidies for sugar to feed our bees in preparation for the winter and also during prolonged inclement periods during the summer should surely lighten the beekeepers' load.

The proposed subsidies for breeders and queen rearers appears also to be a 'Step' in the right direction. Perhaps this is the beginning of a new era where the real value of the honey bee to the human community will at last be fully appreciated.

For too long beekeeping, especially in Britain has been treated as the Cinderella of our agricultural system. In truth, without the pollination of agricultural crops, fruit and vegetables which bees perform as an incidental to their foraging, the agricultural economy and indeed the whole economy of Britain and the world would be drastically and adversely affected — ask your local fruit farmer or seed grower where he would be without insect pollination. In England especially, fruit growers are virtually dependent on the services of beekeepers to achieve the level of fruit set they require to remain in business.

With the increasing use and abuse of agro chemicals and the irresponsible methods of their application, many of the naturally occurring insect pollinators like the bumble bee, wasp, hornet and many of the solitary bees and wasps have been decimated to such a degree that they can no longer be relied upon to perform an effective natural pollination of crops. In many countries where agriculturalists are progressive, and aware of the importance of pollination to the success of seed and fruit setting, no crop development project is considered, without consideration being given to the bee colony requirement to achieve maximum efficiency from the proposed project.

Let us hope that in the not too far distant future the E.E.C. or at least the pundits in our own agricultural policy making machinery here in Britain, will become to beekeeping what Prince Charming was to Cinderella. I would also like to take this opportunity to wish the readership of *The Scottish Beekeeper* a Merry Christmas and a Happy New Year.

STRANGE ON-GOINGS IN MID-FEBRUARY
By A. E. McARTHUR

As is my wont in late winter/early spring, I was topping up my sugar syrup feeders on Sunday 14/2/82, which readers in the West of Scotland will remember as the second warm, bright day for bee flight, which we have had since well before Christmas 81.

The bees were well out, although it was quite late on in the afternoon — this timing was deliberate since, if perchance any sugar syrup had been spilled, by the time the bees had mopped it up the late afternoon 'cool' would then have forced them back 'home' before they could think of looking around for other easy pickings — i.e. robbing. Feeding in this manner when bees are flying has to be done carefully and quickly, checking always that no syrup is leaking from the entrance of stocks already fed.

One particular hive seemed rather more 'edgy' than the others in the apiary. So being always on the watch for out-of-character behaviour I decided, that after I had finished my feeding in the apiary, I would open the hive and examine the combs.

Using a little smoke, the bees were subdued, then I hefted the hive — the Smith lends itself to being lifted bodily — I was appalled to register how light the hive felt. The colony was then opened for examination and to my astonishment I found three frames with large patches of worker brood — there were also adult drones on the combs and emerging drones were noted struggling free from their cells in the middle comb of the three brood nest frames. The hive was virtually empty of stores, although the bees were alert, active and obviously healthy — but they were hungry. They had made great inroads into the two solidly crystallised bags of sugar (1 kg) at the frame tops, but it was quite obvious that without some help — even feeding sugar syrup was not going to be of much use to this colony if the weather turned cold again before Mid-April — which it most certainly will. There was nothing else to do but augment the scant stores in the hive, and a frame of honey and pollen was promptly removed from each of three better provisioned hives in the apiary. These were placed in and around the brood nest of the three frames with sealed worker brood, the frames closed up and this colony given a gallon of heavy syrup (5 lbs sugar with water added to make a 1 gallon solution of sugar syrup!)

This colony is now secure till the sycamore blooms, but meantime it will be fed regularly on a weekly basis as are the other colonies. You may be tempted to comment — "An outrageous piece of beekeeping (mis)-management" — I would add that all colonies are fed their 10lbs. of sugar in syrup before the winter sets in, so the bees were well provisioned for the dormant period. I would make the point that bees in "single brood" overwintering management are vulnerable and, unless the beekeeper is diligent and prepared to take emergency action, needless loss of colonies will inevitably result — especially where British Standard (BS) brood boxes are used. The Langstroth with its larger comb area is more suited to 'single' management than most B.S. hives — with regard to overwintering.

To return to the prolific colony which started the spring build up too soon. Obviously, the queen and the bees got their "biological clock" mixed up. They must have reckoned that after the extreme cold of December/January that the relatively mild weather heralded the arrival of the active season, and active — or rather hyperactive, they became.

Management of the sort just described is not to be undertaken lightly, but I believe that by making decisions and acting judiciously the beginner and not-so-beginner beekeeper can become more "expert" and confident in handling his charges and dealing with the problems which beset them.

This colony will not be included in any breeding programme since the queen -obviously has - blood in her which is not ideally suited to our seasonal pattern.

THE MONTHLY ROUND BY RAMBLER MAY 1983

Here we are in May already. How rapidly the season progresses once the catkins are past their peak. The Germans call May der Wonnemonat' — the month of joy! For Beekeepers in Scotland, May is indeed the month of joy. May heralds the blossoming of the first major nectar-bearing plant, the sycamore tree, whose dull green tassels are eagerly visited by our honey bees for their rich harvest of greenish brown pollen and for its nectar which is elaborated by the bees to produce a dense, highly viscous, dark brown honey which is rather difficult to extract from the comb when cold. This honey, which has a pleasant taste but unfortunately leaves a slight burning aftertaste on the palate, is best extracted immediately on removal from the hive. This is the month where many of us are caught on the wrong foot by the bees, if April has been mild allowing many flying days to the colonies, i.e. early swarming. Where the Spring has been favourable to early bee flight and also where the colonies have been 'Spring-fed' with sugar syrup since late March or early April there is a good possibility that they may build up ready to 'go' inside the first ten days, of May. The nine-day inspection cycle for swarm preparation checking should be started not later than the end of the first week in May in 'normal' years where the bees have been able to fly frequently in late April, i.e. a good Spring! Where the weather has been cold and unfavourable for bee flight in April the spring build-up will usually have been delayed thus buying the beekeeper time. Another worthwhile task to be undertaken at this time of year is the marking and clipping of queens. If this job is done at the start of May or even very late in April when the weather is fine, the rest of the beekeeping year from a management viewpoint will be the proverbial piece of cake. The queen is usually reasonably easy to find in a colony around this time, making marking relatively easy. This same queen at the height of the swarming season unmarked in a 'heaving' colony, especially where the queen is of one of the desirable black strains will elude even the most sharp-eyed and experienced beekeeper for a long time. Where such a colony has already started building queen cells making it imperative that the queen be found, especially where more than one colony is known to be under the "swarm impulse", the facility to find and remove the queen quickly is not only time saving, but also much easier on the nerves. The inability to find a queen in a hive preparing to swarm will invariably result in the loss of the swarm when it ultimately issues, since the procedure of cutting down the queen cells in a hive preparing to swarm has limited worth. Where the weather is fine during this cell cutting regime the bees will merely use three day old larva, have a sealed cell ready three days after the last 'cell chopping event' and be in the air while the idle beekeeper is kidding himself on that he has at least a whole week before he need worry about the colony swarming. The main reason for the persistence of the myth of swarm control by cutting queen cells down is that if a short period of unfavourable weather occurs immediately after the cell cutting operation sometimes the bees are discouraged from swarming. The only really sure method of restraining bees from swarming, without weakening the colony, is to remove the queen and ensure that the bees only have one sealed queen cell in the hive after a minimum of eight days have passed since the removal of the queen. If all sealed queen cells are cut down at the time of removing the queen, then the advice given in the previous sentence is easily put into effect. But only one queen cell must be left after the second cutting operation.

JUNE

The June gap! A phrase to strike terror into the hearts of beekeepers. When the late Spring and early Summer blossoming period is finished, unless the beekeeper is in a particularly favoured area where the bees have access to good clover, raspberry or oil seed rape, the bees will be faced with a chronic dearth of nectar sources at this time. The beekeepers in the East of Scotland around Perthshire and Angus are particularly fortunate in that the raspberry growing area of the Carse of Gowrie is within relatively easy reach. Now as if to spoil the beekeepers in these areas the rapid increase in rape plantings is occurring here as well. So, the June gap is normally only an interesting phrase for the beekeepers of Angus. The growth in the rape acreage now offers the possibility for beekeepers in 'June gap' areas to migrate to the oil seed rape fields in late May early June. Rape is being planted experimentally, also west of Glasgow near Erskine this year — hopefully if it is well covered by bees and a good seed set is achieved the farmer concerned will promulgate his success and perhaps herald massive rape plantings in areas other than the east. If the rape can be relied upon as the clover was many years ago perhaps in the foreseeable future the 'June gap' may well be relegated to mere historical importance — like the Isle of Wight Disease! The work schedule for June is more or less similar to that of May; swarm control inspections, and feeding where necessary. In a case of real nectar shortage in June it does no harm and a great deal of good to feed colonies. Especially where the bees will have access to lime in July, feeding at this time can be an excellent management ploy. In a nectar scarcity, the bees will cease to rear brood. Whereas if they are fed (for a strong stock 4 lb sugar in syrup/week!) the colony will react as if it were enjoying a nectar flow and continue to rear brood. Using the rule of thumb - six weeks from egg to forager and estimating that the lime blossom will 'peak' at around the 15th to 20th July ('these dates are valid for West Central Scotland) then every egg laid from 1st

June onward to the 14th June will be of critical importance for the lime flow. If it materialises! This is the exciting aspect of beekeeping, pitting your wits against nature by using your skill to take the calculated risk—not chance! By encouraging the colonies to continue expanding during June by feeding, and if the lime secretes, the sugar feeding costs will be recouped many times over. If you don't feed and the colony populations stagnate during late June, the bees produced on the sycamore flow being relatively old, will die quite rapidly when they start to forage on the lime and they won't be replaced. Then the beekeeper will lose out, since the hives will be well below optimum honey gathering strength at the critical time. A good indicator for a successful lime flow is a wet period two weeks in June — preceeded by warm weather in late May, early June. If the bees get flying weather during the lime blossom period after such a situation, make sure you have plenty of spare supers and shallow frames! If, however, the weather has been cold during late May and early June and if around mid-June the weather turns very warm and dry, then the lime blossom will invariably be arid.

JULY

This is the month to re-queen, if the job has not already been done, where colonies are headed by queens which have already seen two winters. A young healthy queen of proven ability given to a colony being prepared for the heather, at the start of July, will ensure that the colony builds up well on the lime, privet and willowherb flows to bring the bees to spanking condition for the heather flow. A young queen at the heather will continue to lay hard and steadily if the conditions are right. By using this management device, three of the main conditions for good overwintering are provided:

1. Young healthy queen.

2. A good population of young bees.

3. A well stocked brood chamber full of high quality ripe heather honey.

(Item 3 will of course be supplemented with a feed of at least 10 lb of sugar in syrup before the end of September).

If the early part of the summer was unfavourable for flying, and if the lime flow is a good one, the nine-day inspection cycle to check for swarming must also be carried out in July. Swarming preparations will most certainly be made if the season is as described. If the beekeeper does not inspect the hives during this period on a regular basis, swarms will be lost. If the beekeeper left combs of honey — either sealed or unsealed — in the super during the June gap, these combs will by the start of the lime flow be largely depleted of honey. Hopefully the lime will secrete well and the loss of the early honey made good. Ideally during the June gap, unfinished combs should be removed from the hive and stored dry and bee-tight until the lime or other late flow occurs.

The colonies, in the interim, should be fed sugar syrup as suggested in the June "Monthly Round". If supers of honey so treated are placed on the hives at the start of the late summer flows, the bees will move up into the supers with a will and the beekeeper will thus harvest more honey than otherwise and the cost of the sugar feeding in June will be more than recouped.

Hives of colonies being prepared for the heather should be checked nearer the end of the month, to ensure that there are no wide gaps or splits in the hive body and to make sure all hive parts are fitting snugly with no spaces between the various parts i.e. floorboard/brood box, brood box/super, super/crown board. Any gaps at these points will allow bees to leak out during transit to the moors. Better a few minutes spent making this check than an uncomfortable drive (in a closed vehicle) with irate bees spewing out of previously unnoticed splits between parts.

It is a good ploy to give the bees a new super fitted with drawn or partly drawn empty combs for transit to the heather. This has the double advantage of giving the bees space to move into during the journey, thus inhibiting panic and a subsequent rise in temperature, which of course can be fatal, causing comb melt down. The drawn comb in the super will encourage the bees to work in the super more readily. A word of caution regarding the use of supers having only undrawn foundation fitted for transit to the heather! — in certain weather conditions, and where the hive population is at the optimum (i.e. bursting at the seams) the sheer weight of bees on this foundation can cause it to sag and buckle or even to collapse altogether. It is good sense to check any hives, which make the trip to the heather fitted with foundation, as soon as possible after the bees have settled, perhaps after a couple of days — but soon !

AUGUST

If the hives going to the heather were well prepared at the end of July, then as soon as the heather bloom shows around 15% open blossom the colonies can be transported "painlessly" to the moors after closing the hive entrances when flying has finished for the day and securing the hive parts against relative movement. It pays to keep a close watch on the progress of the nectar source being worked for — to work by rote is fool hardy. Last year the heather bloomed very early, around 27th July in some areas. The traditional move to the heather on the 12th August was thus far too late. With the disastrous collapse of the weather last year where the wind blew strong and continuously after the 8th August, only those beekeepers who moved to the heather with the advent of the bloom period got some heather honey — but really not a lot relative to the hard work involved in moving — but that's beekeeping — win some, lose some!

Hopefully all hives going to the heather will have had their swarm urges requited by either the natural act or by colony manipulation. Any hive which has not yet shown swarm preparation by August, where the queen is of the previous year,is still very liable to swarm at the moors if the weather is right. Our present strains of bees do not yet seem to be able to respond correctly to the stimuli of the active year and realise that after a particular date the hive resources of population and stores have to be husbanded in preparation for the coming of winter, rather than squandered in a final suicidal act in August, suicidal for the swarm, because its chance of build up to overwintering proportions are minimal, suicidal for the parent stock left with only queen cells, where the selected virgin will have very little chance of mating successfully so late in the year. Despite the foregoing logic, swarming still occurs at the heather. The reason for this "unreason" could possibly be laid at the door of the variety of races of bees from which our present "indigenous" bees derive. Many races of honey bee coming from vastly different ecological backgrounds and whose "biological clocks" were out of synchronous with regard to the British climate were imported into these islands after the devastation caused at the turn of the century, to our native "British Blacks" by the Isle of Wight Disease. This disease was reputed to be a combination of the diseases Nosema apis and Acarine, and it is credited with the demise of the then indigenous British honey bee population. Many leading authorities on beekeeping are now beginning to question the significance of the I.O.W. disease and its contribution to the loss of the "Black".

At the time of the tremendous losses of bee life around 1904-1920 the traditional skep system of bee-keeping was rapidly being displaced by the revolutionary "moveable frame hive".

This new hive gave the beekeeper greater access to the bees than ever before — also greater access to the honey in the hive. Many of the losses attributed to I.O.W. could have been caused or aggravated by starvation resulting from the beekeeper removing more honey than prudent, because of the lack of sufficient information on the over wintering needs of the colonies at that time. The introduced races of bees have been selected by beekeeping management and husbandry rather than by nature and will never, due to this fact, match the life cycle of the original British Black which evolved naturally as our indigenous honeybee. The Black evolved according to the laws of nature by adjusting its population dynamics to suit the environment. Experiments have been carried out with pure strains of particular honey bee races in Europe. *Apis mellifera nigra* from the Pyrenees, where they had naturally adjusted to the late season nectar flows characteristic of that region, were relocated to an area where the nectar flow occurred early, in the season. Similarly *Apis mellifera carnica* a race which is indigenous to early nectar flow regions was relocated to the Pyrenees. Not surprisingly these races did not adjust automatically to the nectar source blossom periods in the new environment. A. m. nigra built up in the typical manner, slowly, reaching full strength too late in the season to benefit from the main nectar sources. *A. m. carnica* built up rapidly in its new location and by the time the late nectar flows came 'on stream' the bees were a spent force. It is small wonder, due to the fact that our bees in Britain are a random mix of gene pools from diverse races of bees, that they sometimes have difficulty in knowing "which way is up".

SEPTEMBER

Hopefully the heather nectar flow has been a good one this year — to compensate for the disaster of last season's total failure of the heather flow. The hives at the moors should be left on location for at least a full week after the heather blossom has ceased to secrete nectar, i.e. the nectar flow is past. This gives the bees the chance to ripen and hopefully seal the last of the new nectar gathered during the closing days of the heather bloom. If the hives are moved back to the 'home' apiaries too soon, raw nectar or unripe honey could be shaken out of the comb resulting in great discomfort and possible death for the colony, due to the bees panicking and causing 'melt down' of the wax in the frames.

There are numerous methods which can be employed to remove the heather surplus. Ideally the honey supers should be removed from the hives before the bees are brought home, especially where the beekeeper works single-handed! Some beekeepers with less than three or four hives at the heather use the "shake and brush" method, each shallow frame is removed with its adhering bees and given a skilful bump with the closed hand to dislodge the bulk of the adhering bees — the bees still remaining on the comb being brushed off using a handful of long grass or the 'classic' goosewing or bee brush. This method can be quick and effective in the hands of an expert, who knows his bees and has a good weather eye, but in no way should the job be tackled by a beginner, especially at the heather, without advice or qualified supervision. If bad judgement is used the bees will win the day and the beekeeper will have to retreat, probably in disorder — but I suppose we've all got to start learning the limitations of beekeeping sometime, somewhere. At all costs — "Bee" protected! The easiest and least painful way for the average beekeeper is to utilise a clearing board — either with Porter escapes (which sometimes clog up with the occasional dead bee!) or better still, a clearing board similar in design to the New Zealand "hole and slot" type, which has no moving parts and is virtually — dare I say it - idiot proof! So long as it is placed the correct way up on the hive. Using any kind of clearing board method entails a double trip to the moor, but in particular years where a 'shake and brush' method or forced draught blower is unsuitable the clearing board will give the beekeeper reasonably easy access to the (his) honey. The colonies being cleared must of course be smoked in the customary manner to subdue the bees before any hive manipulation is attempted — especially at the moors. When the colonies have been re-sited in their "home" apiaries, the pre-winter checks and management , procedures should be started, ensuring that the hives are tilted slightly forward to facilitate moisture drainage, the hive should be checked that it is both 'weather tight' and "bee tight" ensure all entrances are wide open but either fitted with mouse guards or reduced in height to between ¼ inch to $5/16$ inches no more no less, this height will allow access to the bees but exclude even the smallest of shrews. If the hives can be located in an apiary with a southern aspect this will enhance the colony winter survival chances, as well as shelter from the prevailing wind. Apart from the tilt forward for moisture drainage, the hive should be positioned as "plumb" as possible, for two very solid reasons: feeding for over-wintering and comb building in the spring. If the hive is not level, feeding tins or pails of syrup are liable to leak. Autumn feeding should be started as soon as the hives are settled in the apiaries. Feeding should be effected as rapidly and efficiently as (possible and preferably given in the late afternoon — to minimise robbing where spilled syrup might start a "holocaust!"

Feeding for the winter should be done with a thick syrup three kilograms sugar with cold water added to make a gallon of solution is ideal not three kilos added to a gallon of water. This solution made with cold water will probably require intermittent stirring over a period of a couple of hours but the end result will be heavy, clear, palatable food for the bees. In mild autumns feeding into late October is possible but only to very strong colonies. Better to have all hives fed by the end of September until the beekeeper knows what he/she is about, to give the bees the chance to ripen the fed sugar to sugar honey. The colonies should have approximately 40-50 lb. of sealed stores in the hive going into the dormant period.

OCTOBER

Although the bees will still be flying quite freely at the start of this month in particular seasons, October may be regarded as the start of the dormant period. The bees will have started to draw a tighter cluster in the hive and the queen will have reduced her egg laying to negligible proportions. The hives should be given their final check for the winter "bedding down." The entrances should be mouse proofed, the hives should be level, i.e. entrances should be horizontal and the whole hive positioned with a slight forward tilt to facilitate water drainage, either from rain ingress or condensation. Hives fitted with shallow roofs should be weighted to safeguard against their being blown off by high winds. The deep design of roof does not need this treatment. If the colonies have been well fed and placed in a southern aspect, where they can benefit from any winter sunshine, and also sheltered from the prevailing wind little else need be done apart from a weekly check (for bees in out apiaries to ascertain that hives have not been disturbed either by natural or other forces!). Hives in the garden will be reviewed every day by the 'caring' beekeeper! The extraction of honey, centrifuging and pressing should all be finished by the smaller beekeeper by this time. Honey left in the comb till late in the season runs risks of spoiling due to moisture take up, causing sealed combs to weep, down grading their value for cut comb purposes. Cold honey is more difficult to extract, especially where the honey has a heavy body, like Sycamore!

Presses and centrifuges should all be thoroughly washed down with cold or tepid water, never boiling water as this will melt any wax residues and make the wax extremely difficult to remove. After washing and drying the application of a thin layer of petroleum jelly or liquid paraffin to all exposed metal components and bearings of equipment is well worth the effort. All spare equipment should be stowed in such a way that all wooden hive components which may have been soaked with rain before removal from hives is adequately ventilated to facilitate drying. Frames with comb or foundation should be stored in stacks of supers and treated for Wax Moth and made secure against mice. Wax cappings and brace comb pieces should be washed thoroughly at this time and melted down into wax ingots, either for sale to appliance dealers or ready for rendering into homemade foundation. At the first frost queen excluders can now be effectively cleaned. The wax being brittle comes away very easily when the excluders are scraped using a hive tool or some such handy broad bladed tool. For the hobbyist beekeeper October can also mean the start of the beekeeper's dormant period. But for the larger beekeeper October merely signals the start of preparations for next season and a continuing of honey extracting and bottling — hopefully!

NOVEMBER

A quick look at the apiaries once a week for obvious damage or disturbance is all that is necessary in the way of beekeeping management from now until the end of December, unless extremely high gale-force winds occur between visits to out-apiaries. Such winds can play havoc in apiaries exposed to them and a check on these apiaries after high winds is good beekeeping sense. Bees are relatively hardy and can endure even prolonged exposure to the elements — but a colony in an overturned hive exposed to heavy rain, hard frost and snow has greatly reduced chances of over wintering successfully. A quick car journey after foul weather can save bee lives and give peace of mind to the beekeeper, where out-apiaries are worked. Now is the time to make your choice of beekeeping reading material. We can never know enough about our charges. The more we read about them and weigh against what we have observed or learned from practical experience, the more proficient we can become at working our bees and understanding their needs and natural history — to the obvious benefit of bees and beekeeper. Some of the recently published beekeeping books are well worth a read. A new treatise of pollen identification with a comprehensive coverage of the major angiosperms listing the important plant families and relevant species, by Rex Sawyer, titled, " Pollen Identification for Beekeepers," will absorb the reader for hours. The book will also give an insight into the relationship of various nectar-bearing plants. For instance, how many beekeepers know that the sweet chestnut (*Castanea*) is more closely related to the oak and beech, than to the horse chestnut (*Aesculus hippocastrium*) or that the bean and pea belong to the same family as the clovers. (Papilionaceae). The photographic representation of the various pollen grains is par-excellence. The International Bee Research Association, the London based (Slough actually!) clearing house for all beekeeping scientific and literary developments worldwide, publishes information of exceptional interest to bee keepers in general. The magazine "Bee World", is published quarterly and is crammed full of very readable articles. Recent contributions include Dr L. Bailey's latest work on honey bee viruses; exhaustive work on the incidence of stone brood; modern honey hunting in Nepal. Another read of "The Hive and the Honeybee" or Bob Couston's "Practical Beekeeping" would do no harm either!! The awareness of the incidence of endemic disease in our honeybee colonies is also a very important factor in beekeeping. As is the knowledge of how the various treatments for honeybee disease are applied and when. Only by reading the relevant books and questioning our overworked college beekeeping advisers can we hope to even begin to understand the problems involved. In my own opinion a really sound knowledge of the natural history of the honeybee is of much greater value to the beekeeper than the memorising of a multitude of management devices which are utilised without a full understanding of their effect on the bees. By acquiring an 'in depth' knowledge of our honeybees' life cycle, coupled to the habitual use of arithmetic, beekeeping can be a real stress free pleasure. The application of the arithmetic relates to the reading of the fixed cycles in the activity of the hive, viz : the metamorphosis of the different castes; the cycle of egg laid to foraging bee; the period required from egg to sexually mature queen and drone; the time from egg laid to sealed queen cell — the list is endless, but by applying the discussed time scales to practical beekeeping management the beekeeper can not only stay with colony development, but also anticipate events in advance and plan his management accordingly. Bee informed!

DECEMBER

Happy Christmas and a bountiful New Year! At this time of year, believe it or not, the bees are beginning to sense the onset of spring and the queen is beginning to lay a few eggs more each week as January approaches. A crystalized 1 kg bag of granulated sugar (dip the unopened bag in cold water to soak the paper, by the time the paper has dried in the hive the sugar will have crystalized) given to the bees directly onto the top of the brood frame under the crown board will represent a welcome Christmas present to the bees — and could also mean the difference between starvation and survival in some seasons for the colony. It really does no harm to the colonies even at this time of year to remove the roof of the hive and gently raise the crown board. If this is done on a cold bright day and a quick glance is cast over the frame tops, a good estimate can be obtained of the state of stores available to the bees. If the clustering bees are well down into the combs and not readily visible, the stores situation is good. If, however, on making this inspection the bees are found to be clustering right up to the top of the frame, then trouble is brewing and the bees are hungry. They have worked their way steadily upwards eating the crown of stores above them as they went, until they reached the wood of the top bars and can go no further. If this condition is seen in early December, the beekeeper has a problem to solve. The timing of the inspection discussed here is most important if a true result is to be achieved— because on a bright day in winter (they do occur occasionally in Scotland!!), the most provisioned bees will expand the cluster and even swell to the frame tops. The inspection has to be made on a cold day where the bees are not inclined to break the winter cluster. Giving a bag of sugar or even candy to a colony showing these obvious signs of hunger will not normally be adequate. In a mild winter a feed of heavy sugar syrup fed where it is in contact with the bees, i.e. right on the frame tops not over the feed hole of a crown board, might do the trick. But remember the bees then need to fly at some sooner rather than later time to void the excess water imbibed in the syrup feed. The best and easiest remedy for a hungry colony at any time during the dormant period is a full frame of ripe honey laid flat over the cluster and resting on a stick placed at each side bar to give a "bee space" between the clustering bees and the underside of the comb given. This is the time of the year for cleaning queen excluders. The brace comb and propolis adhering to the excluders is very brittle when cold and can be scraped off easily using a hive tool or paint scraper. Out-apiaries should be visited at least weekly during the dormant period to check that the hives have not been disturbed or worse! It is sensible after periods of high winds to visit out-apiaries for an additional check — high winds can wreak havoc on bee hives if they blow from particular directions. I've seen hives where the roof and crown board complete with 'holding down' bricks on top of the roof have been blown for yards after a strong gust of wind struck the hive front on and pressurised it internally. Better safe than sorry!

THE MONTHLY ROUND - RAMBLER MARCH

This is the month where critical observation at the hive entrances can really pay bountiful dividends. Pollen gathering activity can relate volumes regarding the state and quality of population and bee strains in the hives. On bright days where bee flight is possible, copious pollen intake is a real reassurance for the beekeeper, this is an unerring indicator that the queen is laying well and brood is being fed and nursed. For the beekeeper who dabbles or more than dabbles in breeding queens, the foraging behaviour in the early spring months has or should have a particular significance! In apiaries where a number of hives are kept, acute observation will show that particular colonies are invariably first in the field each day, and in all, probability last to "put their feet up" ' in the late afternoon/evening. Also observation for colonies which persistently forage in rain or wind or low temperature is a worthwhile occupation. These colonies should be noted and when and where possible, used for queen and drone propagation. Other characteristics of the strains observed must also be taken into consideration, but these other qualities can only be assessed by the "hands on" method, i.e. internal inspection during the active season. The qualities to be husbanded in a breeding strain are, in my opinion :—

1. A genetic marker (i.e. breed from only black strains or only from the yellow tinged strains, whichever moves the bee breeder!).

2. Good temper (bees which sting on sight are loved by none, bee-keepers included!).

3. Good temperament (bees which zig-zag all over the comb face during examination are sore on the eyes where many stocks are worked!).

4. Good hoarding (goes without saying, after all most of us have a vested interest in the honey returns. Otherwise we'd keep ants!).

5. Stamina (includes longevity, strength, resistance to disease and good overwintering!).

If the above qualities can be combined with the characteristic of foraging in marginal weather conditions, I feel that the bee breeder is working positively with the forces of natural selection — rather than propagating willy nilly and thus perfecting strains which need a great deal of husbanding to get them through poor seasons or particularly hard winters. The keeping of hive diaries is a real necessity for the beekeeper who wants to really know his bees and get the best from them.

March can be quite a dangerous time for the honey bee which in nature, if of a suitable strain, will have adequate stores to sustain its brood rearing programme in early spring. The unfortunate bees which are the charge of beekeepers are more at risk at this time than the feral colon since, if the beekeeper through inexperience or worse misjudges his 'harvest' or take. The bees can be kept in a borderline starvation situation come late February early March. To qualify the attitude to potential isolation starvation in early spring I would say that any beekeeper overwintering bees in well provisioned double brood chamber hives has little to worry about regarding spring starvation in normal winters. The beekeeper with hives having well provisioned brood chambers augmented by a well provisioned shallow crate (no queen excluder present please!) is also in a commendable position. However, the beekeeper overwintering bees in British Standard (B.S.) hives, having only a single brood chamber, is always at risk, good winter or bad! The single B.S. brood box beekeeper has to be on his toes all the time since the depth of the single storey of the B.S. frames on its own is not particularly suited to the storage of the necessary crown of honey above the cluster in winter in moderate to severe winters. As a single brood chamber beekeeper myself, I feed sugar syrup to my bees almost coincident with the gathering of the first pollen in spring. I also am convinced that by supplying even the best provisioned of colonies with a palatable drink, i.e. a light sugar syrup solution in spring the colony is spared much unnecessary work and loss of bee life in the need to forage for water, which has been proved by exhaustive study to be indispensable in brood rearing. It is indeed aqua vitae.

THE MONTHLY ROUND - March

This month embraces the quite dramatic population crash and ultimate core population stabilisation in the honeybee colony. The population depletion at this time of year is due to the loss of the almost senile older bees in the field at a faster rate than they are being replaced by new young bees reared in the late winter/early spring. Most honeybee colonies which overwintered as strong at surplus honey gathering strength in autumn colonies will be noted at this time of year to be covering at most six frames, many will have reduced to around even two or three, but the norm to be expected is four/five frames. This is the core population from which the colony will begin to develop in size, provided of course the queen is healthy, fecund and not a granny too many times over! The bulk of the bees in this core population will be bees produced from eggs laid from early January onward, especially if the late February early March weather was suitable for intense foraging activity which will have taken its toll on the older bees. For the beekeeper working in areas where oil seed rape and/or good stands of sycamore are available to the bees, or where it is the intention of the beekeeper to migrate to these sources of nectar, it is good management practice to feed sugar syrup to simulate a nectar flow complimentary to the hoped for masses of crocus pollen into the hives at this time (in particular years). This feeding has many benefits for the "early-area" beekeeper. The most obvious benefit is of course that colonies in a borderline stores situation will be safeguarded against eating themselves out of house and home. The early liquid feed will also provide much needed water which the bees require for brood rearing to facilitate pollen digestion. The prime benefit is that the queen is stimulated to lay eggs at a higher rate than otherwise and hopefully if the initial colony strength was sufficient to brood a high egg laying rate, by the time the oil seed rape or sycamore are secreting nectar the colony in question will be booming. In a moderately good spring, the beekeeper will get honey returns adequately large enough to well compensate for the energy and expense invested in the sugar feed. The foregoing early feeding procedure in my opinion is of critical importance in single brood chamber overwintering management The beekeeper who generally overwinters on double brood chambers where the top chamber is choc-a-bloc with stores — all other things being equal — will find that his bees will generally build up quite rapidly, naturally in areas where early pollen intake is possible. The bees have an inbuilt feeling of prosperity and security in such a situation, but even here a light sugar syrup feed to supply a palatable drink for the necessary water requirement in the colony will not go wrong. Ideally, liquid feeding should be given to the colonies in the late afternoon as flying is coming to an end. This procedure will help safeguard against robbing starting. When feeding colonies, especially those low in population numbers, it is always advisable to restrict the size of the hive entrance, ensuring that the restricted entrance is adjacent to the position of the colony on the frames in the hive, thus enabling the small population to defend itself against intruders. Sugar syrup feeding, once started, should be continued right up until the bees are foraging actively on oil seed rape or sycamore which constitute the first major nectar sources in Scotland. Mention was made briefly last month about the treatment of colonies in spring, which are found to-be queenless. For the less experienced beekeeper, uniting to a suitable queenright colony is probably the path of least resistance. A useful method, which will not only save the colony but will have it queenright at the earliest possible time for the particular year may be employed by the beekeeper with more than two hives. The queenless colony maybe given a frame of emerging brood (without adhering bees) from a queenright colony in the same apiary. This will help maintain population size. If this is done every other week in spring until around the last week in April and then a frame of eggs and open brood is given, the colony will raise a new queen which will be mated and laying by late May. The timing of the eggs and open brood insertion can be further accelerated if the beekeeper is aware of when the first drone eggs are being laid in the colonies. A warning though — the drone eggs noted should be in a different colony from that which is being used as the donor colony for the queenless hive to inhibit rapid inbreeding which will result in poor colony performance or no colony performance. Inbreeding in honeybee colonies is a phenomenon which many beekeepers are either unaware of or don't care to concern themselves about. Perhaps I'll stick my neck out (again) next month and air my views on this very neglected subject.

THE MONTHLY ROUND - April

I wonder how many beekeepers realise the "powder keg" of potential population explosion that exists in their bee hives in the latter half of this month. Even at the end of April a colony may seem to be not very densely populated, perhaps covering seven or eight brood combs — but most of these combs will be packed with brood in all stages, which on emerging in the first half of the month of May will swell the adult population to quite staggering population numbers. Apparently overnight! Beekeepers who mark or clip, or mark and clip the queens in their colonies are well advised to carry out this operation either at the end of this month, but, normally in a good colony building to honey gathering strength for the sycamore, no later than the end of the first week in May, especially where the now very popular black strains are being husbanded. A black queen is difficult to find in very populous colonies! Where syrup feeding was started earlier in the season it should be continued with until the first major nectar sources are available to the bees — not just blooming, but available! Wind and rain and cold can confine bees to the hive while abundant nectar is present in the field, but not available due to the bees not being able to reach it. Even at this seemingly promising time for colonies bees can still perish more or less in the midst of plenty. Feed where necessary! As soon as the bees show that the nectar flow has started get the supers on and remove the feeders. Depending on how spare comb and supers were stored, the beekeeper will have either an easy or difficult transition from the feeding stage to the honey production stage — i.e. whether he/she incurs wax moth damage or not. The judicious use of certan will protect stored drawn combs from wax moth which is a biological preparation now marketed. A simple method of discouraging wax moth damage is to store supers (and spare brood boxes) wet with the honey residues after extraction. When dry combs are to be stored, by interspersing these with wet combs, seasonal protection will be afforded. The bees will rise rapidly into these 'wet' combs even though they smell slightly or even heavily of alcohol.

Much discussion has been carried on regarding when to put supers on a hive in spring. The best and most sensible time is after any feeding is finished with, but generally not later than the first week in May. Rest assured on two points;— (1) The bees will not put nectar into the super until they are happy with the state of the stores in the brood box: i.e. hungry bees won't store honey in the super; (2) By supplying an external source, the bees themselves will tell you when the nectar flow starts, when the bloom begins in earnest the bees will forsake an otherwise much visited drinking place. It is also sensible at this time to fumigate all spare brood combs with 80% acetic acid to kill any residual Nosema spores and S. pluton cocci lurking in the cells. The interchange of combs between hives is an obvious mechanism for the spreading of disease. Any procedure which limits or eliminates the spread of disease is definitely worthwhile thinking about. Incidentally, treating Nosema using Fumadil B is fine, but spores left in the combs of treated colonies are still 'viable. (note from editor Fumadil B is no longer available for the treatment of Nosema it was found to cause birth defects.) This results in re-infection of the colony unless the treatment is backed up by the fitting of a full set of new frames and foundation or by some form of fumigation or sterilisation of brood combs.

THE MONTHLY ROUND - May

The month when it all began to happen for us. This is the first month where it is possible to harvest honey. Supers should have been repaired where necessary and prepared with at least one frame of drawn comb but preferably with a full complement of drawn comb. If some of these combs are still damp with last year's honey residues so much the better. Combs in such a condition will bring the bees eagerly out of the brood box to take possession and once they have moved into the super they will then be encouraged to store surplus there, providing of course the brood chamber has been transformed into a well provisioned larder before hand. By supplying this first crate of combs drawn it will most likely mean the beekeeper will be unable to harvest cut comb honey from them, but in a reasonable season the bees will soon hopefully require a second super; this second super should preferably be filled with frames of foundation — to give the bees some comb building to do, and also to give the beekeeper his first virgin honey. The second super should be put on top of the first one as soon as the bees are working at least five combs in the original. This gives the bees adequate advance room and could relieve congestion which may lead to swarming perhaps earlier than otherwise. Myself, I like to have all my hives swarmed by the middle of May — in normal years. By ensuring that all the hives have had their swarm urge requited by this time, the management of the colonies is greatly simplified for the rest of the season. The easy way to short circuit the colony's need to swarm (which it invariably will do if it has an overwintered queen heading it! This statement has to be made because in his management at Buckfast, Brother Adam re-queens his colonies in March each year.) is to cause it to advance its spring development using good management. A healthy well provisioned colony fed light sugar syrup soon after the first pollen has been noted being carried into the hive, will respond as in a nectar flow — in mild springs. In very cold early springs the bees will be reluctant to break cluster to take the sugar syrup, unless it is a heavy syrup. By mid-April in a particular spring, a developing colony can be ready for a second brood box (where it was overwintered on a single!)

If a frame of brood is removed from the colony and placed in the second brood box so that it is positioned right above the centre of the brood nest in the lower box (otherwise chilled brood will result), then the bees will be encouraged to move up and take possession of the top box more rapidly than otherwise. And since the frames above the brood nest in the lower box will be warmed by the heat rising from the lower brood frames, the bees will be encouraged to expand the brood nest upwards into the upper chamber. Where colonies are overwintered on double brood boxes it is very worthwhile recovering the boxes at the start of the spring build up. Since all the brood will be in the top box anyway — the lower chamber is invariably free of brood. The bees will soon occupy this otherwise neglected box, when it is set above. For areas which have predominantly late nectar flows (i.e. clover, lime, heather as opposed to sycamore or oil seed rape areas!) this switching of the lower brood box, or the addition of a second brood box to a single brood box colony will enhance colony development and hopefully increase colony strength such that the bees will be at near optimum strength by the time the nectar flow begins. A weekly interval/9-day interval inspection cycle should now be normal procedure. Swarm preparations can occur more or less at any time from early May onward, depending on the weather patterns. Look out for swarms issuing from hives in the 6th-9th day of a prolonged spell of good weather from mid-May onward.

THE MONTHLY ROUND - June

Swarm control or at least swarm awareness should still be the uppermost thought in the beekeeper's mind during early June, especially in areas where raspberry acreage or oil seed rape stands are available to the bees. In less endowed areas which are still subject to the 'June gap' — swarming although always possible is less likely — unless the summer is so ideal that the June gap effect is greatly minimised. In mediocre years where June is not particularly warm, hungry bees can be more of a problem for the beekeeper, than swarming. It pays to check the level of stores in the hives at this time — and where necessary sugar syrup feeding should be undertaken until the later nectar flows commence. Sugar syrup feeding will help maintain the queen in lay — and if one remembers that the bees born in June will have a decisive influence on the number of foragers available for the lime and other sources of nectar in July, to say nothing of the brood — on the effect of the bee populations at the heather, then the importance of June feeding, where necessary, will be readily appreciated. June is a good month to eliminate undesirable combs from the brood box, replacing them with frames of undrawn foundation. During the early internal examinations of hives in May distorted combs or combs with 'pop' holes in them should be moved to the sides of the brood box to allow any brood to emerge from them. When these combs are all empty of brood they can be removed and replaced with the previously mentioned undrawn foundation. Insert this new frame right into the heart of the brood nest — or at least where the queen can get easy access to it near the fringes of the brood nest, not the fringes of the brood box otherwise the bees will just draw it out and fill it with honey (instead of drawing it out and having the queen fill it with eggs!) this forcing more honey into the supers where we can get our ' sticky' fingers on it.

To many beekeepers the idea of extracting honey from brood frames is anathema but often this is not only advisable but necessary since congestion of the brood nest due to honey can occur — would that it occurred more often! Such honey choked frames can be either stored for later use in current nucleus stocks or for later late winter emergency feeding. Where many such honey bound combs are extracted (you will find the honey no different from that extracted from the honey supers!) it is a sensible idea to place them in a spare brood box and fumigate them as a matter of course using 80% acetic acid. By employing such a system as a normal management procedure the possibility of recycling Nosema or even A.F.B. residuals in these brood combs will be greatly minimised — to the benefit of colony hygiene and better beekeeping. The high ambient temperature in summer greatly enhances the fumigation effect. Fumigate combs for about a week then ventilate them for at least 2-3 days to get rid of the concentrated acid vapour. The bees will come to no harm even if residual acid vapour is present in the combs when they are donated to a colony — and no contamination of honey will occur after the stated ventilation period.

THE MONTHLY ROUND - July

This is the month of the lime, bell heather and willow herb — to say nothing of the "dreaded" privet. Some beekeepers even go so far as to recommend the removal of honey supers to prevent privet nectar contaminating the honey already in the supers. From experience we all probably know that despite the overwhelmingly sickly sweet smell given off by privet blossom, the honey elaborated from its nectar tastes of burnt sugar or something equally acrid. However, where privet and lime exist together in the same foraging area the resultant natural blend of honey can be quite pleasantly flavoured. So, for the beekeeper who gets a predominantly privet and biased honey harvest — blending it selectively with a more palatable honey can save the day. I would in no way recommend any beekeeper to remove the supers due to the 'threat' from this honey — first get your honey! As a matter of fact some palates actually like the taste of pure privet honey — I myself have acquired a taste for the stuff — but if the truth be told my favourite honey is hawthorn, which unfortunately is not all that easy to procure — the hawthorn (Crataegus monogyna — can't help showing-off since I got Rex Sawyer's "Pollen Identification for Beekeepers"!) is a mysteriously fickle producer, blooming in late May/early June in most areas.

For the 'heather' man, July is of critical importance since the eggs laid in the comb in the early days of July, up to around the 21st of the month will produce the bulk of the foraging force of bees for the moors, by counting six weeks from egg laid to foraging bee. Assuming the heather bloom period behaves itself, and commences around 7th-12th August (some years the bloom begins as early as 27th July!) and effective bloom period ends around 30th August-3rd September. Then counting 6 weeks back from these critical dates we get the latest date for effective egg laying as around 23/7/84 and the first critical date for the stimulation of egg laying rate as around July 1st. A colony being prepared for the heather can also to good effect be reinforced with two or even three combs of sealed brood around the start of the second week in July. If the queen(s) of this (these) heather hives are current year queens, then by the time the heather blossom opens the colony so prepared will be well able to give a good account of itself — provided the swarm urge of the colony has already been met. If a colony so prepared hasn't swarmed or been "swarm managed", if the weather at the moors is exceptionally good, then the chances of the reinforced colony swarming on the moor are quite high. At least clip your queen if the situation just discussed holds — you have been warned!! Fully drawn shallow combs of clean, white wax of the current year should be selected and prepared in advance of the move to the heather. These combs should be placed in the hives after the removal of the late summer blossom honey and at least a day before the planned exodus to the heather to give the bees the chance to secure the super with propolis — this helps keep the boxes tight — it also allows the bees to clean and dry the newly donated combs which will probably still be sticky with blossom honey residues from recent extracting work.

THE MONTHLY ROUND - August

For the 'heather' beekeeper, August heralds the start of another beekeeping year. Strange though this statement may seem its truth is irrefutable. August gives the colonies the chance to have a final late 'fling' at brood rearing on the abundant nectar and pollen which can be obtained at the moors from the legendary ling heather (*Calluna vulgaris*). The massive amounts of pollen available from ling can be easily established and demonstrated by walking slowly through the blossoming heather tufts; (not to be recommended for hay fever/asthma sufferers!) the clouds of brown pollen billowing from the dehiscing blossom has to be seen to be believed. The dehiscence is also a sure sign that the blossom is secreting nectar and is a reassuring sign for the beekeeper anxious to establish if the heather nectar flow has indeed commenced. On bright warm days during the ling nectar flow, the beekeeper will be left in no doubt about whether nectar is available or not — the seemingly chaotic activity of flight at the hive fronts will tell the full story as the incoming and out flying bees dodge each other in their frantic efforts to procure the available bounty of rich amber nectar. The heady aroma rising or wafting from the hives during the ling flow is unmistakable and gives a positive indication that the hives are indeed prospering. The brood produced at the heather, of course, results in the population of the hive containing a relatively large number of young bees to carry the colony through the winter. If the queen leading the 'heather' colony is in her first year she will be encouraged to 'lay her heart' out, stimulated by the great mass of ling pollen and nectar coming into the hive. A word about bees going to the heather; it is essential to ensure that hives at the heather are headed by queens which, if not of local origin, are of a strain which originated from traditional late blossoming areas and where these queens are expected to lay right up to at least the end of August — tailing off at around mid-September. Many queen bees purchased from certain areas in the south where the bees traditionally have no access to heather, cease to lay effectively around the end of July. These same queens probably originate in areas which by virtue of their geographical location, are in comparison to the northern regions, much farther forward in floral development in the spring. Thus resulting in the queens responding too early to their "biological clocks" and exhausting their food stores before the later northern, first major spring nectar flow from the sycamore begins.   The aware beekeeper will, of course, notice the depletion of stores in such colonies and feed. But then again, the aware beekeeper probably would not have tolerated such an exotic queen in his hive in the first place, so the colony would probably be lost at the hands of the non-aware beekeeper. Ideally, hives should be left on the moors for at least a week after the heather 'flow' is finished. This gives the bees the chance to ripen and seal the last of the nectar gathered. Moving too soon can result in raw nectar spilling from the combs and causing the bees to panic, perhaps resulting in the loss of the hive due to 'melt down' of the combs. If the supers are removed using the clearing board method at the end of the extra week, the hives can be brought back to the winter or home apiary at leisure. In normal years sugar feeding can be carried on with well into late October — but the syrup should be thick, at least 1 kg. sugar dissolved in water to make 1 litre of syrup (not dissolved in 1 litre of water). So, there is no need to panic with regard to feeding bees returning from the heather. Bearing in mind the colony will require around 40-50 lb. (18-23 kg.) of stores (i.e. honey and pollen!) to see it through a reasonable winter, the beekeeper can judge how much feeding has to be done.

THE MONTHLY ROUND - September

The "bee year" has begun again. Yes, begun! The young bees produced on the heather or other late nectar flows determine the quality of the colony preparing for the long haul through the dormant period. A colony possessing a fecund young queen, a good population of healthy young bees and an adequate or hopefully, more than adequate amount of stored food, will winter well. Other important factors to be considered for good overwintering are.

A good sound weatherproof hive situated in a location giving a maximum, exposure to winter sunshine (even on the frostiest day in winter when the sun shines the few degrees increase in warmth on the hive walls can be a major factor in colony survival!) Try lifting the roof from a hive located in a shaded apiary during frost and compare the result with trying a hive roof in an apiary in a sunny aspect on the same afternoon where the sun is shining. The hive roof on the sunlit hive will lift easily, the shaded hive roof will be 'cold welded' to the crown board. Even a few degrees increase in hive temperature can make a dramatic difference to food consumption in colonies. Witness the considerable difference in food consumption in colonies in the recently past mild winter, relative to the massive depletion of stores in the ultra-severe winter of 1980-81 where night temperatures of minus 17 °C persisted for weeks on end.

Ventilation must also be considered in the preparation of a hive for the winter. A full width entrance is very desirable, BUT, the entrance MUST be proof against mice, a wire screen trimmed to allow bees to pass under it, or with a gauge which will allow bees to pass but keep mice out is ideal. There are many proprietary mouse guards on the market. However, the simplest and least fiddly entrance is merely a full width entrance not any higher than $^5/_{16}$ in (8 mm). This height of entrance will keep even the smallest shrew at bay. If the hive is set on its stand such that it is horizontal (i.e. the honey combs are hanging "plumb" or vertically in the hive) and inclined so that there is a slight gradient from back to front, to facilitate drainage of any moisture or condensation from the floorboard, and the roof is made secure against wind and weather, then the colony will overwinter effortlessly.

The hive should also be orientated such that the front does not face the direction of the prevailing wind, adequate ventilation is extremely important but not of the 'forced draught' sort.

An important point which is obvious but bears mention, is that the hive should always have some kind of pedestal (even 3 half bricks will do, one at each side at the front of the hive and one at the rear in a 'tripod' type of disposition!) to give an air space under the floorboard, otherwise the floor will eventually rot.

Feeding of colonies which are light in stores should have been attended to and completed by the end of this month — in some mild autumns, an extension into perhaps the first week of October can be accepted, but ideally the end of September should be the aim. Opinions on adequate winter (minimum) stores vary but there is no such thing as too much. Feed the bees as much as you can afford or as much as they will take down if you only have a few hives. A complement of 50 lbs. of stores should be an adequate minimum amount for stress free wintering — for bees and beekeeper.

The ideal overwintering condition for Scotland anyway is two brood chambers, the top one being chock full of stores. Next best is a brood chamber topped with a shallow super full of stores please be sure to remove the queen excluder before bedding the bees down for the winter! The most hazardous overwintering method is the single brood box method, and really not to be undertaken by the novice. Single box overwintering requires good management and an awareness of the limitations of the single chamber system, Especially where the hive used is one of the British Standard hives like the National, Smith or Wormit. Even the Langstroth is suspect due to the depth of the brood chamber ($9\,^9/_{16}$ in) The choice of hive type is not a question I propose to get embroiled in at this present time — but each to his/her own, depending on needs and preferences. There is an almost fool-proof method to ensure good overwintering in the single B.S. brood chamber, but later!

THE MONTHLY ROUND - October

The hives should now be in winter rig. All feeding of liquid supplementary sugar in the form of sugar syrup should have been successfully completed. Rather belatedly, a comment on the problems incurred in syrup feeding: If feeding is done during the time of day when bees are flying freely, care has to be taken not to spill any syrup. Spilled syrup can cause robbing to start. Also, any hive fed should always be checked not later than 5-10 minutes after insertion of feeder, any tell tale dribbles of liquid running from the hive entrance requires the removal and checking of the offending feeder. Either the feeder is damaged, i.e. has "pin holes" in it, or the lid is not properly "home", or the feeder is 'canted' badly due to resting on brace comb on the frame tops. The plastic pail or friction lid type of honey tin feeder are worthwhile feeders so long as it is realised that they can, and do, leak! The best feeder for winter preparation feeding is the "Miller" which rests on the hive similar to a super — it can be filled with syrup without disturbing the bees at all — merely lift roof and crown board off and pour in the "sweet goodness". The "Miller" is not so ideal for spring feeding because the bees are reluctant to break the cluster and move up to the feeder at this time of year. Better with a pail or friction lid type of feeder placed either right on the frame tops in contact with the cluster or over a crown board above the brood nest, where the gap between feeder and brood frame tops is $^3/_8$ in or less. All spare equipment should be examined now and either repaired or set aside for repair. Spare brood frames should be stacked in their brood boxes with a sheet of paper between each brood chamber to discourage wax moths, greater (Galleria melonella) and lesser (*Achroia grisella*), which can make a desperate mess of honey comb when the moths are in the larval stage. For the smaller beekeeper all the honey handling should be about finished and the extracting equipment washed, dried and put away safely for next season's work. The tinplated honey centrifuge, after cleaning should be lightly coated with petroleum jelly, before storing away. This will enhance its length of life and inhibit rusting. Any honey supers with "wet" shallow frames should be stacked in such a way that they are alternated by any spare shallow crates having either dry drawn shallow comb or undrawn foundation to inhibit wax moth invasion. Queen excluders can be effectively cleaned at this time of year. As the evening temperature drops, the brace comb and propolis residue can be scraped or wire brushed, in the case of the 'Waldron' excluder in preparation for the 'big off' next spring.

For the hobbyist beekeeper this is the time for rest and contemplation and background reading. The background reading bit cannot be too strongly recommended. Too much conflicting knowledge can be confusing sometimes, but better that than little or no knowledge at all. Mention was made in the previous issue about an almost foolproof method of successfully overwintering bees in a British Standard single brood chamber, i.e. Smith, National or Wormit design. For many years now I have been advocating the use of 1 kg bags of granulated crystal sugar given to the bees in mid winter, around mid December. The method advocated was: take a 1 kg bag of sugar (unopened) and immerse it in cold water for about 10 seconds then place the wet bag right on the frame top where the clustering bees can reach it merely by rising up as a cluster with no need for individual bees to separate from the cluster to gain access to the sugar, which they do by chewing through the paper of the bag. By the time the bees reach the sugar the original dampness has caused a crust of solid crystal to form which the bees consume eagerly. For the past two winters I have been using a complement of four bags of sugar given simultaneously as a "crown" of sugar over the centre of the bee cluster, right on the frame tops as usual. This has proved to be an outstanding success, and for many reasons. Recent work done on the Continent on the honeybee winter cluster and reported in Bernard Mobus North of Scotland magazine *'Apiculture'*, has demonstrated the mechanics of bee movement within the cluster during the dormant period. Excess water content in colonies of particular sizes has been shown to be a major problem for the bees, causing them to do certain things like rear brood very early in winter in order to dissipate some of this surplus water in the form of brood food or they are forced to defecate in the hive, with all the attendant dangers of this event. I have found that by giving the bees the advocated 'canopy' of bags of sugar, that the colonies come through the winter really well and apart from eating into the sugar as the winter progresses, for their immediate needs, the bees actually liquefy the crystal and store this 'sugar honey' in the cells immediately under the canopy. Where in late winter the bees might be faced with empty cells up on the frame tops — they now have liquid stores right above and in easy reach of the cluster. The major bonus from feeding crystal sugar in this way is that the bees now have another more positive method at their disposal for the dissipation of any excess water within the cluster — IN FACT any excess water is now more of an advantage than a disadvantage. Try the method and say 'goodbye' to the fears of isolation starvation. I will not say that I now look forward to winter, but all other things being equal, the overwintering of colonies of honey bees now holds no terrors for me. The sugar can be given earlier than mid-December but if given when the bees are still relatively active, it must already be crystalized in the bag. If the bees are "showered" with sugar grains they throw most of it out of the hive.

THE MONTHLY ROUND - November

Winter is deepening, bee flight, except in unusually mild late autumn or early winter weather, will now be a rare occurrence. The waiting time for the beekeeper has begun — waiting for Spring. Management is now reduced to merely checking hives in out-apiaries on a weekly basis, to ensure no tampering or worse has occurred at the hives. It is good practice to check out apiaries after storms or other periods of high winds, even though the previous check was made the day before the high winds. Wind coming from particular directions, depending on the apiary location, can wreak havoc among hives, so it is always best to assume the worst and make that little extra effort which hallmarks the good beekeeper, to ensure all is well with your charges. Letters and reports appear in the magazine from time to time commenting on the mysterious goings on of particular colonies — especially in early spring, when the hives are particularly vulnerable. Colonies are reported just to dwindle away or queens disappear for no reason, or population build-up does not happen as it should. All sorts of puzzled questioning is sounded, but rarely do the beekeepers concerned think of the most probable cause of 'malfunction' namely covert or even overt disease in the hives. Even many highly skilled and experienced beekeepers have this massive 'hole' in their experience. The incidence of the major bee diseases in Scotland is thankfully comfortingly low — but by being so 'thin on the ground' they tend to be rather forgotten when we look for reasons for the strange things that happen to our bees sometimes. The disappearance of a queen from a hive can occur at any time of the year, due to death from injury or old age or disease! Mature laying queens have also been seen merely flying out of hives at times, never to return — perhaps even insects, like people, sometimes go crazy — who knows? By resolving to re-queen hives either each year or every other year, the beekeeper can to a greater or lesser degree safeguard against senility and possibly the fatal effects of lingering disease in queens. Care during hive manipulations will normally ensure the queen is not damaged. However, sometimes she is located on the very first comb out of the hive or on the adjacent comb face, and in a "tight" brood box she can be "rolled" and possibly crushed. It is always politic to heavily smoke the frame spaces of side frames a few seconds before withdrawal. This gives the 'old lady' the chance to move out of danger. Many beekeepers are adamant that they have no disease in their colonies, but in nature anything which lives and breathes, be it animal or plant, is host to some kind of parasite or pathogen or other — even parasites have their predators and pathogens! It is best always to have an open mind. Many honeybee diseases are endemic, that is they are present continuously in colonies but are of low level incidence and never show, only when a colony suffers a setback can endemic disease suddenly become epidemic. The major honeybee diseases we have to contend with in Scotland are Nosema, Acarine, E.F.B. (European Foul Brood) and A.F.B. (American Foul Brood). Of these diseases, A.F.B. is the most dangerous and is to honeybees what Fowl Pest is to poultry or Anthrax is to cattle. A.F.B. is a notifiable* disease and colonies found to be suffering from it have to be destroyed, by law. E.F.B. was at one time also looked upon as notifiable* and infected colonies had to be destroyed. The Bees Bill 1982, however, removed E.F.B. into the treatable category but only under qualified supervision. Terramycin and streptomycin are effective antibiotics in the treatment of E.F.B. Sodium sulphathiazole and terramycin are used to treat honeybee colonies suffering from A.F.B. in some parts of the U.S.A. and Canada, where, due to the method of beekeeping i.e. package importation in spring, and autumn destruction of colonies with the brood combs being stored, dead larvae and all, until they are recycled in the colonies in the following spring. While chemotherapy may be necessary at times in the treatment of bee diseases, there are many good management practices which can be used to reduce disease incidence in hives. A continuous year by year check on bee health in early spring will give a fair indication of the general state of health. Early spring sampling is very important, because in endemic conditions, samples taken later in the active season can show negative results due to the older infected bees dying off as the active season progresses. Residual infection can cause disease to flair again the following winter, resulting in more mysterious goings on in the hive being observed by puzzled beekeepers. Many bee diseases can be, if not completely cured and eliminated at least controlled at endemic levels without the beekeeper having to resort to chemotherapy, by using a systematic regime of fumigation of spare brood combs and hive bodies. To safeguard against the incidence of A.F.B., formalin as a fumigant is extremely effective in killing any dormant and residual spores of Bacillus larvae, the bacterium which causes American Foul Brood. The disadvantage of formalin fumigation is that it is difficult to sterilize capped cells, and stored pollen and honey are permanently toxic to bees. The most effective easy to use fumigant for the average beekeeper is 80% Acetic Acid. Recent work done by Dr L. Heath has demonstrated the effectiveness of 80% Acetic Acid in treating Chalk Brood which seems to be on the increase in certain beekeeping areas. The effectiveness of 80% Acetic Acid is well-known in the treatment against

Nosema and Melissococcus pluton (late Streptococcus pluton) the causative agent of E.F.B, and also in combating wax moth, the acid kills both eggs and adults of both wax moth species. Thus, all stored, spare brood combs if placed in spare brood chambers in stacks, by soaking pads with approximately 125 ml., 80% Acetic Acid and placing these pads, one per brood chamber between the brood chambers, can be safeguarded against five potentially dangerous diseases and conditions of honey bees, if Malpighamoeba mellificae is included in the list, (Nosema, Malpighamoeba mellificae, E.F.B., Chalk Brood and wax moth). It must be stressed that only spare combs not in use should be sterilised, on no account must hives with bees in them be treated with either formalin or 80% Acetic Acid. All fumigated combs should be ventilated for at least a week before being given to the bees. Spring is the ideal time to carry out fumigation. L: Bailey gives a full account of various treatments for honey bee diseases in his excellent book "Infectious Diseases of the Honey-Bee". Every beekeeper should read or have read this important work.

*(you can contact Bee Inspector - Scotland by emailing: beesmailbox@gov.scot)

THE MONTHLY ROUND - December

This month normally heralds the hardest part of the winter for the bees, in the West of Scotland anyway. The frosts which have been reasonably moderate up till mid-December begin to bite deeper each night. This time of year is among the most depressing periods of the beekeeping year. The nights are long, dark and still lengthening this winter has still a long way to run. So long as the bees were well provided for, i.e., at least 50 lbs of well sealed stores, before the onset of the first frosts, and the hives are situated in a sunny spot, sheltered from the harsh winter winds, and not located in a damp, dreary frost trap. If the hives roofs were secured against being removed by high winds and the entrances made secure against the entry of mice and the dampened 1 kg bags of granulated sugar were put in place at least before the middle of this month, then the beekeeper can rest easy until at least early March. He has done his good management bit. Snow drifting around the hive and even blocking the hive entrance should not give any cause for concern to the beekeeper. A "snowed-in" hive in winter can actually be a benefit to the bees since, on bright, crisp days after snow has fallen the bees might be enticed to leave the security of the cluster by the dazzling light entering the hive. The waste of bee life when this sort of thing happens can sadden the toughest heart. Although a moderate mortality rate at this time can also be beneficial due to the loss of the older bees which might have been tending toward dysentery or senility, thus removing a negative factor from the other factors governing the survival of the colony as a whole. However, in general, bees should be discouraged from flight in snow because even healthy young bees can become disorientated by being unable to distinguish up from down due to snow glare. The hobbyist can screen the hive entrances in winter using a canted board in front of the hive in bright snow conditions. The commercial man must take his chances. If the beekeeper can brave the cold at this time of the year to work in the honey house or shed, then repairs to spare hive parts can be carried out at this time. Queen excluders are easily cleared of propolis and brace comb during the colder weather. Don't be tempted to do anything with foundation or drawn virgin comb, the wax is so brittle at this time of year you'll do more damage than good. Reading and study of honeybee lore should be going on apace. A gentle hint to one's spouse may result in that sought-after reference work turning up in the Christmas stocking. A present in the shape of a subscription paid to the International Bee Research Association will give a worthwhile return to the really interested beekeeper. The wealth of information — right at the forefront of beekeeping thinking and practice appears in each and every issue of Bee World, I.B.R.A.'s magazine. A widening of beekeeping horizons can be achieved by encouraging some member of the family to present you with a subscription to "Bee Craft", the magazine of the British Beekeepers' Association — or even treat yourself! Out-apiaries should, of course, be checked on a weekly basis and as previously advised particularly after high winds. A thin twig or wire rod should be inserted into the hive entrance with a sweeping motion at each visit from now on until the Spring to maintain the entrance clear of dead bees and debris, especially where the late autumn weather/early winter weather was not conducive to flying — resulting in a higher population of older bees going into the winter and dying en mass in late winter — accumulating on the floorboard and possibly blocking the hive entrance!

THE MONTHLY ROUND UP - January

The hives at this time of year stand stark in their winter sites without living leaf or vegetation cover or may even be just visible as shapeless clumps under a heavy white mantle of snow. However gaunt the hive exteriors look in January they belie the stirring of movement in the heart of the clusters of the overwintering occupants. As the days slip past and February draws near the colonies begin to sense the onset of Spring. The canopy of crystal sugar or block candy will very steadily be eroding away as the bees begin to increase their food consumption to support the first few larvae being reared. Nearer the end of January, the queen is becoming more active, seemingly anticipating the income of early nectar and pollen from the snowdrops which will usually, in West Central Scotland, be in bloom in late January if the ground is not frost-bound. January is notorious as the month of severe gales in Scotland so the earlier advice about out-apiary inspection after such events is still cogent. So long as the hives are intact with roofs in place and weathertight, the bees will come to no harm. It is quite astonishing how soon the "active season" can be upon us — although at time of writing it seems a long way away. Thoughts should be turning to the plans for the coming season, equipment to be repaired or replaced needs consideration, jars and cut comb boxes for the anticipated bumper crop this year. Does the current management system measure up to the number of hives now being worked? Beekeepers who expanded last season may find that they are short of both the necessary equipment and expertise to meet the new demands in working larger numbers of hives. The rule for adequate equipment is at least 50% more spare comb and hives than the number of successfully overwintered stocks. Thus, the beekeeper with perhaps six stocks of bees should ideally have accommodation for at least a minimum of another three spare, to be able to meet most of the exigencies of the active year. There is nothing more frustrating than having swarms to hive, either your own or somebody else's which you happened to acquire, and not having enough equipment to cope. The beekeeper who expands to out apiaries after being a garden beekeeper will find out, perhaps too late, that his previous methods of beekeeping are just not adequate for the new circumstances. Even the mode of transport may show itself to be inadequate, perhaps the saloon car has been outgrown and an estate or pick-up truck is needed. Be sure of one thing, during the height of the active season when it is all happening everywhere for you, the bees and the ease with which you can tend them will search and find out your limitations. That is a promise! To end on a sour note and perhaps spoil the reverie. This could be the year of Varroa. Just because the beast is maintaining a low profile in the beekeeping Press does not imply that the threat has been lifted from us. Did you have your paper inserts in place? —it still is not too late to slip them into place on the hive floor. It really is quite sad that the horrendous silence from the ordinary beekeeper in Britain is encouraging the Government to ignore the pleas of the governing bodies of the S.B.A., B.B.K.A., B.I.B.B.A., I.B.R.A. and the B.F.A. who have continuously asked for a total ban on the import of queen bees. When Varroa reaches these islands on the back of a New Zealand, Israeli or American queen bee or her attendants, who will we have to blame? — only ourselves and our present apathy. The hobbyist, especially the "let alone beekeeper", will disappear into history after Varroa has become established here. The "involved" beekeeper will really be hard pushed and involved to keep his bees alive when Varroa reaches us here in Britain. Unscrupulous appliance dealers (I have a well-known dealer's catalogue to hand advertising American queen bees!) will continue to put British beekeeping at risk by importing unless we as a body state loudly and clearly that we are angry. What price an Anti-Varroa Lobby by beekeepers the length and breadth of Britain?

THE MONTHLY ROUND - February

February is the month where the beekeeper can first witness externally the condition of his, now increasingly active colonies. The activity at the hive entrance can speak volumes to the beekeeper who knows what to look for as the snowdrop and crocus become more abundant. Whenever the weather is suitable, queen right hives will show an activity which will gladden the heart of any beekeeper in the gathering of the rich golden/bright orange pollen of these blossoms. Large pellets energetically transported through the entrance and the hurried impatient business-like ingress of the early nectar-carrying bees is a sure sign that brood-rearing is increasing satisfactorily in the hive. Any colony which appears to be idling when its neighbours are going energetically about their business should be noted. If lack of foraging activity is evident consistently over a period of days in early February, there is no need to panic. Some colonies are just naturally slow or late developers. If, however, by late February early March where other colonies are 'going strong' in their foraging activity, these colonies are still idling, then the colony(s) in question would seem to have a problem. A number of approaches can be made with suspect colonies at this early time of the year. The most positive, least disruptive, is to lift bees from the entrance on mild days when activity is possible and put them in a match box (wear a veil while doing this, but no need to smoke the hive. Do not, of course, wear gloves!). Collect around 20-30 bees and check for disease. If the bees are diseased, nosema, acarine or affected by any of the virus diseases you must take action. If as we must hope the hive is healthy then the obvious next step is to make an internal examination. Select a bright day in late February/ early March to do this inspection. Smoke the hive in the usual way and after a couple of minutes open it and gently and carefully check the frames at the centre of the cluster. If there is evidence of sealed worker brood, however little, the colony is queen right! Check the level and quality of stores in the hive at this inspection and feed either with spare combs of honey, stored dry from the previous season, or give the bees a thick heavy feed of sugar syrup if necessary. Close the hive and continue to monitor the behaviour of its foragers at the entrance. If no eggs are present and the colony is strong it can still be queen right but the chances of this are not high, a colony like this can be saved if a frame of brood in all stages is given to it at fortnightly intervals. The donated brood will emerge and maintain population levels — if no queen is present the colony will ultimately raise its own queen from the frame of eggs and open brood given around early April. In good springs such a colony could be queen right and thriving by early May. Such are the skills to be learned in beekeeping! Nothing ventured nothing gained! First, though, establish colony health — there is no point in fostering disease which could ultimately be transmitted to other colonies in the apiary. It is always good practice to have more than one stock of bees in each apiary, this gives the beekeeper the chance to compare hive behaviour. Otherwise abnormal behaviour is less easy to detect early enough to take 'rescue' measures. Hive stores can begin to be quite rapidly depleted from this time on due to the increasing brood rearing rate. Around a week after the first early pollen is seen going into the hives, sugar syrup feeding can be started. Even hives heavy in stores can benefit from this early feeding. Bees require water for rearing brood and if this is given in the form of heavy syrup for colonies light in stores and light syrup to give a palatable drink to colonies with good amounts of stores, it can save much 'bee life' in eliminating the need for the bees to forage for water in the more unfavourable spring days. With more winter oil seed rape being planted in Scotland now it is becoming increasingly important that colonies are strong enough, early enough, to exploit this new crop which blooms in May. Early feeding will encourage an earlier build-up to effective honey gathering strength. Beekeepers who have good stands of sycamore around their apiaries can also benefit from this early feeding management regime. The sycamore in May is a very reliable early major nectar source in most years. Well worth the calculated risk of the sugar investment to increase the chance of a surplus from this worthwhile nectar source.

THE MONTHLY ROUND - March

Despite the fact that the weather is still far from sub-tropical, the micro-climate temperature inside the honey bee cluster in our hives is high enough to encourage the bees to rear increasing amounts of brood. A very close watch must now be made on stores. For the more conservative beekeeper another slab of candy right on the frame top under the quilts and crown board will safeguard against isolation starvation and chilled brood. For the more daring beekeeper sugar syrup given at the correct strength relative to colony needs and remaining stores will keep the bees in good heart and ease the burden on the honey stores being used to produce the copious amounts of 'bee milk' as increasing amounts of pollen from crocus and aconites roll into the hives. All queen right hives should, by late March, on favourable days, be extremely active in their foraging activities, any laggards at this time definitely have problems. The measures discussed last month are also relevant for March, i.e., health check and judicious reinforcement. If by this time a hive is hopelessly queenless but nonetheless healthy it may have developed laying workers, easily identifiable by the erratic appearance of 'domed' sealed brood scattered over the brood frames. The eggs laid by laying workers are invariably located on the walls of the cells, not at the bottom of the cell, due to the short abdomen of the worker caste relative to the queen caste. To unite a colony having laying workers to a queen right colony is not good practice unless certain measures have been taken beforehand. The queen of the queen right colony used may be killed by the laying workers! A frame of eggs given at weekly intervals for perhaps three weeks will help restore a natural balance in the laying worker colony where it is still quite strong. After two or three weeks this colony can then be safely united to a queen right colony — a health check having already been carried out! The queenless stock must always be placed under the queen right stock at the union, using the 'newspaper method', i.e., a sheet of newspaper between the two brood boxes of the colonies being united. The bees will chew through the paper and unite peacefully. If the queenless, laying worker colony is healthy but weak, dispersal is the best method of utilising the residual foragers in the hive. On a bright day in early spring subdue the colony. Remove the hive to another location in the same apiary, remove the frames of bees one at a time and give each one a hefty shake or bump. (When holding frame by lug of top bar strike your fisted free hand on the clenched hand holding the frame of bees. The bees will be jolted cleanly off and fly back to the old site.) Later they will disperse to other hives in the apiary to perform some useful work before they go to that great blossom-covered meadow in the sky! A colony having a drone laying queen at any time of the year is a great nuisance. If the "dud" queen is eliminated and a frame of open brood given at weekly intervals, the colony will ultimately re-queen itself. If, however, after repeated attempts to find the queen she still remains elusive, the best method to find her is to give the colony as much open brood as it will be able to cover adequately. Place this brood in a second brood box above a queen excluder (preferably with the adhering bees from the donor colonies, checking carefully of course that the queen of these colonies is not on the donated combs. If in doubt shake all the bees from the 'to be donated' combs of brood!) and place it in the brood box right of the cluster of bees in the lower box. The bees will rise to cover and nurse the donated brood leaving the "dud" queen in lower box and the attendant population greatly diminished. The 'drone layer' should then be easy to locate. If she still eludes capture, remove the hive to a different location in the apiary. Place the brood box with the brood and covering bees on a new floorboard and put it on the original hive stand. Shake all the bees off the frames of the drone layer. The foragers will return to the old location and the 'dud' queen will be "lost". The so treated hive can be managed now as for laying workers — either united to some queen-right stock or left to re-queen itself, as discussed previously.

THE MONTHLY ROUND - April

This month should see the beekeeper beginning to anticipate the possibility of surplus honey in mid-late May. Sugar feeding should be well under way. Any suspect colonies should have been examined by now, and the necessary steps carried out to have the bees of such hives working positively or killed, depending on state of health at the first internal examination.

Any spare brood comb should be fumigated, using 80% acetic acid, to eliminate any residues of nosema spores lurking in the woodwork of the frames or the nooks and crannies of the brood cells. The fumigation can be easily carried out by placing the spare brood combs in an empty brood box. Make the brood box or stack of brood boxes air-tight and place a soaked pad (containing approx. $^1/_4$ pint of 80% acetic acid) on the frame tops of each box of frames to be fumigated. Place the brood boxes being fumigated in a warm corner of the garden or apiary and the April sunshine will do the rest. At the end of a week remove the, by now, dry pads and ventilate the brood chambers for around 2-4 days before using them in occupied hives. A slight residual vinegar smell from the combs will do the bees no harm after the recommended ventilation period. Fumigation using 80% acetic acid will also kill the larval and adult stages of *Galleria melonella* (greater wax moth) and *Achroia grisella* (lesser wax moth).

There are beekeepers who will tell you that they do not have disease in their hives and they may think sincerely that this is so, however the causative agents of disease are always present in hives, they are endemic. Beekeepers who are fortunate to be located in areas where abundant nectar and pollen sources exist and who are expert enough to maintain strong, well provisioned colonies will rarely see any signs of overt disease, unless a massive infection is introduced by the robbing out of a weak diseased colony by a strong hive. Colonies of bees kept in borderline areas with sparse forage or poor climatic conditions are continuously in a state of stress unless the beekeeper is fully aware of the limitations of the area and feeds supplementary pollen and sugar. Stress will encourage the incidence of disease which may or may not become epidemic, depending on the critical limitations of available forage and climate. The message is therefore, maintain all your colonies in a state of prosperity by good management. Good management entails such parameters as best possible forage location; best possible geographic location (i.e., sunny, sheltered, well drained apiary site); best possible management system for the area (i.e., double brood chamber, brood chamber and super or even single brood chamber). By experimenting with parallel systems with a number of colonies over a period of at least three years the best method will soon be apparent to the beekeeper. The best method will be deemed the one which gets the best results for the beekeeper with the minimum of effort compatible with healthy bee stocks and successful overwintering.

Queen excluders should have been scraped and cleaned of propolis and brace comb during the cold weather in late winter. Spare floorboards should be cleaned, sterilised by wood preservative or scorched using a blow lamp and generally made ready for replacing the overwintered floorboard, which will be covered with all sorts of "horribles," like dead and decomposing bees, drifts of wax debris in which wax moth eggs and larvae will be nestling and possibly faecal deposits. All these components are capable of harbouring and encouraging disease — a clean floorboard gives the bees a brand new start and incidentally saves them a lot of exhausting work.

Stocktaking should be carried out at this time, in anticipation of honey surplus. Sufficient shallow frames and shallow crates to give at least two, and preferably three shallow crates, per hive; sufficient spare brood chambers to accept any swarms which may come your way either due to your own indecisive management or somebody else's; sufficient jars and cut comb boxes to cope with at least the anticipated spring surplus. Is your veil still serviceable? Did you maintain your smoker at the end of the year? Did you clean and vaseline your centrifugal extractor at the same time? Time now to degrease the extractor and check the smoker bellows. Have you made a start to rearing your own queens yet by inserting drawn drone comb into at least 50% of your best stocks around the second week in April? Have you thought about marking your queens at the end of the month, before the population explosion occurs in mid-May? A marked queen makes life a lot easier when the colony shows swarm signs and the queen has to be found.

The MONTHLY ROUND UP - June

As the traditional problems in beekeeping management due to the June gap recede with the increased acreages of oil seed rape being planted, other problems in honey handling and marketing arise to replace them. The massive amount of adverse comment accruing from the incursion of oil seed rape into the range of major nectar producing plants has puzzled me for some time, especially since little positive information has appeared in the popular beekeeping press regarding the handling of such a crop as oil seed rape. The trouble with brassica honey is that it tends to granulate very quickly. This granulation is of course not a unique phenomenon. All honey will eventually granulate albeit at different rates. Storage conditions also exert an influence on rate of granulation — the optimum temperature for rapid granulation of honey is 57 degrees Fahrenheit (14 degrees Celsius). Seeding honey will also accelerate granulation. Seeding is a device used by beekeepers and honey packers to ensure a particular type of granulation which will result in a smooth palatable product rather than leaving the granulation process to nature, where sometimes it may be necessary to give the purchaser of the honey a hammer and chisel with every jar! Although the bulk of the early adverse comment on oil seed rape honey originated from the beekeeping fraternity in England, the problem of handling rapidly granulating honey was not a new problem for many beekeepers in Scotland. Beekeepers working in areas where large or even moderate acreages of raspberry are available to the bees, have had to contend with rapid granulation for many, many years. And without a murmur! I feel there is a vast wealth of untapped knowledge for the oil seed rape novice, locked in the lore of the 'rasp' beekeeper. Perhaps these uniquely experienced members of our craft could be encouraged to give a helping hand to those of us who have been caught short with frames and part frames of uncentrifugal blossom honey due to oil seed rape, dare I say it, contamination… The problems arising from oil seed rape are ongoing during the beekeeping year. Combs which contained oil seed rape honey after extraction will be replaced on the hives, giving the bees drawn comb in the supers to accept the nectar from later nectar flows in summer and autumn. Thus, the honey harvests from clover, lime, willow-herb, privet, bell and ling heather will invariably be 'seeded' with oil seed rape. The result of this seeding is of course that these honeys will granulate much more quickly than normally expected of them, making the production of cut comb an extremely hazardous business, due to the uncertainty of the degree of oil seed rape contamination. Lime-and heather honey in the pure state granulate extremely slowly, lime remaining clear for many months, depending on storage temperature ranges. Because of this, lime is an ideal honey for the cut comb trade, however, where it has oil seed rape contamination the whole batch can be ruined if it is not sold and eaten quite quickly after 'boxing.'

Heather honey is another honey which is slow to granulate. Heather honey will remain un-granulated for years in the pure state, but only if it is stored as comb honey. Any contaminating honey in heather comb honey will eventually begin to crystallise out as pearl-like granules at the base of the honey cells. However, with this same granulation resistant heather honey — almost as soon as it is bottled after being pressed or centrifuged (after the use of loosener devices) it will become cloudy, losing its-attractive bubble filled orange hued appearance. This clouding is not granulation. It is the result of acute aeration, synonymous with the foaming characteristic of some aerated waters when shaken. True granulation sets in much later, and invariably takes the form of large individual bead-like crystals, if the granulation is natural, that is, not due to seeding. The granulation characteristic of honey is due to the pressure of one of its two monosaccharides, dextrose, existing in the honey in a supersaturated solution. Like all supersaturated solutions, the dissolved solids, in this case dextrose, will gradually crystallise out. The fructose, which is the other monosaccharide in honey, strangely enough does not granulate despite the fact that the honey in question may become brick hard. Thus, the ratio of dextrose to water in any honey is the limiting factor of rate of granulation.

Another problem related to granulation is that, as the granulation process develops, more free water is released in the honey and thus an increase in the water content can occur. Ripe blossom honey has on average a 17% water content, although in truth the water content may vary between 17-22%. Ripe heather honey and some clover honeys can have a water content near the upper legal limit for honey, relative to the E.E.C. regulations. The greatest problem presented by honey having a high water content is that at particular storage temperatures it will start to ferment, 54 degrees Fahrenheit is a low temperature critical. Above 100 degrees Fahrenheit fermentation will be arrested. However, when honey is stored for periods of time at particular temperatures, chemical changes occur. The E.E.C. regulations impose upper and lower limits for the presence of certain constituents in honey offered for sale. High temperature or lengthy storage can cause a substance, hydroxy-methyl-furaldehyde, H.M.F. for convenience, to increase in concentration. Any honey having an H.M.F. index of higher than 40mg/kg is downgraded and not usually acceptable as saleable honey.

Diastase is an enzyme which exists in honey and this substance has been used for many years as an indicator of the amount of heating which has been applied to honey. Honey with a diastase number of 8 or less is also subject to being downgraded. The latest E.E.C. regulations published by the Codex Alimentarius Commission titled "Recommended European Regional Standard for Honey" describes various tests and procedures which are to be carried out in the establishment of honey quality. Surprisingly enough, despite the preoccupation of the European honey authorities with H.M.F. and diastase content, the most heat sensitive element in honey is actually the enzyme invertase, which is responsible for the reduction of the sucrose in the raw nectar, to dextrose and fructose. So now when a honey appears to be 'borderline' the invertase /diastase ratio, the Kiermeier Number, is calculated and can give a clearer indication of the quality of a honey which may have a naturally high H.M.F. content or a naturally low diastase number.

An aside on fermentation of honey, which can proceed to two very important stages. First stage fermentation will result in mead of dubious quality unless the fermentation was controlled by the mead maker. The second stage fermentation will produce honey vinegar which is quite a saleable product, unless your honey jars or drums exploded during the fermentation processes. So, the beekeeper, even with the best will in the world may eventually finish up with a stored product which could be high H.M.F., low diastase, non-existent invertase, excellent quality, fish and chip honey vinegar. By the way, if you are worried about granulation or fermentation problems, just pop your honey in the deep freeze. Bottled honey will not granulate at such low temperatures. Comb honey must be sealed in an air-tight packaging. Wrapping in air-tight plastic bags is sufficient to inhibit moisture exchange. Happy honey handling — all these problems without even touching on bees legs and bits of wax debris and bubbles usually found in honey, boggle the mind!

THE MONTHLY ROUND - July

This month sees the start of the late summer nectar flows, which constitute a major component in the annual harvest of honey. In "let alone" beekeeping the colonies will invariably build up on the early summer flows and especially where the system mentioned incorporates double brood chambers, will have achieved large populations with not much stored honey, unless the beekeeper is in a particularly good gean, sycamore, whitebeam or oil seed rape area. The later summer flows from the lime, willowherb, privet, clover, bell and ling heather are thus critically important to all beekeepers but especially so to the "let alone" person. All swarm management should now be past tense, the colonies all being at peak population strength for the July nectar flows which can produce massive honey crops if the weather is "flyable" for the bees. Even in seasons where the summer weather is extremely dry and warm over a prolonged period, causing the very drought-sensitive lime to fail to secrete nectar, secondary sources like privet and willowherb can save the day and give quite incredible harvests as happened in the summers of 1983 and 1984, in west-central Scotland anyway!

For the beekeeper who wants to maximise his honey returns, the only way to do this is to be itinerant unless the home territory is one of those god-gifted places where broad-leaved arboreal angiosperms exist in close proximity to the two important ericas, bell and ling and where, according to the legend of Rumpelstiltskin, fox and hare greet each other with "Good day" when they meet. Yes indeed, an all-season location is indeed fairytale life but they do exist! I've heard of some! Anyway, since the secret of honeybee swarming was discovered recently, there is no need for the progressive beekeeper to sustain the, till now, seasonal hiatus in brood rearing in colonies due either to swarming or traditional "post-swarm preparation" swarm control management. Thus, having massive populations of bees available, and by moving to the right area at the right time, effectively exploiting any bright warm or even wet, dull, warm days in July. It is quite surprising how actively bees will forage, even in rain, during a nectar flow where the temperature is reasonably high (>14°C), and the wind speed low (<15 m.p.h.). It is a really worthwhile exercise to acquaint oneself with the available bee forage in the beekeeping territory. By being aware of bloom dates the beekeeper can perfect his hive management so that by timeous intervention his hive populations are optimum at critical times. Reinforcing hives with frames of eggs and frames of brood with due regard to the time lapse from age and stage of reinforcing eggs and brood until foraging age will be achieved (1. Egg laid to foraging bee = six weeks. 2. Sealed cell to foraging bee = $3 - 4^1/_2$ weeks!) will secure good population levels at the right moment. Moving hives of bees is relatively easy, depending on experience and preparation of transport. In hot weather, colonies should only be moved either early in the cool of the morning, timing the journey so that the bees will be at their destination and released early in the forenoon. For daylight moving ventilation screen boards and extra clustering space in the form of a super fitted with empty drawn comb is ideal. The more experienced the beekeeper the more short cuts he can take! Personally, I have found moving bees in the late evening is the least problematic, so long as the extra clustering space is given and the hives are not being moved more than 50 miles or so.

Provided an open vehicle like a pick-up truck or trailer is being employed, then there is no real need to use ventilation screens. Merely make the existing crown board bee tight, secure the hive parts together in your favourite manner, plug up the hive entrance with a wooden block, strip of foam rubber or a suitable length and width of cloth after the last bee is home and heave-ho away. Securing the hive parts together some days before moving gives the bees time to re-propolise the seams between the different parts, making the hive unit even less shock prone. Preparation of the new site should be attended to well in advance, especially where a fence has to be erected and many hives are being moved. If all the hive stands are arranged beforehand, moving bees can really be quite a satisfying experience but, if preparations are done as you go, the business of moving hives can become a tedious and strenuous job of work. Always have the hive set up with the roof already on before releasing the bees — otherwise!

THE MONTHLY ROUND - August

This is one of the most critical months in the calendar of the active bee year, especially for the migratory beekeeper who makes the effort to transport his hives to the moors in anticipation of a settled August, which will hopefully result in a bumper crop of one of the most delectable of all our Scottish delectable honeys, ling heather honey. Moving hives was briefly referred to last month and for the heather beekeeper this will now all be past history — either he will have made a smooth but strenuous transition from lowland to highland or he will have yet another story to tell the grandchildren. As is religiously repeated in this column each year at this time, August heralds the beginning of another bee year, because the colony should produce the bulk of the young bee population which will carry the hive through the winter, at the moors. If the August weather pattern is erratic and generally unfavourable this can inhibit brood rearing due to the unreliability of supplies of incoming nectar and pollen. Abundant pollen and nectar income to the colony is a dramatic stimulus for the queen to lay steadily at a high rate — providing she is not approaching senility which is bad beekeeping practice. All hives going to the moors should be heaving with foragers at the correct age and have a queen not less than one winter old. A current year queen timeously introduced will give the beekeeper a reasonable guarantee of successful overwintering. Although nothing is ever certain in beekeeping except that, in the long term, bad beekeeping practice will eventually result in no beekeeping practice. It is good practice to examine bee colonies at least a week to ten days after they have been set down at the heather. The examination is merely to establish that the hive is still queenright. All that is required is that eggs be noted. There is no need to find the queen. The commercial or multi-hive beekeeper uses his experience in external observation of hive activity to reassure himself that all is well. A good indicator is the intake of large pellets of the dull, browny, tan heather pollen into the hive after about the end of the second week at the moors. No large pellets, means no brood rearing worth talking about and no brood rearing can mean no open brood and no queen in the hive. Where other hives are busily doing what hives should in an apiary and the odd hive is not, then it is worthwhile having a quick internal look in the suspect box. Rumour has it, and so also does bitter experience, that bee colonies at the heather are not the most even tempered.

For the novice this is most certainly the case — and he has the stings to prove it. However, if the beekeeper is of the right stuff, he will be aware of certain drawbacks in opening hives at any time during the active year. Weather is very important in that it should be good! But if this good weather coincides with a nectar flow, then this is the ideal! Now, from experience, we know that bright sunny windless days during nectar flows are not the rule in Scotland, therefore it is not good beekeeping sense to do an internal examination of a bee colony on such days, because of disruption to precious foraging time. As with bee colonies in the midst of the summer nectar flows, so it is with the colonies in the midst of a heather nectar flow. Thus, the optimum time to examine a bee colony transpires to be late afternoon/early evening on a bright day, during a nectar flow when on listening intently at the hive entrance the beekeeper can discern a low, steady, contented hum issuing from the hive. Even at the moors, an otherwise tetchy colony after being subdued if the conditions as described are present the beekeeper will be able to carry out the necessary inspection painlessly and with virtually no disruption to the foraging activities of the bees. At most heather locations it will be necessary to erect some kind of fence to fend stock off the hives. A good gauge of 4" mesh pig wire fencing is ideal and can be rolled up for use year after year. A post spacing of approximately 3 metres is enough to support a moderate sheep invasion attempt. Where no sheep are on the location the beekeeper is still advised to erect a token fence to dissuade deer from using the hives as "head scratchers." Even a roe deer can topple a hive if it gets it right! Hives should be left with the honey supers intact for at least a week after the heather flow is finished. The bees should be back home on their pre-prepared winter sites before the end of the second week in September to facilitate the winter feeding programme.

THE MONTHLY ROUND - September

The heather beekeeper should still have his hives at the moors at the start of this month. Depending on when the heather came into flower and subject to weather patterns the bees should remain on the moors for at least a week after the nectar flow has stopped. Usually the end of the first week in September sees the start of the long trek home. In seasons where the heather starts to bloom late in August and continues to secrete nectar until well into the second and even the third week in September, if an early frost occurs it can bring to a halt an otherwise perfect flow period, even when the heather blossom is still theoretically capable of secreting nectar. This may not be the rule but I have experienced two such frosty events in the past and in each case the nectar flow has ceased abruptly to my great dismay and disappointment. It would be interesting and educational if readers with experience of frosts during a late heather nectar flow would comment on how it affected them, if at all. In some heather areas beekeepers report that their colonies come back from the moors sadly depleted in population, even in good heather years. These colonies must be located in particularly hostile regions or perhaps are headed by older if not aging queens. My problem in the West of Scotland, where the heather is reasonably accessible and not excessively high above sea level, is that my colonies are so populous after returning from the moors that I often feel a bit guilty when I reduce the colony to the winter rig of a single brood box. However, my conscience is salved by the old saw "the best packing for bees is bees," preferably young bees! Most colonies are fed 10 lbs. of sugar in syrup at around the end of the second week in September. Any colonies which feel "extra light" are checked for queen rightness before any feeding is done — if queenless there is no point in feeding, the colony is checked for disease and if clear united to its neighbour by the newspaper method — queenless stock below the newspaper. This is important since if the queenless stock is placed above, the bees of the stock on breaching the newspaper and beginning to forage, passing through the queenright stock could kill the queen — especially if they have been queenless for some considerable time. Better safe than sorry! When the hives are brought back from the moors, if they are grouped according to weight, then the beekeeper can tell at a glance which hives require most sugar syrup. I find it extremely useful doing this because I know at any time during the dormant season and subsequent spring which colonies went into winter naturally well stocked and which colonies had to be assisted. This segregation supplies that extra bit of information on colony/ queen quality when I make my selection for queen rearing in the following spring. I always give my colonies an extra lift in their winter stores later in the year. More about that later but since using this supplemental feeding I have lost very few colonies to isolation starvation, which is always a problem in single brood chamber management in hard winters.

THE MONTHLY ROUND - October

Sugar syrup feeding for the winter should normally be finished by the beginning of this month. It is, however, sometimes unavoidable that feeding of syrup continues into mid-October, especially if the bees had a poor heather or late autumn season where late summer brood rearing was depressed. A feed of sugar given in early September in such circumstances can finish up as late brood unbeknown to the beekeeper rather than winter stores — resulting in the beekeeper feeling quite satisfied that his stocks are well provisioned when in fact they are in imminent danger of "isolation" starvation if the winter is severe. I have often seen references in books and the beekeeping press to the "beneficial" effect of frost during the winter season, the belief being that a wee touch of rime tells the bees "in words of one syllable" — that it's time to get their heads down. I subscribed to this idea in my early days in beekeeping because it seemed logical that if the bees were less active in winter they would not eat through their stores so quickly. Alas like many idea's and beliefs handed down through unsuspecting generations of mankind (at its worst called dogma) the hypothesis does not stand up under scrutiny. In many "hive-years" of beekeeping experience (beekeeper 2 hives x 10 years = 20 hive years of experience. Beekeeper 10 hives x 10 years = 100 hive years of experience and so on) I have noticed consistently that well provisioned hives will come through a mild dormant period using less stores and be much stronger than hives in the same condition in a cold winter. I have experimented with the limitations of overwintering nucleus stocks and have consistently found that in a mild winter even small nucleus stocks (4-frame) come through without any special treatment so long as the colony has adequate stores. Despite the traditional view that dampness is more deadly than cold for a bee colony, I have found time after time that even in extremely wet winters I rarely lose nucleus colonies but in frosty, cold winters I will lose some nuclei, depending on the duration of the cold spell, a ten day period or more of continuous frost will kill weak colonies, where colonies of equal strength and condition in continuous sopping wet winters will survive easily. All things being equal, that is the colony in winter rig in a south facing site well sheltered from the prevailing wind, healthy, with a young fecund queen, adequately provisioned hive, level but tilted slightly forward to facilitate drainage of any internal moisture/condensation and having good drainage. Such a colony of bees will overwinter better in rainy winters than severe frosty winters as long as the hive is weather tight and has adequate ventilation. The legendary killer dampness which does the damage is hive dampness not atmospheric dampness. The dampness factor which is supposed to be the scourge of Scottish, not English beekeeping, I am certain stems from the former widespread practice of layering the hive crown board with carpets, felt, old coats, linoleum, etc., in winter. Such a hive, subject to poor ventilation will become saturated with dampness as the quilts soak up the natural moisture rising from the cluster. The moisture build-up ultimately results in combs becoming saturated with moisture and easy prey to fungal growth. To achieve good winter ventilation of a hive located as previously advised, all that is required is a full width entrance no higher than 5/16" (7.9 mm.), an empty super on top of the brood chamber, then the crown board (preferably a wooden one) and then the hive roof. The empty air space supplied by the super above the brood chamber facilitates a more homogeneous hive atmosphere as the air rises into this reservoir and then is changed as a result of indirect wind effects. This handy cavity can be used to house winter supplementary stores such as candy or the increasingly popular dampened 1 kg. bags of sugar given directly onto the top of the frames on which the bees are clustering in late November/ early December.

THE MONTHLY ROUND - November

The hobbyist beekeeper can now put his feet up for the next three or four months. If he has prepared his bees well the winter period will be a time of mental and physical (beekeeping wise) relaxation. The garden beekeeper has it particularly easy. He can view his charges daily and rectify storm disruption and make the occasional check on the "sugar bag feed" or state of the candy block without great trouble. The out-apiarist hobbyist has a much less relaxed winter. Always there lurks at the back of the mind the worry of disruption to the hives, either by climatological forces or other more terrestrial agents like animal (including human) disruption where hives may be overturned by a cow, dog, sheep or deer or vandalised by malicious, destructive Homo sapiens. I have met beekeepers who do their winter preparation well, ensuring that the hives are virtually "earthquake" proof and who only visit the out-apiaries very occasionally during the dormant period. These men are made of strong stuff, relative to my own attitude. I never feel completely comfortable unless I check my bees at least once a week in winter and I always check the apiaries after storms or periods of high winds. No matter how secure a hive may be storms and gale force winds can wreak havoc. Perhaps a roof and crown board will lift off, a branch snapping from a wind ravaged tree could overturn a hive or displace brood box and super on falling. An otherwise substantial looking stone wall could be blown over just at that critical point adjacent to the hives.

By visiting out-apiaries on a regular (short) cycle an enjoyable following summer with all the colonies intact is not necessarily ensured but the chances of this are greatly enhanced. This I know from bitter experience of the fickle treacherous dormant period weather. Now is the time to ponder the last active season and try to anticipate problems of management in the forthcoming spring and summer. The two major problems which can potentially put the beekeeper out of business — not considering the vagaries of the weather — are disease and swarming. Disease, at least epidemic disease, has not been too great a problem for the beekeepers of Scotland in recent times. We did have the catastrophe of "Isle of Wight" at the beginning of this century which virtually wiped out our native Scottish race of Apis mellifera nigra, "the black". Disease however is endemic in bee colonies, only held in check by strong, healthy, well provisioned colonies housed in sound weathertight hives, located in sunny, sheltered and dry apiaries. The major debilitating or even killer disease in Britain and Europe is nosema, a disease of the mid-gut of the adult honeybee, caused by the micro-organism *Nosema apis* Zander, discovered in Germany in the 1930's by Dr Enoch Zander at the Nurnburg Institute of Beekeeping. *Nosema apis* is a spore-forming protozoan of the order microsporidia. It is not infective in the vegetative stage. It is the spores which cause the disease to be spread through colonies. The spores of *N. apis* can be easily identified under the microscope using a magnification of around 600x. They appear scattered at random on the microscope slide. Where many spores, which look like short fat sausages, are present the infection is severe and probably has already manifest itself clinically. That is, visibly diagnosable in that the colony build up will be slow, static or dwindling, especially in spring when a healthy colony will be thriving as foraging increases and incoming pollen and nectar stimulate the queen to accelerate her rate of egg laying. Nosema disease can cause queen failure, death or supersedure. It shortens the effective life of workers and affects the development of the hypopharyngeal glands which produce the food on which the young larvae are fed, to the obvious detriment of the colony. Nosema can be treated in a number of ways using: (a) Good management practice. (b) Chemotherapy. (c) Fumigation. Good management practice has already been discussed, strong colonies, well fed and sensibly located in sunny exposures. Chemotherapy can be applied using Fumadil-B. (Editor note - Fumadil -B can no longer be used since it was found to cause birth defects). Chemotherapy on its own is wasteful in the long term treatment of nosema. Fumadil-B will repress nosema infection but it will not of itself eradicate the possibility of the disease becoming epidemic in treated colonies in later seasons.

N. apis forms spores. These spores, in severe cases of nosema are defecated in the hive on the comb and on the frames. They can be dormant for up to 30 years and still be viable and cause disease at any time during this period. Thus so long as spores exist in colonies the risk of epidemic nosema disease is always present. Unpublished work done on nosema incidence and spread, at one of the Scottish Agricultural Colleges has shown that the most effective regime to combat nosema is to use chemotherapy in conjunction with the systematic fumigation of all spare brood comb and brood chambers using 80% acetic acid, which kills the nosema spores. Complementary to this work it was found that colonies which were systematically given brood combs sterilised by 80% acetic acid could be maintained free of nosema disease, without the use of Fumadil-B. 80% acetic acid also has an important role to play in the suppression and elimination of many other diseases which will be discussed next month.

THE MONTHLY ROUND - December

Continuing on the theme of disease from last month, Nosema as a significant and potentially dangerous disease was discussed. As a matter of interest a related nosema organism attacks species of bumble bee. The organism is *Nosema bombi* and causes damage to the ovaries and results in sterility of the victim. Much work has been done in the recent past on virus diseases by Bailey and a definite association has been established between certain viral infections and the known microbiological and other parasitic diseases of the honeybee. Nosema has been shown to be associated with bee virus Y, black queen cell virus and filamentous virus.

Acarine is an important and potentially dangerous adult bee disease. Acarine, being caused by the mite *Acarpis woodi*, parasitises the first thoracic trachea of the bee. The mite sucks the haemolymph of the adult bee and where a heavy infestation is present can seriously damage the health of a colony, killing the colony in late winter in cases of severe infection. Acarine disease can be easily diagnosed if present by removing the head and front pair of legs of a suspect bee and examining the thoracic aperture under a X20 dissecting microscope. Dade's "Anatomy and Dissection of the Honeybee" illustrates and describes the procedure in easy to understand terms. With practice, the examination can be carried out with surprising ease. Work by Bailey on acarine disease in association with virus diseases seems to indicate that the damage to bees attributed to acarine during the Isle of Wight disease epidemic really should be blamed on concomitant viral infections such as chronic bee paralysis and cloudy wing virus. It would be irresponsible to imply that over the years since Isle of Wight disease that beekeepers have become more or less complacent with regard to disease incidence in their colonies — especially when the threat of varroasis still hangs over Britain. But due to the low level of epidemic disease in our hives, in Scotland at least, we sometimes tend to forget that poor colony performance could be due to disease however covert. What we should never forget is that the pathogens which cause disease are always present. Disease in colonies is endemic, only becoming epidemic when stress however caused, lowers the colony's natural ability to deal effectively with the particular disease. A very important point highlighted by Bailey's work, reported in "Bee World", Vol. 63, No. 4, 1982, is that the incidence of disease among bee colonies increases as the number of colonies in particular apiaries rises above that number which can naturally support themselves on the forage available. His conclusion is that moderate numbers of colonies in apiaries is a far better prophylactic treatment than chemotherapy - profound thought!

Last month the good management practice of 80% acetic acid fumigation of spare brood comb and brood boxes was discussed. Research has indicated recently that 80% acetic acid kills not only nosema spores, but is also effective against *Melissococcus pluton* (E.F.B.) and *Ascosphaera apis* which causes chalkbrood. Professor L. Heath has discussed this treatment in his work reported in "Bee World", the magazine published by the International Bee Research Association. 80% acetic acid also kills wax moth larvae so it really is a worthwhile treatment if used systematically.

This month all colonies, even the strong, well-fed ones should get their Christmas or pre-Christmas present of at least four dampened 1kg bags of sugar placed right on the frame tops of the frames on which the bees are clustering and in way of the cluster, so that on the milder late winter days the bees can rise and eat their way into the life-giving sugar without having to leave the safety of the cluster. Using this supplementary sugar system the bees not only utilise the sugar for their immediate needs but also tend to store honey made from this sugar in the otherwise empty cells in the heart of the cluster. The sugar not only affords the bees supplementary rations but also helps keep the hive humidity down as the sugar continuously absorbs any excess atmospheric hive moisture. Try it. You will be pleasantly surprised — and incur fewer winter losses —

especially in particularly severe winters. Considering the catastrophe which the bee year 1985 turned out to be, where sugar syrup feeding might be badly judged and inadequate, the suggested bags of sugar will likely mean the saving of a colony otherwise doomed to starvation sometime between late December and mid-April or even earlier in a particularly severe winter which hopefully our bees will not be subjected to in the pending dormant period.

THE MONTHLY ROUND - January

There is a tremendous wealth of interesting and educational beekeeping prose lying, if not quite forgotten, at least dormant in the archives of beekeeping literature — not necessarily historically ancient, albeit historical — what happened yesterday is also history. There is nothing like the stimulus of studying for an examination to bring back to the light of day important and interesting material which has been neglected for years. In Germany, way back in 1568 a beekeeper called Nicol Jacob wrote the first book about bees to be written in the German language. In his book among many interesting facts Jacob described how he could make queenless hives produce "queen bees" only he called them "king bees." It was not known at that time that the biggest bee in the colony was female or that she laid eggs. It was understood even then though that any hive not having a "king" soon perished. Jacob's work lay forgotten until the clergyman, Scherach rediscovered Nicol's procedure and perfected it. Gregor Mendel did fundamental work on genetics using pea plants. His work was forgotten for years and only discovered by American researchers looking for a breakthrough in the study of genetics — Mendel's long-forgotten work supplied the key. It is, however, not really necessary to reach so far into the past to unearth material which is of great interest to the beekeeper. Thumbing through a 1982 issue of "Bee World" recently, an article on the development of insecticides caught my eye. The article was on the toxicity of insecticides, and discussed some of the currently-used materials. A new breed of insecticides was discussed, namely synthetic photostable pyrethroids. Pyrethroid insecticides have been around for many years. They are widely used as household pesticides and are derived from extracts of chrysanthemum plants. Exposure to light reduced the effectiveness of these naturally produced pyrethroids. However synthetic pyrethroid materials were produced in the late 1970's and a number of different types were developed for field application against a whole host of plant predators. In laboratory trials the pyrethroids demonstrated that they were a particularly hazardous material with regard to bees. However, a surprising fact emerged when these pyrethroids were used in field trials. It transpired that these pyrethroids were much less hazardous than expected and were even less dangerous to bee life than other insecticides which had been listed as being a low risk to bees.

Pyrethroid insecticides have the double advantage that they are effective at extremely low application rates, as little as 1Og /hectare and they actually repel honeybees from the treated crop initially for up to twenty-four hours after spraying, thus saving much bee life. Ripcord, a cypermethrin derivative was extensively tested in France on oil seed rape a few years ago with very encouraging results for beekeepers. The type of insecticide which poses the greatest threat to bees is the micro-encapsulated kind, which can be and has been collected by honeybees as pollen in America, where it is cleared for use on such plants as cotton. These micro-encapsulates extend the life of an otherwise unstable insecticide which loses its potency when applied in the field normally. These materials when stored with pollen become "time bombs" when fed to larvae or are eaten by the bees at some later date,. Lastly! WORK FOR THE-MONTH — Check your out-apiaries after storms this month. Scrape your queen excluders when the brace comb and propolis are brittle due to the cold. Check the "sugar bag" feed and replace any bags which have been eaten out. Keep up your winter reading and think about new management procedures which might make your life and that of the bees easier in the coming active year.

THE MONTHLY ROUND - February

Here we go, here we go, here we go! You can all most hear the bees "humming" this challenge to the weather at the beginning of this month as the snowdrop gives them the chance to take in the first promise of early nectar and pollen. The odd bright day in February will gladden the beekeeper's heart as he observes the first active external signs of colony winter survival. With the first indications that bee life has once again surmounted the rigour of winter and the summer of 1985 the beekeeper should be projecting his mind to the anticipation of the active season and how to achieve the best from the bees with the minimum of disturbance to them. Despite having run rings round myself with different management systems over the years, I still find that, although I now work a system which suits my mind, modifications have to be made to suit the variation in the weather patterns of each new season. Sufficient equipment is the main prerequisite for successful variations to management. A four-frame nucleus box for every other hive is an exceedingly useful tool. In ideal seasons the weather turns warm and windless around the end of March. April sees the bees stacking in the pollen from all the salix species, together with flowering currant and gean. By the end of April, the second brood chamber is almost fully laid up with brood in all stages and eggs. The sycamore opens around the end of the first week in May, followed rapidly by winter rape. By the end of May, the third super is filled and sealed and the bees are begging you to get the supers off and give them undrawn foundation in the new supers for the early June nectar flow from hawthorn and late chestnut. You can wake up now! I did say in the ideal season! The run of the mill season usually gives the bees a total of 2-3 days on the snowdrop and crocus. Perhaps a week on the catkins and a couple of days on the flowering currant. The sycamore has invariably been virtually blown to bits by the late spring gales and the bees assisted to covering perhaps eight frames in late April by copious sugar syrup feeding. If they get decent flying weather during May, by the end of the sycamore bloom period the hive will be bursting with bees and the brood box will be filled from end to end with brood in all stages and of surplus honey there will be little in evidence, the bees having used the bulk of the incoming nectar to increase population. In such an extreme (or normal) season the pioneer spirit and some imagination can make a moderate success out of a near tragedy. There is a legend in beekeeping which states that a really strong hive will gather more surplus than four hives with moderate population — and this seems to be not far away from the truth. So, for the beekeeper with perhaps upwards of four hives, in a poor early season the device to use to obtain best results is simply to have pairs of hives side by side. At the end of April/start of May, (having first ascertained that both colonies are healthy) find the queen of one of the hives — the hives have to be of course subdued in the normal way using smoke. Remove this queen to a four-frame nucleus box with enough bees to fill at least one seam between the frames. This nucleus should be removed to another site at least one mile away. Use Brother Adam's device of leaving the colonies to be united exposed to daylight for about five minutes. Then place the queen-right colony on top of the queenless one, place the queen excluder on top and replace the supers. A week later inspect the hive again, ensure the queen is in the bottom box. Place the queen excluder between the two brood boxes and replace the supers. In a mediocre spring (not a disastrous spring) where the colonies are left to their own devices little surplus will accrue, but in colonies treated as described the upper brood chamber will be well filled with honey as the bees fill the cells vacated by the emerging brood — I promise! Some beekeepers have a taboo about extracting from brood combs — but providing the bees are not completely stripped of honey, the honey in the second brood chamber is fair game. And tastes no different from the honey taken from the supers. Brother Adam's method of uniting colonies by exposure to daylight works well in the spring but later in the season always use the "newspaper method" or derivative of the method when uniting bees.

THE MONTHLY ROUND - March

From late February through until the bees have access to the major early nectar sources such as sycamore or oil seed rape in early May, our honeybees are at risk from starvation unless the beekeeper fed generously at the end of the autumn, i.e. late September or even early October in some mild seasons. Many beekeepers are generous to a fault with their bees. Others can be faulted by their avarice when it comes to the question of winter stores. Ideally, especially where British Standard hives are used, double brood box overwintering is the optimum, so long as the top box is well filled with stores of pollen and honey. The most precarious overwintering practice is that employing only B.S. single brood chamber. If the beekeeper is not "all about" he can lose his bees in an unfavourable spring. Using dampened 1 kg. bags of granulated sugar as described in the autumn articles gives the bees in a single brood chamber a stress-free winter, all other things being equal, However, a feed of light or heavy sugar syrup from mid-March onward will do no harm and a lot of good. Care has to be taken to prevent robbing. No syrup spillage in the apiary; no leaking feeders; entrances closed to a length that the bees can defend easily if the colony is weak. Feed light syrup to colonies which are well provisioned (1 kg sugar : 1½ litre of water), heavy syrup (1 kg sugar : ¾ litre of water) to colonies which are light in stores. As always with spring feeding, especially with colonies which are light in stores, once started it is necessary to continue feeding until the first major nectar sources are available and are being worked by the bees. All stored queen excluders should have been scraped clean of propolis and bridge comb by this time. Spare floorboards should now be ready, cleaned and repainted for the switch at the first major internal inspection, which should always take place before the middle of April. Colonies which are noted to be foraging less actively than others or not at all, should be checked for queen rightness as soon as possible. Around a week to ten days after pollen has been seen being carried into hives in abundance is a good yardstick for starting necessary inspections of hives which are suspect. Pick a mild even dullish, but not wet day and use just enough smoke to keep on top of the bees. If the suspect colony has no sign of eggs, take samples of bees, giving the hive a frame of emerging brood, with no adhering bee, from one of your strong hives.

If the health check proves O.K. give the colony (if it is reasonably strong) a frame of brood every other week until mid-April, then give it a frame of eggs and brood in all stages. The bees, will raise a new queen from this and the colony is saved rather than being a dead loss. The health check and moderate strength (covering at least three frames) are an initial must if the colony is to be saved. Do not be tempted by either indolence or apprehension to delay the first internal examination of any hive, especially the suspect ones, beyond the end of April. A stitch in time…!

THE MONTHLY ROUND - April

Catkins, dandelion, gean and flowering, currant blooming at this time of year are enough to gladden the heart of any beekeeper — especially if the weather is favourable for bee flight i.e. temperature greater than 10°C and wind speed less than 15 mph. Drone rearing should be well established in most colonies if the late March weather was in any way spring-like. The appearance of occupied drone cells should start the warning bells ringing for the beekeeper — that the bees have anticipated the onset of the reproductive period. Beekeepers have known for generations that after due time in their summer development that swarms would begin to tumble out of their colonies as the bees repeated their annual urge to reproduce. Although the swarming of bees can be expected during the late spring and summer — the deducing of exactly when any particular colony will throw its swarm is virtually impossible. The only safe method of anticipating the issue of a swarm is by internal inspection of the hive to check for occupied queen cells, eggs or larvae. The classic "nine-day inspection system" has stood beekeepers in good stead for many years but this method of management has resulted in progressive beekeeping becoming one of the most labour intensive forms of animal husbandry bar none. But come closer and listen! Swarming can be anticipated without colony inspection systems. Not only can swarming be anticipated but it can be easily and surely completely eliminated by one uncomplicated manoeuvre. Experimental work using practical management, published recently in a book which shall remain nameless, by a Scottish author who shall likewise remain nameless, *(Editor comment -The Book Eric is referring to is his own)* has demonstrated that by re-queening an as yet un-swarmed colony, not with any old queen, nor even with a year old queen, but on the contrary with a current year queen, that any colony so treated will most definitely not swarm in that particular season — provided the current year queen is fecund, healthy and undamaged in any way. By using this procedure the labour intensiveness is removed from the "progressive management" of bees at a stroke. The savings in time, effort and reduction of aggravation in beekeeping will have to be weighed by the beekeeper against the cost of buying a new queen, or the effort involved in the rearing on one's own early current year queens, which the book also discusses. Most beekeepers, myself included up until recently, will have suffered the nagging doubt about whether a colony will "go or stay" during the beekeeper's holiday period in the summer, only to come home and find that the worst has happened and the errant, partly anticipated in a nightmare, swarm has gone taking with it all hopes of good honey harvest. Use the method recommended and have a "swarm-free" vacation, returning home to a colony or colonies which instead of being depleted are heaving with honey-laden little darlings. Around the end of the first effective nectar flow is an ideal time to "relay-re-queen" but any time up to the last nectar flow of the year is good — but always prior to advanced swarming preparation! Swarming, reproductive swarming that is, is now a thing of the past — believe me!

The first supers of the season should now be being prepared for the hoped for early nectar flows from sycamore and the dreaded (for some) oil-seed rape. The middle of April is not too early to put supers on the hives where gean, sycamore and oil-seed rape are available. A good management device at this time, if the beekeeper works with a brood chamber and shallow crate to house the brood nest, is to smoke the bees down into the deep box and insert a queen excluder under the shallow crate, which will of course have quite a lot of brood in it. As the young bees emerge the bees will come up and fill the vacated cells with honey - much faster than they will an added empty crate with fully drawn combs. By smoking heavily from the top of the hive before fitting the queen excluder, the queen will be among the first of the bees to scuttle to the safety of the deep box — if she was in the super to start with. Check around four days later for eggs in the shallow. If none are present, the queen is safe below, where she can remain until being replaced by the "relay queen." Do you have enough jars and labels? Have you tried the $1/2$ lb jar yet? With the price of honey rising steadily, the economic cost of the 1lb jar is fast reaching the "price threshold" where customer resistance will increase. An important factor often missed by the beekeeper is that the purchasers of honey don't eat it in the same prodigal quantities as the beekeeper — so despite the reduced size the price is more acceptable to such a buyer.

THE MONTHLY ROUND - May

This month sees the dramatic increase in hive population that the beekeeper has been waiting for. The slabs of brood produced in the massive "pollen deluge" in April are now beginning to produce the new young bees which the colony desperately needs, faster than the wastage rate of old bees lost in the field. The beginning of May is a time for optimism in beekeeping — the as yet unknown weather pattern for the following months has still to unfold and the hopes for settled foraging weather are high in the beekeeper's mind. Sometimes things work out perfectly as happened in the 1984 summer season. Mostly we have to contend with less than optimum conditions where rain, wind, cold or any one from three can degenerate to any three from three as happened in the early part of May, 1985. 1985 will go down in the beekeeping annals as the "Year of the Wind" up to the start of August anyway. Optimism in beekeeping is the key — tomorrow will be better. This attitude makes beekeeping in Scotland bearable in most years. If the beekeeper has ensured his hives are queenright around March, by external observation and confirmation of large pollen pellets being carried into the hive during the early foraging period, there is no reason why the colonies should not be at honey gathering strength for the late spring flows from sycamore and oil seed rape. This is traditionally the first month where swarming can be expected — near the end normally. Perhaps a few thoughts (controversial?) on the honeybee reproduction methods would be acceptable here. The honeybee perpetuates its survival by the phenomena of swarming and supersedure. In the reproductive swarming act one or more casts leave the colony after the departure of prime swarm containing the original colony queen. This event results in the formation of at least two, and probably more colonies, which will ensure the propagation and continued survival of the species. In the supersedure act, swarming does not occur. Instead the bees produce another, single, selected queen from one of only two to three queen cells. This new young queen once mated, works in parallel with the older queen which is failing, for some arbitrary length of time, until the older queen finally disappears from the colony, disposed of by the bees at some yet to be established biological signal.

The question has been asked many times. Why swarming? Why supersedure? Having thought not a little about the honeybee reproductive cycle, I begin to wonder if these related but seemingly contradictory and contrasting events cannot be relatively easily explained. Many years ago in Germany, Dr Ferdinand Gerstung demonstrated to the world at large that the honeybee colony was a complete biological organism like any other animal: squirrel, rabbit, cow, man, etc., the three honeybee casts: queen, drone and worker constituting the complete organism. He even postulated that, being a complete organism, some cohesive force similar to the blood circulatory system must exist in the colony. He was of course correct even although the proof took many decades until Dr Colin Butler discovered "queen substance" the pheromones which impart cohesion and equilibrium to the honeybee colony as they are distributed in the colony by trophallaxis. Considering the honeybee colony as a complete animal removes many obstacles to its understanding. Each swarm or cast may then be viewed as a child of the colony, a birth in fact. In the animal kingdom many species reproduce by either single or multiple birth, normally the smaller the animal the more individuals are produced at birth. Mice may average 4-6 young; cats and dogs can have like numbers of offspring. Multiple birth seems therefore to dominate although single births among these species are not unknown. However, among the larger animals like horse, cow, elephant and even man, single birth seems to be the rule, however, even here multiple births of two, three and more rarely four or more may occur. Research has shown that multiple births especially among our own species are not hereditary but occur randomly in the population, however rarely or often. If the swarming and supersedure phenomena in honeybees are considered as multiple and single birth events, similar to many other animals, with the multiple birth phenomenon.

THE MONTHLY ROUND - June

For the beekeeper who has used the Relay Queen method by mid-June (i.e. replacing the old queens, yes even one winter old queens with a current year queen) the swarm threat will now be past history. He can now devote time to perhaps increasing the number of colonies worked, space and transport permitting, or he may now look forward to the summer holiday away from home, without the nagging doubt of lost swarms. The Relay Method of swarm inhibition has been discounted by some "experts" before ever trying the idea. I concede that new ideas always meet with resistance by many when first mooted. It is usually the highly respected pundit who makes the most noise in resisting any new concept — not necessarily in just beekeeping by the way! The expert has a reputation to safeguard and who do these upstarts with the different, fresh ideas think they are anyway? Darwin, Mendel, Churchill (who was the only politician to see through Hitler's "last" territorial demands). Janscha; (whose marvellous discovery of the queen honeybee mating secret was shouted down by Spitzner, a prominent vociferous and dogmatic beekeeper of that time. Janscha died as a young man at 34, before he could press home his observations which were published in 1770. Spitzner's criticism and jealousy set beekeeping back many years until Huber proved Janscha correct — but he, Huber, got all the credit!

Many Scottish areas, especially in the west, still suffer from that dreaded condition the June gap. Again, for the beekeeper going on holiday in June, experience of poor weather (for the bees) has shown that bees can go hungry even in mid-summer. Never fear, help is on its way! Last year (I refuse to use the term last summer because it wasn't) after the sycamore blossom period was over, bees in the west of Scotland fared pretty dreadfully. For many years I have been advocating placing wetted 1 kg. sugar bags on the frame tops of the brood chambers of colonies going into the winter. Last June I accelerated this procedure and placed not part dampened, but thoroughly soaked, soggy 1 kg. bags of sugar on all colonies in the out apiaries (take care the bag does not come apart under its own weight). These bags soon solidified! All the colonies so treated to this "hard tack" survived. I know of beekeepers who neglected their bees during the months after May and lost many colonies due to starvation. Such unorthodox practices can be the deciding factor in colony survival in seasons of dearth. The "going on holiday" in June beekeeper, previously a high risk animal can now if he wishes pop off to the sunny Caribbean with a clear mind. The new queen will maintain his hives in tip top non-swarming condition and the sugar bags (even placed on the top bars of the super) will safeguard against starvation. The bees take much longer to dispose of a 1 kg lump of solid sugar than a 1 kg sugar feed in syrup. Where the colony is very strong why not try two bags? Even if the weather turns into a heat wave while the beekeeper is away, and the bees start to store nectar there is little danger of contamination of the super honey because the bees will work the nectar sources in preference to the lump sugar. If there is any slight contamination due to sugar honey this will be found in the combs directly under the sugar bags, which the beekeeper can use! I stress here, I do not advocate these sugar bags as a standard practice, only where the beekeeper is going to be absent for more than a week in a period known to be sparse in nectar should the bags of sugar be given when surplus honey is being worked for, or in summers of the 1985 ilk!

For the non-Relay Queen beekeeper the most effective method of swarm control is the nine day inspection method. Using this method the beekeeper can at least keep pace with the developments in the hive and be aware of when his bees are preparing to swarm. What he does after noting swarm cells is very much an arbitrary question, most beginners merely panic! If the queen is marked, the beekeeper's problems are minimised. Removing the queen and cutting down all the sealed queen cells is the path of least resistance. Do not cut all the queen cells down when removing the queen. Leave the unsealed ones. Using the nine day cycle, these unsealed cells will not "hatch" before the beekeeper returns for the next inspection. At this later inspection all the queen cells remaining in the hive should be cut away leaving one beauty to produce the queen for the next generation. The queen cell left should be as close to the centre of the brood nest as possible; should be as close to the centre of a frame as possible so that it has never run the risk of being chilled. Any queen cells built close to drone comb should be rejected, any near the bottom bars or the top bars of the frames likewise rejected. Experience will guide the beekeeper's hand after a few seasons of small-scale queen rearing. Excellent queens can be raised even from extended worker cells, where the tip of the queen cell merely shows as a-"nose" projecting through the face of the surrounding sealed brood. Try one, you'll be pleasantly surprised! The much practised complete "gutting" of queen cells in a colony while leaving the queen present can result in a swarm in the air four days after the slaughter of the innocents. You have been warned! Continuous cutting out of queen cells ultimately demoralises the colony and it ceases to forage!

THE MONTHLY ROUND - July

This month produces the first of the late summer nectar flows and the last of the significant arboreal nectar sources, the lime tree or linden. The privet also blooms coincident with the lime, as does willowherb. Lime honey and monofloral willowherb are extremely fine honeys having quite different appearances and flavours. Lime is a clear honey with a subtle hint of green tasting lightly of menthol. This honey in incidentally a very slow granulating honey making it really ideal for sale as section or cut comb. As cut comb it requires to "bleed" a bit by leaving the pieces of cut comb on a wire grid over a drip tray to allow the cells breached by the cutter to drain, otherwise a pool of liquid honey will form on the cut comb box bottom, detracting from the pristine appearance of the merchandise presentation. Willowherb honey is almost water white, as is privet honey. Similar in appearance these honeys may be but their flavour and physical character are as different as chalk and cheese. Willowherb is light-bodied honey and has an extremely delicate, almost sweet wine like flavour, privet on the other hand when cold has an almost chewing gum like consistency and tastes rather like burned sugar. The scent of the privet blossom which is very heady if light in the air but, where the scent hangs heavy in the air, the perfume tends to be almost overpowering and sickly sweet, is totally different from the flavour of the honey. The bees work all these sources eagerly and despite privet's bitterness as a monofloral honey, blended by the bees with the other two nectars and allowed to age for about three months, it produces a very palatable honey. The privet and willowherb are particularly resistant to drought — in total contrast to the lime which will rarely secrete nectar regardless of how fine the weather is during its flowering period — unless the weather is wet for at least 7 - 10 days before the blossom breaks. If the weather preceding the lime blossom period is very dry the lime will, unfortunately for bee and beekeeper, bloom arid. In areas where privet and willowherb exist in abundance in lime areas the beekeeper, due to their flexibility, is reasonably well assured of a honey harvest (providing the bees can get out to forage), ranging from mediocre to "Help! I've run out of supers!" In a good late summer nectar flow the foundations for the next major nectar source exploitation are laid, i.e. the heather. Populations will reach a second peak around the second week in August after an exceptional late summer nectar flow—providing that the beekeeper gave the bees adequate super storage space for incoming nectar. If adequate super space is not given, the bees will congest the brood box with stored honey to the detriment of the production of brood resulting in, not a population explosion at the heather but rather a population implosion as the older foragers waste in the field in greater numbers than they are being replaced. Complementary to providing adequate super space, if the beekeeper removes any deep combs from the brood chamber which contain only honey and pollen and replaces these with new undrawn foundation, the bees will draw these out and the queen will be encouraged to lay in the new combs. If empty drawn comb is given this will be used for honey storage and the beekeeper still faces the population implosion at the heather. Colonies which have older queens in the hive during a late summer, heavy nectar flow will be likely to store honey in the cells drawn from donated foundation despite the beekeeper's efforts. However if a current year queen is present she will respond to the prosperity and the undrawn foundation and lay her heart out, to the advantage of the heather beekeeper. Many beekeepers are faced with a great dilemma around the start of August, especially when the lime blossom is late and the heather early. The problem, to lift the bees from the lime which is still secreting nectar quite heavily and go to the moors or wait until the lime is finished then move. Heather honey is a highly prized delicacy and rightly so. Some connoisseurs reckon that after tasting pure heather honey everything else is just toothpaste. Notwithstanding, a bird in the hand is worth two in the bush, and the timing of the commencement of the lime blossom secretion is the key to the solution of the dilemma. Leave all the strong hives at the lime and take all the lighter colonies and nuclei to the moors, but not before the heather blossom is showing at least 15% (check heather sprigs and make rough counts of blooming heads to unopened heads — it's relatively easy). Visit the heather sites at least every four days thereafter or get someone close to the moors to check and as soon as the hives on the moors begin to forage hard, gathering large pellets of dusty brown ling pollen, get the lime honey off the strong hives and away to the moors. Any unsealed lime combs can be sent to the heather and be sealed at the moors.

THE MONTHLY ROUND August

It has been virtually impossible to get it right this year in the timing of events in honeybee development due to the devastating weather patterns experienced last summer and autumn and the absolutely grim weather during April to late May, 1986. Writing a column, perhaps two months in advance of publication, by its very nature incurs a degree of inertia regarding reaction to current events. The recent problem has not been a case of how do I manage my bees for best results? It has been a case of how can I keep my bees alive until hopefully the weather will change and give them the chance to gather sufficient pollen to stimulate some colony growth? Whoever heard of feeding sugar syrup to the bees in May? Pollen dearth has shown itself this year, for the first time for many beekeepers, as the most critical factor in honeybee propagation and survival. Colonies which continued to die out during April and early May did so primarily because the necessary pollen reserves were not present in these colonies and the presence of nosema didn't help either. The older bees just dwindled away, not to be replaced by the young bees produced traditionally in late winter, from late December onward. The lateness and inclemency of the spring took a massive toll on the older surviving bees as they valiantly strove to glean what pollen they could from the rain and wind devastated crocus. By the time the catkins blossomed many colonies had already perished where they were located in areas devoid of adequate crocus coverage. The failure of the bulk of the catkin pollen sources, usually so abundant that the homecoming bees look like tiny airborne yellow dusters, due to wind and rain, again, further decimated the colonies in either borderline areas or borderline conditions in the hive. The importance of late winter brood rearing in colonies has been unequivocally established this year. Critics of colonies exhibiting this natural phenomenon ought now to take stock and ameliorate their views. An article on winter brood rearing was published in *The Scottish Beekeeper* in September, 1982. It took rather a lot of stick at the time. In retrospect it appears that the author of the piece had a point! It will have been noted by almost all beekeepers who suffered losses in spring this year that the bees did not die of honey shortage. Hives were discovered, on being opened, to have adequate honey within reach of the bees which were huddled hopelessly in a pitiful cluster numbering around 50-100 bees. This honey was virtually devoid of pollen, never mind covering the usual plugs of pollen sealed in the storage cells. If July was a settled flyable month with long mild windless days then the colonies of bees still surviving should have been able to build to strong enough proportions to do well at the heather. If the weather is flyable. If the weather in August is a repeat of 1985 then beekeeping in Scotland could become an item of curiosity in the history books. Heaven forbid! The progressive beekeeper, if mid to late June/early July were settled weather periods, will have made strenuous efforts to make good the colony losses incurred during the previous dreadful beekeeping period. If the methods of queen rearing described in "The Swarm Trigger Discovered" (*Editor – Eric promoting his book*) are employed, the beekeeper can easily rear any number of queens with the minimum of effort and without the need to resort to either fiddly gadgetry or "boiling" colonies of bees, which would deter even the strongest heart on an unfavourable day at the critical time for cell harvesting. A useful ploy for the beekeeper who missed the chance to breed new queens for replacement purposes would be to split a colony into two units, transfer the queenless unit to a new apiary and introduce a bought in queen to it and feed each unit copiously. This device could be used even up to the beginning of September if the beekeeper has the necessary skill. The long term benefits of splitting a thriving colony of course would have to be weighed against the short term sacrifice of honey from that colony. The choice is yours. Make it a good one! Considering the failure again of the early summer nectar sources due to rain and high winds in many areas, beekeepers who are not normally inclined to migrate their hives to the moors for the heather could offset this deficit by "having a go" at going to the heather. If the lime is secreting well coincident with the start of the effective heather blossom (15% of bloom open) it is as well to stay with the lime flow (a bird in the hand...) until it has passed its best. Then get the lime honey off and replacement supers on, and off to the moors — if the timing will get you there before the half-way point of the heather bloom period. Do not make assumptions about the heather blossoming period. The heather does not always bloom according to the calendar and can vary from as early as July 27 to as late as August 21. Visit the moor weekly and assess the bloom stage by counting the blooming heads on sprigs checked at random around the proposed site for the heather apiary. Perhaps the idea sounds a bit loony but believe me, it is quite an easy procedure and by so doing tells the beekeeper exactly what the state of play is. I cannot guarantee a crop but if the homework is done then the rest is up to the weather man. To my mind, if the beekeeper can input a degree of control to his beekeeping by trying to anticipate events and if his anticipation is rewarded by success there is little to compare with the glow of mild satisfaction that comes from working with the compass of Nature and achieving at least some success. You'll kick yourself all over the place if you have bursting hives of bees sitting in the garden apiary during an August heatwave. Oh, the joy of such an event! The heatwave of course!

THE MONTHLY ROUND - September

The resilience of the honeybee species was aptly demonstrated this year in Scotland. In particular areas honeybees were still struggling for survival up to early June with much major feeding being resorted to. The weather was up to this period definitely not bee weather. Quite suddenly and timeously the weather settled and the final two weeks in June were almost tropical — and astonishingly swarming began to occur in colonies which were particularly advanced relative to the norm. I survived the winter with 44 stocks from the 55 over wintered. These survivors were fed sugar syrup complementary to the still present residual solid sugar from the initial four, one kilogram bags placed on the colonies in mid-December '85. Ten colonies were developing rapidly on the meagre pollen income gathered on the few days when bee flight was possible during April and May. The other colonies were in various stages of development from good, not so good, to poor. By the second week in June the ten colonies were on double brood chambers and bursting with bees. The rest were well behind but growing in population. Many of these were reinforced by frames of sealed brood from the more prosperous colonies which were still being fed heavily to encourage the queens to continue laying at their high rate. At the end of the first week of the fortnight of "Heat Wave" all the ten colonies had queen cells. The queens were removed to nuclei, with sufficient bees (2 seams) to maintain the queen. The queen cells were harvested into mating nuclei made up by judicious removal of outside combs with bees, pollen and honey from the less forward hives. These nuclei were transported between apiaries to ensure that the subsequent queens were not mated by related drones. It was noticed that in the course of the weeks between mid-May and late-June that some colonies which had been growing steadily suddenly became queenless. The queens just disappeared from an otherwise thriving colony. These colonies were split and utilised for further queen rearing purposes by introducing frames of eggs from the other colonies of the later developing but still queenright thriving hives. The timing of breeding this year was drastically out of step due to the general colony weakness in May coupled with the atrocious weather. However hopefully the young colonies will prosper well on the late summer nectar from clover, lime, privet and willowherb, and the devastation wreaked on our beekeeping will soon be redressed and all the colonies will enter this coming winter with new current year queens. I feel that the queens which came through the summer and winter/spring 1985/86 suffered terrible hardship and for that reason their suitability for overwintering successfully must be suspect — I could be wrong but I do not intend to take any chances. The current year queens will be ruthlessly culled after the heather and hopefully the selected survivors and their future progeny will exhibit the qualities which bees in Scotland require to have for long-term survival of the species: thrift, stamina, "borderline condition" foraging ability, good overwintering, low swarming characteristic (in my opinion linked to "queen stamina") resistance to disease. The foregoing conditions are important for bee survival, other characteristics are merely beekeeper convenience factors or "cosmetic" such as: good temper, quietness on the comb, body colour and form, freedom from excessive propolis use, good comb- building. Any strain which combines all of these qualities is worth propagating and should be notified to the Beekeeping Departments of the Scottish Agricultural Colleges for controlled propagation purposes. Since in the aftermath of the disastrous 1985 season and the lateness of the summer in 1986 strains of honeybees which survived the holocaust and exhibit all the desirable qualities mentioned are really good genetic material for the Scottish geographical region. The beekeeping year began last month in August and hopefully if the heather or early autumn blossom did not fail (as it so often does) the bees should have been given a flying start for the fast approaching dormant period. Colonies at the moors should remain there for at least a week after the heather flow ends to give the bees the chance to ripen the last residual nectar gathered as the heather blossom becomes arid. The home/winter apiaries should have been prepared in advance for the bees return. It is quite surprising how rapidly bramble, thistle, willowherb and nettles can encroach on previously clear sites. The removal of the heather honey can be done either before or after moving the colonies from the moor. As a single handed operator I usually remove the supers, using a clearing board, before the colonies leave the heather sites — staggering some 50 — 80 yards across all kinds of rough ground carrying upwards of 80 lbs of hive, bees and honey is not quite a walk in the park — especially after the first ten hives! The colonies should be allowed to settle down in the winter apiaries for a week after the return. Where a number of hives are worked, and if a mental note is made of the various hive weights, by setting the colonies down in order according to weight the beekeeper can assess the sugar feeding requirements quite accurately. In the following spring a comparative assessment of the relative hive weights can give a good indication of the overwintering qualities of particular colonies. In seasons where the late nectar flows were good, winter feeding may be safely started early in September. In poor late seasons early winter feeding can result in late brood rearing, leaving colonies with a good population of young bees going into the winter, but with the potential of a crisis in the stores later in the winter/early spring. Ideally around the end of October a colony should have no less than 40 lbs of stores. Additional supplementary rations can to good account be given in the form of four, 1 kg bags of sugar, which have been dampened by immersion in cold water. These should be placed on the frame tops in direct contact with the cluster — under the crown board using an empty shallow crate to house the sugar. More about this device later. As a foot note many virgins due to mate in late June failed to return from their mating flights — due to swallow and swift activity — thus frustrating many weeks of planning and necessitating a start from virtually scratch — always having queen cells available during the active season can save much time in making good this kind of loss — merely having to graft a sealed cell into the failed nucleus can save almost two weeks in achieving mated queens.

THE MONTHLY ROUND - October

This year was not particularly suited to successful queen rearing, unless the beekeeper was fortunate enough to have mature virgins in abundance around late June, when the weather turned almost sub-tropical for a time. Mostly what seemed to happen was that the most forward colonies swarmed during the later part of the mini-heatwave. Many of these swarms were lost and the subsequent virgins left in the parent stocks failed to mate due to the return of the extremely unsettled weather, with large numbers of colonies being found to be queenless in late July/early August. Surprisingly, my experience of queen rearing in the 1985 and 1986 seasons comes out in favour of 1985, in which I produced many more new queens than in 1986. A major factor in this, despite the dreadful late summer, was that the month of May in 1985 was reasonable and the final week of that month produced a fine spell of weather. The Colonies were strong in early May 1985 and since I aim at producing the bulk of my new queens, mated and laying by early June, I was eminently successful, repeating the success of previous years for late May/early June queen production. This year, 1986, was a totally different ball game due to May being a complete write-off, with the colonies generally so weak that queen rearing at that time was virtually impossible. By early June 1986 a significant number of stocks were doing well, having been fed sugar syrup to complement the lump crystal residues of the winter 1kg sugar bags still present in the hives all through May. These colonies were utilised for queen rearing and the production of mating nuclei, and by the end of July, a full set (one for each colony) of new vigorous 1986 queens were laying their hearts out, many having been mated during the heat wave in late June/early July. The major problem I encountered this year was that many virgins did not return from mating flights, and I am convinced this was due to the "coronet" of swifts circling my mating apiaries, enjoying a right royal meal. May queen rearing does not present such hazards and I find early queen rearing more efficient for this reason, and many others. Despite the fact that there were only about four or five days which were suitable for queen mating flight in my area between early July and the start of August, all my colonies will enter the winter with the current year queens which I prize highly for the heather harvest and overwintering. July was a moderately successful honey month despite the lime tree, which can produce almost legendary nectar flows in particular years, only secreting moderately this year. The weather pattern of high winds and rain, coupled to lower than seasonal average temperatures also made its contribution to the less than bumper harvest from the July nectar-bearing flourish. However, this autumn the colonies will enter the dormant period in much better condition than last year, when pollen shortage made brood rearing extremely difficult for many or even most colonies.

Sugar syrup feeding should be finished with by the middle of this month at the latest, especially where colonies with less than heaving populations of young bees are concerned. Opinions vary relative to the optimum amount of winter stores a colony should have available to it going into winter, however 40-50 lbs of honey and pollen should be considered as a minimum for stress-free wintering for the colony — and the beekeeper. The hives should now be checked finally for weather-tightness. The hive should be set as level as possible, with the entrance horizontal and at a slightly lower level than the rear of the floorboard, to facilitate the drainage of any moisture which might gain entry to the hive. Measures against the entry of mice and shrews should have already been undertaken. There are a dozen different methods, some fiddly, some very fiddly. I find that using a floorboard with a 1/4" - 5/16" high entrance, saves me a great deal of work and worry. This entrance is left full width all winter, and only reduced in width in spring as the bees become active again, in the case of a weak colony. An occasional sweep across the entrance with a stiff twig or wire will ensure against the entrance becoming blocked in cases where winter mortality is very high. In years where late autumn weather is mild the beekeeper will note that winter mortality is light, but in years where the late autumn weather is unsettled, winter mortality evident on the floorboard will be quite high or even excessive in many colonies. This is due to the older bees dying in the hive as winter progresses, instead of dying in the field as would happen in a mild autumn.

THE MONTHLY ROUND - November

For the hobbyist, honey extraction will now only be a happy memory and all equipment will have been washed, cleaned and carefully stowed away for the hoped for harvest next summer. The equipment being marketed now is invariably high density polythene which one can "wash 'n' wipe." However tinned equipment should be rubbed down with a cloth impregnated with vaseline or light edible oil. Special attention should be paid to the ball bearing supporting the centrifuge spindle and frame. This bearing should be thoroughly dry and well oiled with vaseline as should the gearing and handle. The honey press should be washed down and all the heather honey residues removed from the "anvil" slats. If the press frame is washed lightly with a solution of water and bicarbonate of soda, this will inhibit any tendency for moulds to form where honey residues on taking up moisture can form unsightly spotty growths.

At the first frosts queen excluders can be easily scraped clean as the residues of wax and propolis become brittle and less adhesive. Management work with the bees, although "low key", must still go on, especially where the beekeeper has out-apiaries. A visual check should be carried out on a weekly cycle at least, to ascertain hives are still intact and undisturbed. The garden apiarist can perform his low key management quite painlessly, running his eye over the hives on a daily basis from now until spring. Powdery debris at the hive entrance or on the ground at the hive entrances is a normal occurrence as winter progresses and the bees "open more tins of honey". However, debris which has a much coarser appearance could spell trouble in the form of mice or shrews — if the beekeeper has not fitted mouse guards or ensured that the full width entrance is not higher than 1/4" — 5/16" which will give good ventilation but keep the smallest shrew at bay. An occasional flick across the entrance with a stiff twig or wire will ensure the entrance is clear, allowing the bees to fly on any mild days in December or January. For peace of mind it is always advisable to visit out apiaries after storms or severe gales, lest a hive has been overturned or a roof has been blown off — better safe than sorry! The beekeeping "social season" is now well under way in that the annual programme of winter meetings will have started and beekeepers will be swapping yarns and discussing events of the summer, when all sorts of adventures and misadventures occurred. I had not a few myself— one event in particular which could bear investigation gave me cause for pause. I like to get all my queens, which are always only one winter old in spring, marked before the first population explosion occurs — this year, due to the poor weather in May, the usual population explosion was a mere "pop" and it didn't occur in the west until well into June. I marked my queens at the start of the second week in June. At one particular hive I had difficulty with the cap of my "Tippex" bottle and placed the comb, which the queen was on, against the hive just "operated on." On lifting the comb to mark the queen, I was flabbergasted to discover that the queen was nowhere to be found. I searched the grass around where the comb had been placed but in vain. I was shattered! Having suffered my share of the colony losses in Scotland in the spring, this year, this additional loss was all I didn't need. Two weeks later I decided to dequeen two of the best performing hives in the apiary and one of the queens I removed was taken from the hive against which I had leaned the comb from which the unmarked queen had disappeared. The removed queen had of course been marked two weeks previously and she was quite easy to find. I returned nine days later to the two dequeened hives to harvest the surplus queen cells which I knew from experience would be present. One hive produced six cells, one of which was left for the colony to re-queen itself from. The other colony had produced no queen cells and — you've guessed it! Eggs were present where no eggs should have been. I instinctively thought of laying workers, however the laying pattern was too regular and the eggs were where eggs should be, singly and right at the bottom of the cell. Day old and other young larvae were also evident - Why? I examined the brood combs carefully and, sure enough, there was an unmarked black queen in prime condition on one of the combs. After recovering from my astonishment I marked her white and closed the hive up. I am very happy that honeybees don't read our books — otherwise this errant queen would have been killed. The phenomenon is worth following up. Who knows, perhaps a two-queen colony can be established much more easily than previously believed. I wonder! The honeybee species has many secrets yet to yield.

THE MONTHLY ROUND - January

This is the traditional month for gales (plus or minus two weeks). Hives in exposed sites or in locations under mature trees, or in the lee of a high wall of venerable years are quite at risk due to these high winds of late winter. Roofs and crownboards can be lifted off bodily by direct wind force into the hive entrance acting on the underside of the roof of the hive in exposed sites. Hives sited under the type of trees mentioned can be damaged or knocked over by wind-damaged branches or a fallen tree. Hives located in the lee of a wall could suffer the same fate if the wall succumbs to the massive pressure a gale force wind is capable of exerting on it. That was the bad news! Always check your hives in out-apiaries after such severe weather conditions! The good news is that at the end of this month in favourable years the snowdrop will bloom, providing our bees with their first opportunity to renew their depleting level of pollen stores as the queen steadily increases her rate of lay from a few eggs per week in late December, to the massive 1500-2000 per day at the height of the season in summer. For those of us who used the "sugar bag canopy" method the bees will by now have taken up residence in the cavity they have chewed into the underside of this crown of sugar, feeling probably much like Hansel and Gretel when they encountered the gingerbread and candy house in the forest in the famous Englebert Humperdinck fairy tale. Background reading on things beekeeping should be well underway and if this is carried out diligently then hopefully we will all enter the 1987 active season as more competent beekeepers than before. We beekeepers in Scotland are extremely fortunate in that we have our own very comprehensive library of books on apiculture in all its aspects maintained for our convenience in the Moir Library, which is housed within the Central Library in Edinburgh. A browse through the recently updated library catalogue will surprise if not astonish the reader at just how wide ranging in interest the library really is. Almost every important book on bees and beekeeping published in the English language in recent and not so recent times will be found there, to say nothing of foreign language books and many translations of this foreign language literature. For those interested in the historical aspect of beekeeping world-wide, there are bound volumes of many beekeeping periodicals from all over the beekeeping world from which may be gleaned the changing attitudes and changing patterns of the art of beekeeping over the years. Only by the objective exchange of ideas and opinions can progress be made in any field of endeavour. And beekeeping is no exception! The best place for new ideas on beekeeping is in a book, which hopefully will enjoy a wide exposure where the ideas can be scrutinised, openly discussed and ultimately put to the practical test by controlled, objective experiment. There is a tremendous amount of pleasure and satisfaction awaiting the "explorer" beekeeper who is prepared to ask his bees "questions." To get anywhere near to the correct answers, the question has to be thoughtfully "worded." Often we give our bees answers before we have even given them a chance to look at the question. e.g. a particular colony is deemed to be vicious by a beginner, (answer!). Is it really? Did the beekeeper use sufficient smoke? Were the bees short of stores? Was the barometer falling at the time of the hive inspection? Was the beekeeper crushing bees while withdrawing combs for inspection? Was the beekeeper standing in front of the hive obstructing the flight path? (questions!). Bees don't winter well on heather honey, (answer!). Were the bees healthy at the end of August? Did the bees rear the necessary young bees on the heather pollen? Did the queen get damaged in transit to and from the moors? How old was the queen? Was the overwintering site suitable, viz sheltered from prevailing wind; placed in a southerly aspect; was there good air drainage; (i.e. not sited in a damp hollow). Did the bees have sufficient stores in the first place? (question!). Those are admittedly very crude examples but there are still many mysteries to be solved by beekeepers about their bees, and the more one reads the more questions one is encouraged to pose. Try it! And then write to the editor to assist him to further the art of beekeeping in Scotland through the experiences and ideas of a wide enquiring readership of *"The Scottish Beekeeper"*.

THE MONTHLY ROUND - February

Although spring is "just around the corner", sometimes when we turn that corner we get a fright because although the calendar says "time for the snowdrop; time for the crocus; time for the catkins" Mother Nature is sometimes a bit tardy in catching up with the wind and rain, blown "pages" as they fall from the wall calendar (we've seen it in the movies!). However, hopefully our bees will be able to profit from the early bloom this spring. To date we have experienced three consecutive poor spring periods, 1986 being about the poorest spring since 1984. The old adage of one year in three is the meteorological norm for favourable years in beekeeping — according to the law of averages we are overdue a good spring this year. I'm saying nothing about the summer!

This is the first month when we can begin to feel optimistic that our charges have survived the dormant period. If the bees get the chance to forage on the crocus which will usually bloom at some time in February, we will be able to determine, not only that the colonies are alive, but also if they are queenright. This is where having more than one hive on a site is extremely useful since, in the early spring, when pollen is available, colony activity can be compared. Any colony less active than its neighbours which are foraging well for pollen will show up easily on comparing colony behaviour. This hypothetical less active colony on its own will tell the beekeeper relatively little until it is probably too late to take action, since the beekeeper can only assume any colony movement as normal if no comparison is available. A colony gathering large yellow-orange pellets of pollen during the crocus bloom period gives the beekeeper an almost fool proof reassurance that all is well and that the colony is queenright. However, a queenright colony with a drone-laying queen can also give the (dare I say it?) less observant beekeeper the reassurance he is seeking from observing colony behaviour, in that pollen will also be being carried into the hive. Closer examination will indicate that the pollen loads are smaller than those of bees foraging from colonies with a vigorous, fecund queen and that the bees are working less hard. Colony strength also can be easily gauged by observation of colony foraging activity in spring where queenright colonies are foraging from the same apiary. The density of foraging activity re-numbers of bees entering and leaving the hive, varying in degree from feverish activity to a mere thimbleful of bees moving is a real giveaway. The quality of the queen may also be gauged by entrance observation. A colony where the great majority of the foragers is gathering pollen early in the spring, is a good indication that the queen is a vigorous early layer. A colony which is exhibiting a minority of foragers gathering large pollen pellets relative to its neighbours could be telling the beekeeper that the queen is failing, or that she is a late developer. By observing such a colony later in the active season, it could possibly transpire that it is relatively more actively gathering pollen than its neighbours, indicating that it is indeed a potential heather bee. Thus, from a few minutes knowledgeable observation per hive in early spring the beekeeper is potentially able to read volumes about the queen state, population size, queen fecundity and perhaps the seasonal characteristics of the queen. Most beekeepers are eager to work their colonies in good weather in midsummer, few however are prepared or even dare to enter a colony before May. This delayed inspection can be fatal for colonies which fall queenless during the late winter. Since the beekeeper, on finding a colony queenless in early March, has the option of uniting it to an adjacent queenright colony. Ideally, samples of bees should be taken from both colonies and checked for nosema and acarine or any of the virus diseases which are now known, due to the excellent pioneering work done by Dr Bailey in recent years. A colony, so managed, can effectively reinforce a weaker, queenright colony and accelerate its development. Queenright box on the top! A classic device employed by Brother Adam at Buckfast for uniting bees (in the spring only) is merely to subdue both colonies using smoke in the usual way then expose both colonies to daylight by removing roof and crown board for a few minutes, then place the queenright stock above the queenless one without using newspaper. The bees unite quite happily. For the more daring beekeeper, a strong, queenless colony can not only be saved by timely intervention but even husbanded until it has re-queened itself. More about this next month!

The timing of entering a hive early in spring poses almost insurmountable problems for many beekeepers. The timing however, is extremely easy to get right. Simply observe colony activity in spring, and around a week after a colony has been observed foraging strongly for pollen, that colony can be entered with care and with minimum smoke. Know what you are looking for though!

THE MONTHLY ROUND - April

This month is perhaps one of the most critical months in the apical year since the brood, on which the early summer prosperity of the hive will depend, should already be present in the hive in the early part of the month, ticking away like a biological time bomb, to produce the necessary population explosion in the hive near the end of the month, and allow the colony to make the best of the late spring/early summer major nectar sources. The number of eggs which a queen bee lays, under normal circumstances will determine the amount of brood in the hive and ultimately the maximum population which the colony will achieve. However, not all the eggs laid by a queen bee hatch and produce a new bee. One extremely critical factor, often overlooked by the beekeeper, is the lineage of the queen bees in the hives, despite the fact that the lineage characteristic of a queen governs the ultimate quality of the colony as an effective honey gathering unit. Inbreeding is a most important phenomenon in animal husbandry and this includes bees. If heterosis (hybrid vigour) becomes reduced in a colony due to inbreeding the degree of severity of inbreeding will show in the ability or non-ability of the hive to build up in population. If a queen honeybee mates with too many related drones, many of her eggs will fail to hatch or will produce diploid males, which the bees eat, and the population of the hive will not increase normally, if at all. On the other hand, a queen which mates with only unrelated drones, all other things being equal, her hive population will build up steadily due to all the fertilised eggs as well as being viable also producing worker progeny. The accepted gospel among many beekeepers of only breeding from the best queens or queen in their apiary is in fact an extremely non-scientific and potentially damaging procedure, in the medium to long term in beekeeping husbandry. And to date the only circumstance which has saved many beekeepers practising the "best queen" procedure from catastrophy, in my opinion, is the fact that there are enough unrelated colonies of bees around their apiaries, feral and husbanded, to keep the gene pool reasonably heterozygous. Consider the case of a beekeeper running perhaps 300 colonies, for argument's sake, on an island remote from other colonies, including feral colonies. In year one, he starts with 300 unrelated bee colonies. In year two, he selects the best colonies to rear his next generation of queens from, say six colonies. At the end of year two, he has re-queened all his stocks with the daughter queens of the selected six, reducing his gene pool from 300 units to six at a stroke. In year three, he again selects from the best six queens and, by the end of year three, all of his stocks are headed by the daughter queens of his selected six, resulting in the dramatic reduction from a 300 hive unit containing a quite massive gene pool, to a 300 hive unit whose gene pool is now so greatly limited that a dramatic population crash is imminent, due to the enormous reduction in heterosis and the rapid increase in homozygosity at the sex locus due to inbreeding which results in the production of many non-viable fertilised eggs which do not hatch and which also results in many fertile eggs producing diploid male larvae which the bees eat. Most organisms are diploid, that is the male and female of the species possess the same number of chromosomes: Homo sapiens has 46. However, the honeybee and some other organisms are different in that the male of the species only possess half the normal diploid number of chromosomes. The queen bee has 32 chromosomes, but the drone being haploid has only 16 chromosomes. This phenomenon, and the fact that queen bees are parthenogenic, has great significance in the breeding of the honeybee. Present day scientific bee breeding techniques owe much to the pioneering work of Dzierzon, the discoverer of the parthenogenic phenomenon characteristic of the honeybee. Incidentally Dzierzon had a long, bitter and acrimonious struggle to convince the beekeeping world that his observations and conclusions were correct in his day. But that is another story! You may say that the example of the 300 colonies was a masterpiece of hyperbole, and I would have to disagree, because, apart from the "remote island" location of the example, the other aspects of the tale and its "logic" were reported quite recently in one of the leading beekeeping magazines of these islands. One of the most efficient methods of retrieving genetic viability in an apiary is to systematically dequeen each hive at some suitable time during the active season and allow the bees to rear another queen from the queen cells subsequently built after the removal of the queen, bearing in mind that a colony so dequeened will invariably build more than one cell. The beekeeper must enter each hive again around 8-10 days later and cut out all the superfluous queen cells, leaving one only. If even only two are left the colony will swarm if the weather is right! The next most efficient method of returning genetic viability (not necessarily economic viability) is to allow the colonies to swarm and re-queen themselves naturally. A compromise method I have used for many years is the selection of the best 50% of my stocks for queen-rearing purposes, and the early production of drones by the insertion of drone comb into selected drone breeding stocks in late March, into the middle of the brood nest. These drone-rearing stocks are fed copiously with sugar syrup. This ensures at least reasonable control over the quality of the daughter queens reared annually and invariably to be mated and laying by at the latest early June. By a procedure of early queen rearing and re-queening, I can manage my stocks to achieve full strength for the late summer and autumn nectar flows without the bother of having to contend with swarm management as well. During the swarming season, I collect usually 10-20 swarms from the Glasgow area within a 20-mile radius of my home apiary each year. These swarms are quarantined in an out-apiary and checked for disease and good qualities. Perhaps 50% of the genetic material of those swarms, each year, is inducted into my colony complement. Thus the heterosis characteristic is continuously "topped up." As an aside to the foregoing and the reference to the contribution to "unrelated" matings made by feral colonies: after varroasis becomes established in these islands, our feral colonies will gradually die out, since it is undisputed scientific fact that any honeybee colony infested with varroasis, if untreated, will die out at some time later (around the fifth year after initial infestation. If, as I believe, the feral colonies alive in the past made a major if unnoticed contribution to the maintaining of heterosis in "out" husbanded colonies, then the medium to long term prospects for viable beekeeping in Britain must be in some jeopardy, unless beekeeping philosophy at "grass roots" level undergoes a dramatic reappraisal. Time will tell!

THE MONTHLY ROUND - May

This month is the first month in which it is possible to harvest honey. The oil seed rape and sycamore, weather permitting, which bloom in May can return quite considerable crops of honey if the bees are strong enough to take advantage of them. In "let alone" beekeeping, the possibility of achieving surplus honey from those early sources is a lottery to say the least, most colonies building population on these sources. By the end of May many hives will be chock full of brood and bees and not only moderately provisioned with stores, but also in swarm mood, especially if a long spell of good weather occurs near the end of May. The progressive beekeeper who is aware of the arithmetic of honeybee biology can, by judicious management, ensure that his colonies, or at least some of them are at not merely honey gathering strength but on the contrary are at surplus honey strength for the peak of the bloom of those early nectar sources. By virtue of being progressive the beekeeper will have a good idea of the approximate blossoming times of his nectar sources and can thus work toward them. The biological arithmetic mentioned, involves the knowledge that a six-week development period elapses from worker egg laid to mature forager bee. And that from sealed cell to mature forager a period of approximately four and a half weeks elapses, the progressive beekeeper will also be aware of the fact that a strong colony of bees (35,000 - 40,000) will produce more surplus honey than four moderate strength colonies (ca 15,000). A colony of bees with a population of around 9,000 bees will, relatively speaking, increase its population size more rapidly than colonies of lesser and also greater size. Bearing in mind also that any worker eggs laid after the end of the third week in April will produce mature worker progeny too late to be effective for the sycamore bloom, assuming that the sycamore period ends at the end of May (West of Scotland). By knowing the approximate blossom periods in the particular area of operation, the beekeeper can modify the timetable to suit the area. Thus, using the foraging logic, two strong hives are worth eight moderate strength hives and twenty strong hives are worth more than eighty moderate ones. I have used biological arithmetic for many years now to great advantage. Working with 60-80 colonies single handed, overwintering in safe remote out apiaries, the colonies are all over-wintered on single brood boxes with the supplemental bags of sugar. Syrup feeding is commenced in early March as soon as the colonies begin to forage in earnest for pollen. Colony status is noted at the commencement of syrup feeding and the relative rate of population increase in each colony is monitored thereafter. Around mid to late April, colony assessment is made and, using the four to one rule previously discussed, the strongest 25% of the colonies is selected and moved from the overwintering sites to the sycamore apiaries. (To date, oilseed rape has not approached within effective range of my operation - when it does, management will be modified to accommodate it). Over the following few days a similar number of colonies of moderate strength are moved to the same sites and placed, one each, on top of the hives previously moved into the sycamore apiaries. The colonies are allowed to settle for a few days, then about a week before the sycamore is due to open the pairs of hives are united. One queen of each pair is removed to a nucleus box with a frame of eggs and open brood, from the lower of the two colonies being united and sufficient bees and stores for the nucleus to support itself. To unite, the Brother Adam spring uniting procedure is used. The colonies are subdued using the normal smoking technique, exposed to daylight for at least five minutes (it will probably take at least this time to find the queen to be removed) and then the top brood box placed directly onto the bottom one, making a two brood chamber unit. The newspaper method should be used if such a procedure is carried out later in summer. In spring/early summer, uniting without the use of newspaper may be done with impunity. The hive is then closed up. It comprises the bulk of the bees and brood of both colonies, giving a sizable work force for the early nectar flow in a unit which can be managed with a quarter of the effort required to manage the original colonies and which will eventually guarantee a heavy crop of honey, if the weather plays its part. I now have my bees in a condition where the number of colonies is back to the original, but where the very strong hives are situated where they can do the most effective work. The other hives are worked with two ideas in mind:

Build up for the lime;

The rearing of new queens.

As I have often stated, it is totally unnecessary to use 'boiling' stocks of bees for the rearing of quality queens. A well-provisioned, well-populated 4-frame nucleus stock of bees can be utilised to produce a surprisingly large number of new queens, provided the arithmetic factor is understood and acted upon. If the beekeeper is prepared to "bite the bullet" and either de-queen or split moderate-sized colonies relatively early in the season, it is possible to have new queens mated and laying by late May/early June. Depending on the number of colonies possessed, the beekeeper can plan his schedule to suit his needs, either going for a total re-queening of all colonies or a percentage re-queen, always bearing in mind the perils of rapid inbreeding. Aim to re-queen from the best 50% of the stocks possessed. Thus, the ten-hive beekeeper would select the five best colonies and use eggs from these to produce queens to head the ten colonies. For example, using the six strongest hives as the projected honey gathering unit, unite them as described previously to produce three really powerful stocks for the very early nectar flow in spring. Make up three queenless 4-frame nuclei and remove a frame of eggs from each of three of the best colonies. Insert these frames of eggs into each of the queenless nuclei. Each nucleus should now have two frames of stores with the frame of eggs and open brood between them. The fourth frame should ideally be a fully drawn comb, empty of stores, but with most of the cells filled with water. The absence of water has been shown by research to be a decisive factor in successful queen rearing. Around three or four days after making up the nuclei, all the sealed cells should be destroyed since these have been around mature larvae. The queen cells have to be harvested ten days after the nuclei were made up in order to avoid the emergence of virgins in the nuclei. At least a day prior to the cells being harvested, three 3-frame queenless mating nuclei should be made up, preferably from colonies in an out-apiary, and brought to the mating apiary. These nuclei should consist of two brood frames of stores and one central frame with at least a 2" x 3" patch of brood to hold the bees in the nucleus. This brood comb should have a slot cut in it into which the queen cell with its heel of comb will be inserted. After the selected queen cells have been harvested, all remaining queen cells in the original queen-rearing nuclei except one should be destroyed. When the new queens have been successfully mated they can be transferred to the strong stocks as required. Subdue the stocks to be worked and remove the old queen in a matchbox. Cut a slot at the corner of a brood comb at the edge of the brood nest. Having already also secured the new queen in a matchbox, slip the new queen matchbox into the slot in the comb, ensuring that the box is open enough for the queen to be fed but not wide enough to allow her out. Close the hive up. Check a week later for eggs in the colony and close up — no need to find the queen. The old queen can be retained in a nucleus or disposed of. Strong colonies so treated before they have started to make swarm preparations will not swarm in the current year. There is a slight snag which most beekeepers will encounter in the procedure of harvesting queen cells. Traditionally, brood foundation is wired and queen cells coinciding with these wire strands will be difficult to cut out. A small pair of sharp wire clippers, obtainable at any ironmongers will overcome this problem. I myself, having for years sought economies and shortcuts to traditional beekeeping, made my own deep foundation which I do not wire. The legend that unwired foundation will sag in use, is indeed a legend and I have never had any of my hundreds of unwired combs sag on me, even during the few good days we are gifted during particular summers. I also regularly extract from deep combs and rarely have trouble with sagging combs, especially after the first cycle of brood rearing has occurred in them. The residual cocoons strengthen the combs so effectively that the procedure, in my opinion, is superfluous. It of course goes without saying that records of queen lineage must be kept religiously and accurately in any queen rearing operation, large or small. This will help facilitate stock improvement and inhibit rapid inbreeding.

THE MONTHLY ROUND - June

For the majority of beekeepers, the first flush of the summer major nectar flows are over. The sycamore bloom will be finished as will that of winter sown oil seed rape. Apart from the fortunate minority of beekeepers whose bees still have access to abundant clover pasture and spring sown oil seed rape, the rest of us must patiently bear the period which has become a legend: "the June gap." Strangely enough the gap associated in the mind with June can and does occur randomly during the active season as a natural phenomenon. Not so much a gap due to a dearth of nectar bearing plants as rather a dearth of suitable fine days to allow effective bee flight, causing a gap in nectar income to the hives. In borderline weather conditions during the bloom period of major nectar sources, even if the bees do not produce surplus honey, they are usually able to sustain their own needs. In borderline weather conditions during a period when only secondary nectar sources are blooming, such as in June, a critical gap can arise, necessitating the feeding of sugar syrup to colonies until the late summer bloom is available to the bees. However, having said all that I have said about the June gap, I have to confess that in the light of experience the June gap does not really exist. If the beekeeper casts the mind back to that halcyon summer of 1984, when the sun came out on Easter Sunday and shone virtually without halt until late July, there was no June gap that particular year. The bees continued to prosper and nectar kept coming in, thus confirming that although no obvious major nectar source is blooming, there are adequate minor nectar sources available in the month of June, which require favourable weather conditions to allow the bees to exploit them. What we really need is the return of the clover to give the bees and the beekeeper the facility to achieve bumper harvests reminiscent of the days when silage wasn't even a twinkle in the stock farmer's eye. Come to think of it, we beekeepers have it in our power to regenerate the clover bounty of yester year if we would purchase even $^1/_2$ kg of suitable clover seed and sow a few seeds in likely places where it could become a well established and abundant weed — a weed being defined as an out of place flower.

In particular seasons, swarming may be expected in late May or early June. There are many ways to control and deflect swarming. I will not labour here the relay re-queening method which, in my opinion is the most effective method when carried out correctly. Where a colony is found to have reached the advanced stage of swarming, where sealed queen cells are present, the most effective method of control is to remove the queen with enough bees to fill the seam between two deep combs and transport her and her bees to another apiary in the queen holding nucleus and cut down all the sealed cells. It is good management to leave all the unsealed cells. This helps inhibit a surge of further queen cell building which would otherwise result if all cells were cut down.

Having cut all the sealed queen cells out, the hive is closed after removing the queen. It is now only necessary to return to the colony the following week and reduce the hive to one sealed queen cell. All other queen cells, unsealed as well as sealed, should either be harvested and given to queenless nuclei or ruthlessly destroyed. One cell, and one cell only, must be left in the colony. If even only two are left, the beekeeper may as well not have bothered with the previous cell-cutting procedures because the bees will swarm with the first virgin out, unless prolonged unfavourable weather occurs around the time of the emergence of the first virgin. If the queen is left in the colony where the cell-cutting procedure is used, the bees will construct replacement queen cells from 2-day old larvae (i.e. 5 days old from egg laid), have a cell sealed in four days from cell-cutting operation and, if the weather does not turn unfavourable, the colony will swarm while the beekeeper is kidding himself that he still has five days in hand considering the application of the 9-day swarm control cycle. The onset of unfavourable weather will inhibit the swarm from issuing and, in my humble opinion, this factor has contributed more to the apparent success of cell cutting where the queen is left in the hive than any other factor and as such makes the cell cutting method extremely suspect and unreliable. If the beekeeper stays on the ball and persists in cell destruction on a 4-day cycle in a settled weather period, the bees ultimately become demoralised and foraging becomes either depressed or ceases altogether. All of us have our own preferred method of swarm control varying from total unpreparedness, resulting in panic, to sensible anticipation using methods which either deflect or eliminate the swarm impulse. If you find a reliable, simple method, tell the world. A caution here, don't be afraid to resort to feeding your bees in summer, especially in a long unsettled spell of weather. Better safe than sorry as the last two summers have shown!

THE MONTHLY ROUND - July

Mention was made fleetingly to the control and management of swarming last issue. It might be of interest to discuss some of the more popular traditional methods used to dissuade honey bees from swarming; To be able to take effective measures to inhibit swarming the beekeeper must at least know the state of his bee colonies week by week (preferably day by day) and the most popular and positive system which gives this facility is the 9 day Inspection Cycle. This is based on the biological fact that the time lapse from egg laid to sealed queen cell is approximately nine days. It is also a biological fact that a colony of bees will normally swarm on the first fine day after the first queen cell is sealed. The "9-day Inspection' Cycle" has to form the basis for any swarm control method which is based on management procedures after eggs have been noted in queen cells. This 9-day cycle in beekeeping is one of the main reasons why progressive beekeeping is so labour intensive. Any mangagement procedure to inhibit swarming, which is effective and eliminates the need for the 9-day cycle should be promoted vigorously for the benefit of beekeeping in general. I know of only two such methods where, once the procedures have been carried out correctly, further swarm control management is unnecessary. One method is queen removal before any swarm signs have been exhibited by the colony and the removal of all queen cells subsequently built, except one a week after the queen was removed. This procedure will result in the colony requeening itself and producing surplus honey. But more importantly the colony will not make any further swarm preparations that year. The other method is of course the Relay Re-queen Method which, if carried out correctly, will achieve the same end result with the added advantage that there is no interruption in the production of brood.

Now to the traditional swarm control methods, most of which are effective most of the time, but some of which are only suitable for the beekeeper with less than ten hives. **Snelgrove's Method** is a very effective system for the garden beekeeper and it is worth doing one's homework with this method which entails removing the bulk of the young bees and open brood above a special screen board which is fitted with double "ports" by which flying bees can be deflected back into the parent stock by intelligent manipulation of the hinged entrances. The method is labour intensive. **Taranov Board Method** is also very effective but extremely labour intensive and time consuming and may result in confrontation with irate neighbours. A sloping, elevated board is placed in front of the hive, leaving a gap of around 2 inches between the edge of the board and the front of the hive. The bees are shaken from their combs onto the board. The flying bees return to the hive on the wing; the young bees run up the board and form a cluster under the lip of the leading edge of the board in the presence of the queen. This cluster of young bees is the equivalent of a swarm and can be housed adjacent to the parent stock without loss of these young bees. This method is used only after swarm preparations are noted in the colony. **Shook Swarm Method**. This is a method which depends on judicious shaking of bees into another hive, thus depleting the parent colony population. It is done to best effect when a spare queen is available for either the shook swarm or the dequeened parent stock. Either the shook swarm or the parent stock must be removed from the parent apiary, lest the mature foragers reunite with the stock on the original hive site. **Caged Queen Method**. This system is very effective if carried through correctly. It is however quite labour-intensive and totally dependent on the 9-day cycle for its success. The queen is found and caged in a queen excluder cage when swarm preparations are noted in the colony and the cage is wedged under the bar of one of the brood frames. The queen cells in the colony at this time are ruthlessly destroyed. The following week the subsequent queen cells are again destroyed. On the next visit (i.e. 2 weeks after the initial caging of the queen) the queen is released into the colony. Sometimes the queen is killed; sometimes the bees resume swarm preparations but mostly they accept the queen again and resume normal business. **Demaree Method**. This entails the separation of all the brood except one frame of sealed brood. The separated brood is housed in a new deep chamber and placed above the original chamber over a queen excluder. The original chamber, containing the queen, the frame of sealed brood and a full complement of empty, drawn comb is intended to ease the congestion. Queen cells built in the upper chamber have to be destroyed the following week or the colony will swarm. **Pagden Method** is quite similar to Demaree except that instead of placing the brood chamber, into which the bulk of the brood has been placed, on top of the chamber containing the queen, this box with brood and no queen is moved to a new position on its own floorboard in the parent apiary. The mature bees return to the parent hive, depleting the population of the queenless stock. The queenless stock will build queen cells and these must be reduced to a single cell the following week. Care is required with this method, otherwise chilling of brood in the queenless colony will result. **The Heddon Method** is similar to the Pagden but entails the positioning of the queenless stock with the entrance at right angles, adjacent to the parent hive for a couple of days. The queenless stock is then moved to the other side of the parent stock, being placed again with the entrance at right angles to the parent hive entrance. Two days afterwards the queenless stock now well depleted of bees, is removed to another apiary to re-queen itself There is a danger of chilling brood using this method. A common factor to be noted in many of the traditional methods of swarm control is that an attempt is made to separate the brood and younger bees from the queen and the older bees.

The beekeeper of yore knew instinctively that there was some rational in this but could not make the correct connection because scientific beekeeping still had a bit to go. Colin Butler supplied the solution when he discovered the queen substances and demonstrated that the older bees did not pay particular attention to the queen, the pre-forager stage bees being the age group most interested in the queen and queen substance. The critical importance of the queen substances have been demonstrated unequivocally, both their presence and their absence in the hive. Thus it is not the actual physical separation of the young and older bees in the colony which is necessary for the success of a swarm control method, but rather the need to maintain a sufficient supply of queen substances to satisfy the increasing numbers of pre-forager age bees at a time when the overwintered queen is beginning to decline in her ability to secrete the necessary quantities of these queen substances. The obvious solution to the problem of declining queen substances supply is to renew the source of the supply. Another queen! But not just any other queen. A current year queen, flowing over with the necessary queen substances, and full of the joys and energy which only youthful vigour can impart, is the key. Even a year-old queen is entering middle age by the end of her second summer. After the age of 35, how many of us could compete on equal terms with an energetic teenager? (For deeper treatment of the swarm prevention methods mentioned, refer to Couston's "Principles of Practical Beekeeping" or Cumming and Logan "Beekeeping — Hobby and Craft", both obtainable from the Moir Library).

THE MONTHLY ROUND - August

The beekeeping year begins again! The brood raised in this month could be the most critical factor in the successful overwintering of a honeybee colony. For this reason alone, it is sensible to ensure that by late July each colony has a queen not more than two winters old heading it. To capitalise on this young queen, it is good management to migrate to the heather — for two good reasons:

- a potential harvest of Scotland's premier honey.

- a dramatic stimulus is given to the colony to continue to rear brood due to the massive income of the tan coloured pollen of the ling heather.

A colony entering the dormant period with a high proportion of old bees is at a great disadvantage, due to an inordinately high winter wastage rate resulting in a very low core of population in the hive around mid-March. The smaller the core population in spring, the slower the colony can build, even if it is well fed. The queen can only lay the number of eggs which the colony can brood. So, lay a good foundation for the spring next year by doing the right thing now. Moving to the moors can be effected quite easily if advance preparations are made to ensure the colony is in "migration rig." Using a pick-up type of vehicle for hive transportation makes the job relatively easy and eliminates the problem of bees leaking from a hive or hives being transported in a closed vehicle, gradually festooning the car windows as the escapees increase in number. Worse still, a sudden jolt, due to a pothole or emergency stop, can cause hives to split at the floorboard or super and allow thousands of irate *Apis mellifera* to stampede to freedom, causing all kinds of problems for the unhappy beekeeper. On an open-backed vehicle such an event, although not to be recommended, is relatively speaking of no great consequence, especially if the hives are being moved late at night or very early in the morning. The escaping bees will eagerly spill out of the "split" but, as soon as they feel the wind on their faces, they will remain on the hive surface and eventually percolate back into the hive again. It is useful to have the smoker lit and ready at all times when moving bees. At unloading, a quick blast of smoke can make all the difference between getting stung to bits and merely getting stung. Despite the problems encountered in moving colonies, it really is a most rewarding procedure when the weather in August turns trumps and the bees get the chance to show their mettle on the moors. It is good management practice to examine bee colonies at least a week after they have been moved to the heather. This examination is merely to establish that the colony is still queenright. All that is required is that eggs be noted. This will be indication enough that the queen survived the migration to the moors. For the beekeeper who is not inclined to disturb the colonies at the heather, observation of the colony activity at the end of the second week on the moors will give a good indication of the colony condition. If the bees are taking large pellets of the dull, tan coloured heather pollen into the hive at this time during foraging then all is well — brood is still being reared and fed. If no pollen is going into the hive while other adjacent colonies are carrying pollen this can mean a number of things, mostly unpleasant for the bee and the beekeeper:

1) Hive is queenless.

2) Hive has a scrub or aged queen.

3) Hive has a drone-laying queen or less serious.

4) Hive has an early queen which is not suited to late nectar flow areas and due to this has ceased to lay.

If 4) is the case, do not use her for breeding purposes and re-queen in the following summer with a "local strain" queen. The traditional optimum type of day to make an internal examination of a bee colony is during a nectar flow in the early afternoon. Being a fervent non-traditionalist, I threw that particular piece of counter productive information away many years ago. In Scotland, good bee days are a premium and to disrupt a strong colony of bees on a perfect day during a nectar flow, especially at the heather, is to me the height of folly. Bees at the heather are not noted for their good temper. However, when a colony has to be examined at the moors — or even at any time during the active period — if the examination is carried out in the late afternoon/early evening, the beekeeper will be pleasantly surprised to find out just how amenable a colony of bees is to being opened up after an excellent day in the field when the bulk of the foraging is over for the day. The bees are so busy with their stomachs already full of nectar being evaporated and processed into honey that they barely seem to notice the beekeeper's intrusion. Try it! You'll be pleasantly surprised — and get more honey in the process. It is sensible to erect some kind of fence to protect hives against overturning by sheep, cattle and even deer in some locations. Hives make excellent scratching posts for deer, and grazing cattle and sheep can easily turn them over. A weekly visit to the heather site is worthwhile and can catch an exposed hive before rain and wind can wreak havoc on the bees. The timing of migration to the heather can present problems to the beekeeper, especially if hives are located in areas where the lime tree is still secreting nectar when the heather starts to bloom. Having been presented with the problem many times, I have devised a method which allows me to hedge my bets. From bitter experience I have learned that it is easy to misjudge the start of the heather nectar flow, despite the heather beginning to bloom. I now do not move hives to the moors until at least 15% of the blossom is open — this state can be judged quite accurately by randomly checking heather sprigs at various distances from the intended apiary site. Merely doing a rough count of open to unopened heads can give a surprisingly accurate estimate of the state of play. If the 15% bloom state is reached while the lime is still effective, I clear the lime honey supers off 50% of the hives and lift this 50% of strong colonies to the heather, leaving the other 50% on the lime, along with the nucleus stocks. As soon as the lime nears its final date, I repeat the removal of the lime supers from the remaining strong hives. All hives going to the moors, apart from nucleus stocks, which are also migrated, are fitted with supers of drawn comb where possible, but where at least the two centre shallow combs are drawn to get the bees into the super and storing honey as quickly as possible. Any un-ripened lime honey can be placed on the colonies at the moors and if the weather is unstable the bees will finish the job of ripening and sealing using heather and lime honey. Let us hope!

THE MONTHLY ROUND - October

Sugar feeding should be completed by the middle of this month at the latest. Preferably though especially in the case of colonies covering less than eight frames liquid feeding should have been finished by the end of September. Opinions vary regarding the minimum amount of stores, pollen and honey, a colony should have going into winter, but few experienced beekeepers would accept less than 30 lb. of stores. The average beekeeper in the past has probably never considered effecting control over the amount of pollen contained in an overwintered colony, although for centuries we have tried to ensure that liquid stores always attained a particular safe level. The poor early springs of the seasons 1985, '86 and '87 have demonstrated quite dramatically the critical importance of adequate supplies of pollen in the colonies. The devastation wreaked on population development during the dreadful non summer of 1985 was the reading of the first lesson in pollen importance. The second lesson was read during the prolonged adverse spring weather in 1986 where colonies were still dying and dwindling up until early June due to pollen shortage curtailing and inhibiting brood rearing. The third lesson, in spring this year, should have driven the pollen point completely home. Colonies were struggling to maintain population levels and the normal late March/early April population stabilisation and gradual increase, where natural wastage of the old field bees gradually became balanced by the new bees emerging in increasing numbers from the eggs laid in late February/ early March, did not materialise. The greatest population losses and dwindling occurred primarily in colonies which had not had access to the late nectar and pollen from heather and ivy in September, '86. The poor weather in July and early August inhibited much pollen gathering and colonies subjected to these conditions went into the dormant period, low in pollen again! Population stabilisation occurred around mid - April, but population increase did not occur at that time. To quote Lewis Carroll's Queen of Hearts in "Alice in Wonderland", the honey bee populations were "running very fast just to stand still." The emerging new bees were just about keeping pace in numbers with the wastage of the almost senile field bees. The most critical spring day in the beekeeping calendar in most of Scotland occurred on April 12th, '87, almost at the tail end of the massive "catkin pollen deluge" period. This was the first day where the bees could really forage well in windless, sunny and warm conditions. The pollen intake of each colony was a sight to gladden the beekeeper's heart. Hundreds of yellow "clusters" were witnessed "scrumming" for access to the hive with their precious burdens of life giving protein. Real colony growth only began around May 2nd/3rd as the masses of brood reared on that, from 12th March '87 continuous avalanche of pollen which lasted until about 30th April '87. The 21 day development period from eggs to emerged adult bee was very discernible during this period. And it was the sensible beekeeper who supplied his bees with adequate amounts of sugar syrup at this time. Up until 12th April '87, very little sugar syrup was taken down by the bees. After 12th April '87 the beekeeper of sharp hearing could almost hear the "vortex effect" of bath water "going down the plug hole" from the stronger hives as the bees took the sugar syrup down! So, pollen availability in the hive is a major critical factor. Last month (September 1987) a procedure was mooted, tongue-in-cheek, to secure adequate pollen for the spring build up need of our colonies. For the beekeeper who does not relish trapping pollen but who wishes to have some control over the pollen content of his colonies there is a relatively simple and very effective method of doing this.

Most beekeepers working their colonies during the traditional nectar flows in their area or out apiary area will have noticed that the combs immediately fringing the brood nest are normally well provisioned with pollen usually packed dry into the storage cells. If the beekeeper removes one of these pollen combs from each of his colonies on a weekly cycle he will quickly accumulate a quantity of pollen in the comb suitable for feeding back to the bees at the correct time in the following spring. To store this pollen safely and guard against spoilage, all the beekeeper has to do is take a lesson from the bees and pickle the pollen under a layer of honey or even warmed refiners partially inverted syrup the Tate and Lyle variety from a tin (not sugar syrup!). Each comb so treated, ensuring that no pollen is exposed to the atmosphere should then be placed in a suitable polythene bag and stored vertically in a safe, dry location until next spring, to be inserted at the edges of the cluster in early March the following year, using little or no smoke.

A cool evening is the ideal time for insertion since the bees will have all night to remove any spillage of excess honey or inverted syrup thus inhibiting the possibility of robbing starting. Four such combs given in two cycles, two combs in early March and two combs in late March/early April will go a long way to eliminating the problems which have recurred in the recent past seasons and which resulted in either population crash or inability of colony to increase population. For the 'disease conscious' beekeeper the marking of hives and removed frames of pollen will ensure that the same frames are returned to the colonies from which they were removed. By being able to donate pollen and sugar syrup, the two nutritional components of the honey bee, the beekeeper has it within his compass to really manage his colonies optionally. As a matter of interest, research done in Russia by S. A. Rosov has demonstrated that 1 kg (2.2 lbs) of pollen will rear 8,500 new bees. Moreover, in the same experiments Rosov proved that 1 kg. of honey would rear 3,500 new bees. His research was carried out at the Institute of Bee Culture in Moscow, not quite a "Mediterranean environment"! Rosov's work and results are relevant for the Scottish climate and indicate the potential of pollen feeding to bees in early spring.

THE MONTHLY ROUND - November/December

The hardest part of the winter is yet to be endured by the bees and beekeeper alike. For the beekeeper who does not use the $1/4"$ high full width flight entrance, all mouse guards should already have been fitted to the hives. If your bees are hived in double brood chambers or even a deep and a shallow and both boxes are chock full of stores, then the remainder of the winter should hold no terrors for such colonies, provided that the queen has not seen too many summers and there are plenty of young bees in the hive population. All the tools and equipment should now be clean and well preserved for the next active season. The extractor should have been washed out, using cold water (don't use very hot water or this will melt the wax residues and coat the extractor surfaces, inviting fungus formation) dried and then lightly coated with vaseline or vegetable oil to inhibit corrosion. The honey press should also be washed in cold water into which liberal amounts of bicarbonate of soda have been dissolved. By doing this fungal growth will be greatly hindered. Queen excluders are easier to clean in the colder weather and by careful light scraping, taking care not to damage any of the slots, brace comb and propolis will break off effortlessly. This is the month when I enter my seasonal diatribe about placing 1 kg bags of granulated sugar directly on the frame tops of the combs on which the bees are clustering. Four to six 1 kg bags of sugar, previously dunked in a bucket of cold water for about a count of ten, to dampen the paper of the bag, then placed as a canopy over the cluster will give the bees supplementary stores which will crystalize rock solid and into which the bees will chew or lick over the winter period. Believe it or believe it not, this canopy of "K" rations is as effective as a full super of honey and removes much of the hazard of single brood box wintering. The need for the sugar bag canopy in double brood box overwintering is not quite so pressing. If this canopy is given around mid-December, by the time the bees have breached the paper of the bag around early January, the sugar will be rock hard. The beekeeper should now be habitually placing the insert of screened paper on the floorboard as a varroa check. If this is done nation-wide it will give us all the reassurance that we haven't got varroa epidemically in any of our hives when the spring analysis of the insert is carried out by D.A.F.S.* in Edinburgh. I use the term epidemically, because by the time varroa mites show on the insert, if left to nature the colony in question will already have had varroasis for at least two

* Editor note - D.A.F.S. no longer exists. It is now the Science and Advice for Scottish Agriculture (SASA) Roddinglaw Road, Edinburgh, EH12 9FJ that floorboard scrappings should be sent to clearly marked ' Bee disease'.

years and will be on the verge of a varroa mite population explosion which will severely debilitate and ultimately kill the colony if it is not treated. Do you know how to diagnose and treat varroasis? Do not lose sight of the fact that we in Britain will not escape varroa in the long term. We are at present engaged in a "Phoney War" with it. Get to know your enemy so that when you find it in your colonies there will be no panic, only dismay, but backed by the ability to set about the parasite without delay. Much work has been done on combatting varroasis worldwide and the situation is not now so desperate — so long as the correct measures, biological as well as chemotherapeutic, are taken in time. If varroa is not detected and thus not treated by the fourth year of infestation, the colony is virtually beyond saving and has probably infected all the other colonies in the apiary and immediate vicinity.

Now is the time to make your choice of beekeeping reading material. We can never know enough about our charges, the more we read about them and weigh against what we have observed or learned from practical experience, the more proficient we can become at working our bees and understanding their needs and natural history — to the obvious benefit of bees and beekeeper. Some of the recently published beekeeping books are well worth a read. A treatise on pollen identification with comprehensive coverage of the major angiosperms, listing the important plant families and relevant species, by Rex Sawyer, titled "Pollen Identification for Beekeepers" will absorb the reader for hours. The book will also give an insight into the relationship of various nectar bearing plants — for instance how many beekeepers know that the sweet chestnut (*Castanea sativa*) is related to the oak and beech, rather than to the horse chestnut (*Aesculus hippocastanum*) or that the bean and pea belong to the same family as the clovers (Papilionaceae)? The photographic representation of the various pollen grains is excellent. The International Bee Research Association, now based at Cardiff University, the clearing house for all beekeeping scientific and literary developments worldwide, publishes information of exceptional interest to beekeepers in general. The magazine "Bee World" is published quarterly and is crammed full of very readable articles. Contributions include Dr L. Bailey's latest work on honeybee viruses; exhaustive work on the history of the development of artificial insemination of queens. Another read of "The Hive and The Honeybee" or Bob Couston's "Practical Beekeeping" would do no harm either! The awareness of the incidence of endemic disease in our honeybee colonies is also a very important factor in our beekeeping. Only by reading the relevant books and by questioning our peers can we hope to begin to understand the problems involved. In my opinion, a really sound knowledge of the natural history of the honeybee is of much greater value to the beekeeper than the memorising of a multitude of management devices which are utilised without a full understanding of their effect on the bees. By acquiring an in depth knowledge of honeybee life cycle, coupled to the habitual use of simple arithmetic, beekeeping can become a real stress-free pleasure. The application of the arithmetic relates to the reading of the fixed cycles in the activity of the hive viz: the metamorphosis of the different castes; the cycle of egg laid to foraging bee; the period required for sexually mature queen and drone; the time from egg laid to sealed queen cell — the list is endless. By applying the discussed time scales to practical beekeeping management, the beekeeper can not only stay with colony development, but also anticipate events and plan his management accordingly. Bee informed! The question of caste differentiation between queen/worker has been simmering away steadily for many years. Many theories have been postulated and discarded. The quality of nutrition is undoubtedly a factor, as is warmth and adequate nursing by the bees. Some years ago a scientist in America, M. Haydak, postulated that water could be critical. Not much notice was taken of his idea. Recent work with premature babies caused me to look at Haydak's work again this summer. The findings in the case of premature babies, born before a particular critical age the lack of oxygen caused brain damage due to the respiratory system being under developed. The medics decided to make up this oxygen deficiency by supplying the baby with a controlled oxygen-enriched environment to the horror of the doctors involved, although the children did not suffer brain damage, some babies became blind. The problem was closely scrutinised and it was discovered that the percentage level of supplementary oxygen held the key — too much oxygen and blindness occurred; too little and brain damage resulted. The dividing line between the oxygen criticals was discovered to be extremely fine. If the orientation of the queen cell is compared with that of the worker cell, the difference is quite remarkable. The worker cell is orientated at a particular angle upwards. The queen cell is built pointing downwards or angled downwards relative to the midrib of the comb. The drone cell is angled upwards and is of course much larger than the worker cell to accommodate the larger insect. The queen could equally well be reared in such a cell, elongated to suit her greater length. It isn't. Why? Research work in swarm control done by inverting brood boxes bodily was reported in a recent American Bee Journal. Inverting the brood boxes caused the queens in the open cells to die. This result stimulated a controlled experiment whereby batches of queen cells were inverted day by day up to the sixteenth day which is of course the day the queen emerges. The most significant result of this experiment was that in each case where the queen cells were upturned while still unsealed, all the larvae died. However, after the sealing of the cell, inversion had absolutely no bearing on the wellbeing of the developing queen and virtually all of the queens emerged successfully. Why? If one takes a test tube and immerses it in water in a vertical position, open end up, on submerging it will of course fill with water. Now take the same test tube and invert it open end down and immerse it. The tube will not flood. The trapped air prevents the water from rising beyond a particular level.

THE MONTHLY ROUND - January

As a West of Scotland beekeeper, I have no great love of the month of January. It can be the hardest month of the winter for the bees. The hardest frosts always seem to occur in this month and fall prolonged. A spell of weather with below freezing temperatures at night and blue skies during the day when the sun can warm the hives will not harm a strong, well-provisioned healthy colony. However, a period of hard frost, where the frost remains also during the day, if prolonged over more than a week, can wreak havoc on even quite strong colonies where stores are scant in way of the cluster, despite the fact of full store combs outside the bounds of the cluster. To a colony held fast on empty combs during a prolonged spell of hard frost, these just-out-of-reach combs may as well be on the moon! They are inaccessible. Or are they? The beekeeper who understands the limitations of overwintering bees in Britain, Scotland in particular, realises the importance of bridges for the bees to give access to otherwise remote stores.

The feeding of slabs of candy should have a greater purpose than merely to contact a small fraction of the area of the brood chamber frame tops. The candy should be given with the knowledge that it may encourage the bees to move reasonably freely across the frame tops in the cavities chewed out of the candy mass. This facility to fetch and carry stores from otherwise remote combs on the milder days could mean the difference between survival and a horrible, untimely demise for the colony. In my own case, I long ago gave up the chore of making candy for my bees, once past 15 colonies candy making can become an almost, full-time occupation. Managing 60 - 80 hives can tip the scales and make the almost full-time become full-time.

As anyone who has been tempted to read this column for any length of time will now well know, sugar bags are my winter supplementary stores device. By arranging a canopy of 4 - 6 x 1 kg bags of sugar with the bags touching each other a solid crystalline ceiling eventually forms over the winter cluster, performing the same life-giving function for the bees as candy. The bees use the crystal sugar as required and can move around at will in the 'honeycomb' of passages which they chew out of the solid. Even the empty shell of the paper bags affords some shelter from the hostile atmosphere of the hive outside the cluster in winter. The bags can quickly and easily be checked for content during the dormant period and with experience, the beekeeper will learn just how damp to make the donated bags relative to the particular point in time of application.

Ideally, the bags should be given to the colonies around late December. The bags merely being 'dunked' into a bucket of cold water to dampen the paper. By the time the bees start to use the sugar it will be rock solid and eminently edible and stable. Sugar bags given later in the winter should be punctured to allow water to actually dampen the sugar. If a punctured bag is placed in the afore mentioned bucket and observed, as it starts to sink, it should be lifted out and 'plunked' into position on the hive frame tops in way of the cluster. The job is best done right at the hive being serviced — a few seconds delay could have the bag collapse in your hands yugh!! You have been warned!

From all the written criticism I received on propounding the 'Sugar Bag Trick' I began to feel as if I was the only beekeeper capable of carrying out the operation successfully. However, on meeting beekeepers face to face (the silent majority!!) I have been greatly encouraged to hear them relate the ease with which the operation has been successfully carried out by them! A word of warning however — bags of sugar are no use to bees dying of hunger — such bees need honey and fast. Candy has limited effect also in such a case. If your candy or bags of sugar are on the hives by late December at the latest and your bees have honey around them at that time then they are in no danger of 'isolation starvation' in the later winter period.

We are now aware of the problems caused to bees in a poor spring period where the early pollen sources like crocus, aconite and catkins fail. Sugar syrup, candy nor sugar bags will avail a colony suffering from chronic pollen shortage. In poor springs it is advisable to feed pollen/ pollen substitute as a complement to sugar.

Mice are a great problem at this time, and the use of some form of mouse / shrew defence is critically important. For the busy beekeeper the full width entrance not more than 5/16 " high is the ideal - a stiff twig or wire swept across the entrance around late January and at intervals thereafter will ensure a clear entrance for the bees. For the indulgent hobbyist there are a million and one good to indifferent anti-mouse devices (ouch! Who threw that!) which may be used.

Any stored, surplus brood combs which have been out of service for more than one season should already have been fumigated by 80% acetic acid on pads. If they haven't — be it on your own head if the mass of tangled debris in the frames in the spring horrifies you. Wax moth is a greater danger than most beekeepers like to think, the last three poor seasons have demonstrated to many — who laid up spare brood comb after colony losses — just how expensive wax moth damage can be.

With a bit of luck by the end of this month, the first snowdrops will be blooming — always a welcome sight for the beekeeper.

According to statistics it is a virtual impossibility in beekeeping to get three good seasons in a row — let us hope that it is also highly improbable that we will get a fourth poor to dreadful consecutive season, this coming year.

Using the Moir Library, from which you can borrow a vast range of books on apiculture, the beekeeper in winter should never be bored, and if the reading material is absorbed into the beekeeper's general experience a better and deeper insight into the craft of keeping bees will be the result.

THE MONTHLY ROUND - February

The bees should already have indicated their survival of the darkest part of the winter by early February — the tiny piles of capping dust at the hive entrance tells the story that the colony within is stirring and brood rearing should-be underway on a small scale by this time. With luck the crocus will bloom early this year — always a good sign! The beekeepers anticipation of a good year ahead rises with the increasingly animated behaviour of the bees with each day where the temperature lifts above 10°C.

The early pollen intake should be complemented by feeding sugar syrup, thick (1 kg: 0.5 litres water) for a colony light on stores, then (1 kg sugar: 1^{1}/$_{2}$ litres water) for a colony with adequate stores. The water in the feed is as important to the bees as the sugar in many cases, because of the need to metabolise pollen for brood feeding. Beekeepers working for O.S.R. or sycamore are well advised to feed syrup from mid-March — to encourage build-up to effective honey gathering strength for mid-May. The beekeeper in the later areas perhaps working mainly for clover, lime and heather should merely ensure that there is sufficient candy or crystal sugar still in the hive to forestall any possibility of the colony running short of stores. In the light of the past pretty devastating seasons however I feel that by stimulating colonies regardless of whether the area is late or early, colonies will be encouraged to develop as rapidly as possible as early as possible. These colonies may then be split to supply the necessary 'increase' in colony numbers to begin the long hard road to pre 1985 beekeeping levels. It is 'guesstimated' that a mere 20% of colony numbers are alive today relative to the numbers of colonies alive in 1984.

The activity at the hive entrance on the 'flyable' days during this month can tell the observant knowledgeable (i.e. well read!!) beekeeper volumes. Large pollen pellets in the corbiculae (pollen baskets) of returning foragers confirm the vigour of the queen of that particular colony. Any colony noted 'idling' when its neighbours are hard at work should be noted. If the laggardness persists over a period of a week relative to the industry of its companions the colony should be examined internally for 'queen rightness' at the first opportunity in early March — on any bright day — using the minimum of smoke.

It is as well this month to start taking stock and noting any shortages in equipment and other needs like jars, labels, frames, foundation, queen excluders. This is the time to make up new frames and render the wax which has been hopefully soaking away in tubs of rain water over the winter. Ingots of pure wax laid aside, for use with the wax mould which every association worth its salt will have, can soon be turned into foundation, saving the beekeeper much expense. Excluders should all be scraped clean by this time. Do it now!

The fumigation of spare brood comb should have also been done. If it hasn't, do it now, or you could really regret dragging your feet. One bad experience with wax moth will show you why!

Spare floors should now be being prepared for the annual spring clean, scraped and painted with the right cuprinol and well aired prior to fitting will give the bees a flying hygienic start to the year. Think, Varroa. If we've got it and we detect it quickly enough we can treat it and save ourselves from the pending "Apicultural Holocaust". Don't attempt to make up foundation into frames yet, unless you have heated premises — or the wife lets you use the living room. Foundation is very brittle in cold weather and will split and crack with the impact of the frame nails being driven into the top bar wedge! Care should also be exercised when moving supers with undrawn foundation in them, a bump can result in hours of work showering out of the frames.

Comb which has been used for brood rearing is much tougher and will stand a bit of handling in cold weather but even this comb has its limits because the wax anchor under the wedge is subject to the same fragility as new foundation

— this I have found out through bitter experience.

It is a good idea to paint the hive stool a bright colour (not green!) and this time of year is as good as any to do it!

Measures to be taken for colonies which fall queenless, have a drone layer or laying workers will be described next month and these will range from the 'adventurers' to the standard practice. Colonies which are strong but queenless in late March can be saved intact if certain procedures are carried out — but more about that later.

THE MONTHLY ROUND - March

Spring is advancing, the real danger time for bees is imminent, 'depleted' stores can soon become 'exhausted' stores as the brood to feed increases in number. This stores depletion is a constant threat for the bees in 'let alone beekeeping', especially in a spring weather situation where access to early crocuses has been available to the bees. A heavy deluge of crocus pollen can encourage the queen to increase her laying rate beyond the capacity of the remaining stores, and where a sudden collapse of the weather confines the bees to particular combs, they can die in the midst of plenty due to 'isolation starvation' by consuming the honey immediately accessible to the cluster within its environs. Any honey in combs outside the fringes of the brood nest might as well be on the moon as far as the bees are concerned, because if the weather turns really cold after brood rearing has become established and the bees are unable to break from the cluster they will not be able to reach the stores on the outer combs.

Nothing in beekeeping management should be done by rote. Each season is different! The trick is to either consciously try to anticipate events or at least respond to the weather patterns as they happen. Always bearing in mind that the weather is the most important factor and like natural queen mating out with the control of the beekeeper and homo sapiens in general — thank goodness!!

There are years when the spring rushes in like a breath from Paradise and the bees gain easy access to snowdrop, crocus, aconite and willow and pound surplus pollen by the proverbial ton into the storage cells. If the beekeeper is located in an early area where his/her bees have access to gean/sycamore and, or O.S.R. in such a spring as just described it will pay handsomely to anticipate that May will also be fine, and by feeding sugar syrup from early March to complement the masses of pollen coming in, colonies can be husbanded to quite formidable honey gathering strength to coincide with the early major nectar sources mentioned.

However, the recent spring norm, has been far from idyllic and the main problem has not been that of synchronising peak population with peak nectar source bloom period — but rather maintaining bee life in the sometimes forlorn hope that the 'killer' weather pattern would change. In such a spring, from bitter experience, the beekeeper must turn his mind to pollen or pollen substitute feeding to complement the candy or crystal sugar which should still be available to the bees. In many of my own hives crystal sugar given as a canopy of 4-6 bags in December 1984 was still providing "life support" up to early June 1985.

Many beekeepers have difficulty in either knowing when to start feeding sugar syrup, or change over from candy/crystal sugar to syrup. It is now a well established fact that when bees begin to rear brood in earnest they require water. Brood rearing begins in earnest in spring when the bees begin to bring in abundant pollen. Thus, the time to begin feeding sugar syrup is soon after the predominant number of foragers returning to the hive are observed to be carrying pollen.

By hefting his/her hives from each side in turn the beekeeper can get a fair idea of the level of stores still remaining from the dormant period. Where a hive is reassuringly heavy, but not merely due to its construction, the sugar feed can be given as a palatable drink perhaps in the proportion 1 lb. sugar dissolved in 1 pint of water. Where the hive seems to be rather light feed in proportion 1 lb sugar dissolved in water to make 1 pint of syrup. Where any doubt exists in the beekeeper's mind regarding the stores situation feed the thicker syrup.

Varroa inserts, where fitted should now be removed and checked.

It is good beekeeping practice early in March to flick across the entrance with a stiff wire or twig, to ensure that the entrance has not been clogged up with debris and dead bees. This procedure is especially important in a spring which follows an autumn/early winter when the weather during late September/October was unfavourable for bee flight, in such a case the bulk of the older bees which would have flown from the hives never to return, will still be in the colonies entering winter and die in the hive instead of dying in the 'field' as happens when the weather is flyable during late September/October.

Close external observation of hives in the early spring, late February/March during favourable weather can tell the observant beekeeper much about the condition of the bees. The most reassuring sight being that of each colony during crocus or willow blossoming time where the bees are bustling into the hives loaded down with pollen pellets of enormous size. In such a colony the queen is alive, well and laying her heart out. The other end of the scale is a colony in which the bees are aimlessly wandering around the hive entrance when other colonies in the same apiary are carrying abundant pollen. Such a hive should be noted and observed closely, just in case it is a classic late developer but if no pollen activity comparable with its neighbours occurs in the course of a week, there is nothing to lose by having a quick internal look to check for brood or eggs on the first bright days in mid March. If the colony is queenless and weak it may be as well to destroy the bees which in all probability could be diseased, they will most certainly be old and of little use as foragers. On the other hand if a listless colony is opened up and found to be queenless (this is where marked queens are worth their weight in platinum!) but still quite strong and still having sealed brood, indicating that the queen has only recently "keeled over", the beekeeper then has at least two possibilities for 'salvaging' the colony: i) Uniting or, ii) Fostering.

If option i) using the classic 'Newspaper Method' is followed it is imperative that a disease check is made. A sample of adult bees should be checked - using your local association microscopes check the bees yourself for Acarine and Nosema. To be doubly sure any brood in brood cells which look suspicious, i.e. watery/sunken cappings or perforated cappings should be checked for A.F.B. and E.F.B. This check is a more specialised process and it is recommended that where there is doubt cut the offending brood out on a 2" X 2" section and pack it off to SASA. Do not use a polythene wrapper, brown paper is ideal.

If option ii) is followed — recommended for the slightly more experienced than 'beginner' beekeeper who has more than two hives. Check hive queenless. Subdue best hive in apiary. Enter hive. Remove a frame of open brood from this hive. Split cluster in queenless hive. Insert frame of open brood into queenless colony (having first shaken all adhering bees from frame before removing it from its colony!). Insert replacement frame (preferably containing stores!) into donor hive. Close hive up. Close recipient hive up. The donated open brood will help maintain colony morale in the queenless box and also ultimately replenish some of the wastage as the older bees die off If this procedure is carried out on a two week cycle, in a favourable spring, this colony will not only have requeened itself by mid-May but will be in a condition where it can develop normally and hopefully produce a crop of honey in a reasonable summer season. A disease check in option ii) will also not go wrong!

THE MONTHLY ROUND - April

Let us hope that this is the year in which the pattern of late springs is broken. Every spring has been poor since 1984: yes, even spring 1984 was poor to say the least. It was April 23nd, Easter Sunday, when the rain stopped and the sun came out to shine in an almost unbroken run right through to late August.

Every season has its problems for the bees, and it is difficult to try to imagine which of the seasons is the most critical for them. I myself am of the mind that, although winter is a time of great vulnerability for bees with regard to their defencelessness against predators and exposure to the elements due to natural or unnatural catastrophe, the late spring period is the most trying time for bees when the weather is unfavourable for bee flight, especially when traditional beekeeping practices are followed, i.e. "Let Alone" or the feeding of merely sugar syrup or candy.

Recent past bitter experience has demonstrated the weakness in such procedures. In seasons where the spring weather is flyable, sugar syrup on its own is fine. The queen's egg-laying rate accelerates with the abundant pollen coming into the hive, and the bees can utilise the water and the sugar to supplement their natural stores as demands on the hive reserves increase to meet brood-rearing needs. But in continuous adverse weather the bees not only require to be fed, they require to be fed a "balanced diet of pollen as well as sugar in its various forms, viz. candy, solid crystals, sugar syrup, etc.

Although times can be difficult for bees in summer and autumn, sugar feeding can maintain bee life until the weather changes for the better. In winter so long as the beekeeper has done his/ her homework and fed copiously where needed, and also rendered the hive mouse, weather and "earthquake" proof, there is no great threat to bee life. Even a hive turned over by whatever agency is not necessarily doomed if the beekeeper resets it back to rights within a reasonable time scale. However, in poor spring weather of the order of those in recent years, normal "progressive" practice will not pull the bees through. The beekeeper now needs to re-learn the critical importance of combined pollen/ pollen substitute and sugar feeding, or incur massive losses among his colonies as has been the norm since 1985. Who knows, 1988 may see us back to the halcyon days when statements like: "There is an abundance of pollen available to our bees in early spring in Scotland", or "The bees always get out to the crocus and willow for the early pollen" will again have a ring of truth about them.

Let's assume that the spring be a "normal" one. What should we expect? This is the time of rebirth in the colony. The brood which will constitute the early summer foragers on the sycamore and O.S.R. should be being reared during early to mid-April, bearing in mind that a time lapse of six weeks occurs from egg lay to mature forager. With the advent of catkin, dandelion, gean and flowering currant to gladden the heart of the beekeeper and excite the bees to peak activity, the mass of pollen going into the hive is indeed a reassuring sight. Drone rearing will be well underway especially if late March was fine, allowing the bees to "access" the last of the crocus bloom. The beekeeper can take it as read that when he finds drone brood in his hives in early spring, his bees are "feeling good" and enjoying prosperity. The advent of drone rearing also has its slightly ominous side for the beekeeper, because this is the colony's first tentative step toward swarming. However, the progressive beekeeper need have no fears about losing swarms if he/she has been reading *The Scottish Beekeeper* Magazine for more than a year.

It is now definitely time to be thinking about the preparation for supering the stronger colonies, especially where O.S.R. and sycamore honeys are worked for. By giving extra room early, the bees have the chance to spring-clean and explore the additional space before the hoped-for "flood" of new nectar begins to stream into the hive in early May. A useful device, for the not too harassed beekeeper, to get the bees into the super in quick time, is to put the super in place without using a queen excluder. About a week later, smoke the bees down into the brood box from above — the queen if up in the super will be among the first to dash for cover "below."

Remove the super, put the queen excluder in place and replace the super. The bees will soon rise again to re-occupy the honey "room'."

Have you anticipated your equipment needs for the year? Labels, jars, frames, foundation; stocks of these items should be checked and ordered in good time. There is nothing more frustrating than being short of equipment when it is urgently needed on the instant. That is usually the time when almost every other beekeeper in the land is short of the same items, and the dealers are sold out. We are due a 1975 summer; this could be the year, so don't get caught short as some were in 1984, when by late July, almost every spare piece of equipment capable of holding bees and honey had been thrown onto the hives and filled by the bees at an almost embarrassing rate. One beekeeper I know very well was so hard pressed for equipment in his preparations for migrating to the heather that he moved his bees to the moors early, so that any heather honey he got would not be "contaminated" by the late July flower honey deluge of that year. The heather in 1984 was not particularly "write-home about-able": the moors were almost bone dry. Moral — first get your honey: never run away from a nectar flow, but never!! The summer of 1985 brought beekeepers back to reality with a jolt, generating visions of the truth of the old adage about the Lord Mayor's procession.

The end of the third week of this month is the time to begin thinking about starting queen rearing in order to be in the position to requeen the strongest hives with the current year's queens. This will obviate the nail-biting and anxiety associated with the traditional methods of swarm control in the coming summer — provided of course the spring has been favourable or the beekeeper has been feeding pollen and sugar during March.

THE MONTHLY ROUND - May

This is the month when the aggravated active season management begins for the 'traditional' beekeeper. The progressive traditionalist will be committed to scrabbling through the hives on a 'nine-day' inspection cycle to check for swarm signs until the end of July, even into August in particular years. The 'traditional' traditionalist beekeeper will merely await the issue of swarms and hopefully be on hand to take them. The 'nine-day' system is an excellent management ploy, it is also extremely labour-intensive. Even using this method the beekeeper on seeing queen cells or eggs in queen cups invariably behaves as if the finding was totally unexpected and panics — because the event although subconsciously expected had never been consciously anticipated by forward planning, and the beekeeper has no clear idea what the dickens to do! Cutting down the queen cells is the easiest procedure *and* the most risky — especially when the queen is not removed at the same time. This procedure only works well if the weather becomes unfavourable and unsettled coincident with the cell cutting operation. If the weather remains warm, windless and dry after the cells are cut down the hapless beekeeper has lost a swarm four days after the cells were destroyed and a good five days before the next nine-day inspection falls due.

The device of clipping the queen's wing(s) is a commendable ploy for the progressive traditionalist, especially when out-apiaries are maintained. The beekeeper has of course to know the limitations of the 'clipped queen' method. The system only buys time, it does not in itself prevent swarming. Where the beekeeper misses the swarm issue critical date the clipped queen method will prevent loss of the prime swarm emerging with the clipped queen. The queen falls and invariably becomes lost. The cluster on forming senses the absence of the queen and ultimately returns to the hive — to await the first virgin emerging. If the beekeeper does not return to 'manage' the stock in question, the bees leaving with this virgin comprise the original frustrated prime swarm as well as the bees which 'in nature' would have formed the first cast, and this constitutes a lot of bees. The effectiveness of a hive to gather a surplus of honey after losing such a swarm is almost nil.

There are as many ways of trying to stop bees from swarming as the skinning of the proverbial cat, but a management device common to most of the more effective systems is the marking of the queen, so that she can be found on 'demand'. During the height of the swarming season the beekeeper trying to find an unmarked (especially black) queen among clumps of greasy, listless bees has a most unenviable and almost impossible task before him/her. A glint of colour from the depth of a bee-festooned frame can quickly draw the beekeeper's eye when seeking an illusive queen in a populous 'swarm intent' colony.

I use, and have depended wholly on, 'Relay Re-queening' to control swarming for a number of years now, having tried virtually every popular method of swarm control and found them lacking. The 'Relay Re-queen System' allows a dramatic reduction in labour input in management and virtually stress-free beekeeping for the active season after late May/early June — depending on the season.

By rearing new queens early in the season and replacing the over-wintered queens of the strong colonies with these new queens as they become available, the re-queened hives will be inhibited from swarming in that season — in my experience! By mid-June in most years the 'Relay Re-queen System' allows relaxed, stress-free management of quite large numbers of colonies. The basis of the Relay Re-queen system depends on an amalgam of a method used at Buckfast Abbey and of a major discovery by Dr. Colin Butler. At Buckfast Abbey all colonies are re-queened by mid-March every year using young over-wintered queens; swarm control still has to be exercised. The discovery by Dr. Butler was that queen bees secrete pheromones — queen substance — which maintain colony equilibrium; these pheromones are 9-oxodec-2-trans enoic acid and 9-hydroxydec-2-trans enoic acid. It is a well-documented fact that the absence of queen substance results in the colony building queen cells and the worker ovaries developing during the active season.

My logic anticipates that at some point between an abundant, sufficient quantity of queen substance and zero queen substance a 'critical' quantity level is reached which causes an imbalance in the colony and triggers the 'swarm mode'. Continuous strenuous expenditure of energy, mental or physical, results in a fall-off of performance. In *homo sapiens* athletes get 'stale', students face insurmountable mental blocks; an 'out of condition' phase occurs.

The queen bee, in laying increasing numbers of eggs as the season develops, not only lays almost her own weight in eggs each day but must walk vast distances in concentric circles seeking empty cells in which to lay these eggs — pretty strenuous stuff! She must ultimately go out of condition and in so doing fall off in ability to produce the necessary amounts of queen substances — triggering queen cell building and laying worker ovary development and ultimately swarming! The consistent swarm-free results accruing from the practical application of the foregoing logic seems to confirm the idea, because the new vigorous current-year queen is capable of producing the required queen substance for colony equilibrium well in excess of the lower yet-to-be-established critical levels.

There is an interesting aside to my postulation about critical levels of queen substance:

There are many species, subspecies and races of bees. The *Apis* species alone has many mystifying, variable and wonderful qualities: *viz. Apis dorsata,* the giant honeybee which builds a single comb 4-5 inches thick and up to 4 feet in radius; *Apis florea,* the dwarf honeybee which builds a single comb barely the size of a man's hand, *Apis indica (cerana)* the original host of the dread *Varroa jacobsoni.* The continent of Africa offers even more delights to the interested beekeeper from the Tellian bee in the north to the notorious *Apis. m. adansonii* in the western regions. However, right at the tip of South Africa there exists the unique Cape bee. *Apis m. capensis.* It is she that I wish to focus on because experimental and purely practical beekeeping observations have demonstrated that *A. m. capensis* is a very low-incidence swarmer compared to the other *Apis* species and races. Analysis of the relative amounts of queen substance has shown that *A. m. capensis* queens secrete far greater amounts than other species and races. The relative quantities are *A.m.mellifera* — 37.25%, *A.m.capensis* — 84.83%. A marvellous phenomenon *of A.m.capensis* is that the race is parthenogenic in a totally different way *from A.m.mellifera.* In *A. m. mellifera,* where laying workers are present in a queenless colony the eggs laid produce drones; in *A.m.capensis* these same laying workers in a queenless colony produce workers and also fertile queens. The great disadvantage of *A.m.capensis* is that when the colony becomes queenless 'civil war' breaks out in the colony and the older bees attack and kill the younger bees. No one theory seems to be able to explain this fatal phenomenon. After a period of around eight days the colony stabilises; this coincides with pseudo-queens (laying workers) which can produce as much queen substance as true queens, beginning to lay fertile eggs. But not before about one third of the colony population, mostly young bees, has been killed. It would appear that it is rather dangerous to transport exotic bees between countries — the devil you know...! as always. I will stick my neck out as an interested observer of the suicidal behaviour of *A.m.capensis* in the newly queenless state. From the analysis of the relative queen substance levels which shows a massive level of production in *A.m.capensis*, it appears obvious that there is a demand for a high level of queen substance in the colony. When the queenless condition occurs the queen substance quantity available diminishes rapidly, resulting in an enormous 'thirst' in the older bees. The younger bees which are the main victims of the attacks produce varying quantities of queen substance and are killed as a result of the rush of parched of 'queen substance' bees to get access to the meagre and insufficient supplies. It is significant that the 'killing' ceases at the time the pseudo-queens begin to lay. These pseudo-queens have been shown by laboratory experiments to secrete as much queen substance as true queens.

Sources: TRIBE, G. D. Plant Protection Research Institute, Rosebank; S.A. Bee Journal (1981) 53 (5) 10-12.

CREWE, R. M. — Dept. of Zoology, University of Witwatersrand, S.A. Bee Journal (1984) 56 (16) 20

THE MONTHLY ROUND - June

June, although regarded as a 'high summer' month can be a most difficult month for bees in particular areas, especially in less than ideal weather conditions. When June is 'blazing', the proverbial 'June gap' is bridged by the easy access the bees have to the many secondary nectar sources available in this month. However, if the weather be windy, wet or cool the bees are denied access to these minor nectar sources and can soon find themselves in dire straits — unless the beekeeper is aware of the pitfalls of beekeeping in June, which lie between the abundant flows from sycamore and oil seed rape in May and early June, and the hopefully copious flow from the lime, privet and willowherb from July onward. The 'raspberry and clover' beekeeper has normally fewer worries than ordinary beekeeping mortals in June. I can hear you say it — 'Cripes, he's at it again'. However, undaunted I intend to advocate a device especially useful to the queen-rearing beekeeper using nucleus stocks. Around the end of the first week in June thoroughly soak a 1 kg bag of sugar (holes in the bag!) until the contents are quite damp (watch the bag does not collapse!) and place this bag or bags on the frame tops of the brood frames on which the bees are located. This sugar quickly crystallizes and the bees subsist quite happily on it all through the 'June gap' if necessary. Growing, strong colonies can also benefit from such management. The trick is judging the 'sogginess' of the sugar bag — if the soaking procedure is done alongside the hive being manipulated the work can be carried out effortlessly.

Traditional beekeepers should still be working the 'Nine Day' swarm control system which will be necessary even up to mid-August in some years. By mid-June anyone practising the 'Relay Re-queen' method correctly will be at the fishing or playing golf or bowls with a clear conscience.

For the heather beekeeper June is an important month especially the final two weeks. The brood produced in the last two weeks of June and the first two weeks in July is of critical importance to the hive population at the heather, because using the legendary apicultural arithmetic the beekeeper knows that six weeks elapse from egg to foraging bee. The classic heather bloom period is from August 4 to the end of the first week in September. August 4 minus six weeks is around mid-June. The effective bloom period of the heather is approximately four weeks! During the four-week period mid-June to mid-July at a modest laying rate of 1000 eggs/day, 30,000 new bees for the heather are the potential for a well-managed June stock. By judicious management the beekeeper with five or more hives can really exert his Influence on the harvest potential of his/her bees. Consider hypothetical 'booming' colonies around the end of the sycamore flow. In reasonable May weather these colonies will be prospering well with surplus honey and masses of brood in all stages. Undoubtedly there will be the odd colony of the apiary complement which is not quite 'booming'. By projecting the mind ahead to July and the start of the second phase of nectar flows and anticipating probable population densities at that time, the colonies can be managed in such a way that virtually every hive is at optimum size for the critical time during the July flows. The brood in the colony in mid-May will have no effective role to play in that colony during the sycamore/oil seed rape flows (winter rape!!), thus in recognising this the beekeeper can remove perhaps two/three sealed brood combs without adhering bees at this time and donate these brood combs to less advanced colonies — providing there are sufficient bees in the recipient colony to cover these new combs. The massive population lift which this 'device' can give to a mediocre colony has to be seen to be believed. In so saying though, reasons have to be sought as to why particular stocks are less populous than others. A word of warning, do not immediately think, "I'd better replace this queen with a queen reared from my best stock." By all means replace this suspect queen from a 'best stock' queen, but preferably another quite distant beekeeper's 'best stock' queen, because breeding from the best in a localised situation has dangerous implications for inbreeding. In my opinion, once Varroa has become established in Britain, inbred bees will become a major problem for beekeepers practising blindly 'best stock' selective breeding. Again, in my opinion the role played by the feral colonies around our apiaries has cushioned us against rapid In breeding for many, many years. After Varroa (A.V.) in the third or fourth year after its Incursion into the U.K. we will begin to lose our ferals and with them the benefit of an enormous latent gene pool. The loss of this gene pool will be of critical importance to many beekeepers. Another disadvantage which Varroa will bring is that such management procedures which give the beekeeper good control over populations as described above will not be possible without incurring the great risk of causing a Varroa population explosion in a previously 'clean' colony.

I am convinced now that it has been established that Varroasis is virtually epidemic in America, that Varroasis will be found in British colonies this summer — if we look for it! The American infestation has been present, according to their own experts, for around three years. The Southern States incurred the initial infestations probably from random Brazilian Bee incursion or some stupid breeder bringing queens from South America into the country illegally.

Yes, it is illegal to import bees and queens into the States, but Britain has been 'sucking in' all kinds of dangerous apical material for years and years. It needs no great stretch of the imagination to extrapolate to Britain the problems now being experienced throughout America - due to, by the expert's admission, migratory beekeeping and the traffic in queen bees between South and North America. We have brought in our share of potentially dirty queens from these same American breeders. I hope I am dreadfully wrong — but I feel I am dreadfully right!

THE MONTHLY ROUND - July

This is the-time when the big second major nectar flows are available to the bees. Lime, clover, privet, willowherb and, for the lucky or determined few (who migrate!), bell heather. It pays handsomely to manage colonies for optimum population size in anticipation of these late summer sources. The only real reasons for failure to achieve maximum possible populations in the selected colonies (not all colonies will be at optimum honey-gathering strength, i.e. nucs, either recently established or used for breeding purposes, etc.) are (1) loss of swarms; (2) poor weather; and (3) bad or non-existent forward planning. Items (1) and (3) are both related and under beekeeper control or should be. They were discussed in the June issue. The weather, thank goodness, is something which we as yet have not managed to get our nasty little fingers on! Although certain attempts have been made at affecting Cuba's rainfall pattern, negatively, and on the continent in late January "snow making" machines were employed at some ski resorts, as well as at Calgary for the winter Olympics.

The beekeeper, on recognising that a colony of bees can be "managed" like cattle or sheep or poultry, has "come of age." In putting this recognition to work the beekeeper suddenly realises just how much positive power lies under his/her hand. Despite doing everything almost right however, the beekeeper can still achieve a massive failure — due to the weather either being adverse or becoming adverse at the critical periods. However, if the beekeeper gets it right and the weather complements his/her efforts, the satisfaction of having worked within the limitations of nature and succeeding, beats lying 6 at bowls or hitting a perfect 20 ft putt! Having prepared well, even if the weather fails the beekeeper can say with satisfaction — "at least I went down fighting"! The beekeeper who doesn't prepare even in an ideal weather situation will do as well as he/ she deserves, usually badly! Be a beekeeper, not merely a "bee haver", because after Varroa, there will be few "bee havers" around after year four!

In preparing for the migration to the heather the colonies should be headed by a queen not more than two winters old, have bees right across the brood chamber and at least two drawn "virgin' combs in the centre of the shallow crate. The remaining combs should also ideally be drawn comb.

If foundation only is fitted, it must be well secured in the frame, otherwise the bees on moving into the super to cluster during the journey to the moors will pull the foundation from its anchorage — with disastrous results. A stimulating management ploy is if moving hives fitted primarily with foundation to move without a queen excluder. This gives the bees, especially in a really powerful stock easier access to the extra clustering space they need while travelling. After the bees have had around three good days working the heather the beekeeper could then put the queen excluder in place. This will serve two purposes — (1) a check can be made to see if any foundation in the shallow has collapsed, (2) the bees will be well into drawing the foundation in the super due to the absence of the queen excluder. I used the word *could* deliberately because from long experiences at the heather with and without excluders I have come to the conclusion that in most seasons excluders are not necessary at the moors. The queen, although laying quite well, seems quite happy to remain in the brood box.

If colonies being moved to the moors are prepared even two/three days in advance of the move, this gives the bees the chance to propolise the parts again, making the hive just that little bit more secure. In many years of migration to the heather, I have used all kinds of transport including a Mini- Minor saloon. Triumph Herald 1200 Estate, Ford Cortina Mk II estate; these vehicles are of course closed and can become quite "bee-filled" if hives leak. The best transport is a pick-up type vehicle. I have been using the Mazda 1-ton pick-up type for many years, being now on the third vehicle. This type of vehicle allows a maximum lift of 20 Smiths in one bite and makes for stress-free migration. I will not go so far as to say the shift is "strain-free", because the colonies have to be on-loaded from the summer sites and then off loaded at the heather. Even using a shoulder harness to take the hive weight, a Smith hive in "heather rig" can feel like a similar sized block of lead, especially after the fifteenth hive. I move all my colonies single-handed and will admit that migration is the most labour-intensive period of my beekeeping. Colony management is child's play relative to moving bee colonies in such numbers. The operation is greatly improved for me by the use of a spring-loaded staple gun which is used to secure the hive parts. Complemented by propolis and the pick-up facility, preparations for migrating are relatively painless.

THE MONTHLY ROUND - August

What will the heather bring this year? The unspoken question rings around the "heather beekeepers" mind as preparations are made for the annual pilgrimage to the moors. The important factor is, of course, that beekeepers indeed make the effort to go to the heather. Anybody who likes honey and has savoured the taste of the orange, translucent, thixotropic substance which the bees harvest from the ling will rarely need a second reminder that heather time is here again. As I have often remarked in this column, the ideal transport for shifting bees to pastures new is an open-backed pick-up type of vehicle — a trailer comes a poor second to this kind of transport. With a pick-up moving bees is, I'll not say easy, but less fraught with problems than any kind of closed or partly closed vehicle. I remember a story about a heather move using a horse box, and because the interior is dark relative to the open top of the half door, when the horse box hit a bump and hives split the resulting exodus of bees through the opening was reminiscent of the tail of Halley's Comet — according to eye witnesses.

Some take elaborate measures in heather site preparation and erect firm stands which can be used year after year, others use sheets of corrugated iron cut to size to set the hives on. I myself favour three half bricks, one at the back and one each side of the entrance at the front. Half bricks are much more efficient and cost effective than whole bricks (provided you get them from a rubble dump) because three whole bricks broken in half by dropping them on the edge of a good sized rock will produce two hive stands. After use they can be stowed for the following year in a nearby ditch or along the line of a handy dyke. Bricks have one decided disadvantage though. They crumble — like us! — as they get older and a few replacement bricks are needed each year. Apart from that I think they are ideal.

At most heather sites a fence is essential to keep stock at bay. Cattle stampeding through an unprotected apiary will please neither cow, farmer nor beekeeper, to say nothing about the feelings of the bees. Deer can be a problem as can sheep, using the hive as a scratching post. I used to feel a great need to check the colonies at the heather at the end of their first week to ensure the queen was OK — looking for eggs but not necessarily sighting the queen. However, with the increasing passage of time I find that time passes increasingly fast and I have less time to carry out this procedure. I now make a point of picking a good day around ten days after the move, when the bees will be flying freely, and I check colony activity at the hive entrance merely looking to ensure that each colony has its share of pollen carriers. I learned also long ago not to trundle along in front of the rows of hives — this can cause a massive reduction in foraging activity due, I am certain, to the bees sensing danger and staying at home to repel invaders into the hive.

Observe closely next time you move among your bees at the moor and the phenomenon will become readily apparent. A pair of binoculars from a distance of twenty yards will tell all to the considerate, concerned and efficient beekeeper. Any idling or non-pollen carrying stock can be noted — but not worked on immediately. It is good beekeeping practice to wait until the bulk of flying is over in late afternoon or early evening, before disturbing the colony, and, from the previous statements — the whole apiary!

If a colony is found to be queenless, it can either be united to its neighbour using the newspaper method and the colony reduced to a single brood chamber at the next (late afternoon!) visit, or it can be re-queened, depending on the beekeeper's whim or need.

When I was younger and full of the joys of spring, I used to bring my bees back from the moors at night. I decided some years back that I had had enough of stumbling about in the "stygian" and falling into haggis traps while staggering perhaps 50-60 yards with a (delightfully) heavy hive of bees. I now leisurely remove the colonies in batches during the daylight hours at the end of the nectar flow and enjoy a stress-free migration to and from the moors. The end of the first week in September is usually time enough to start bringing the bees home — for those of you who still feed sugar syrup prior to the winter, the sooner the bees are home and syrup feeding started the better, although in a mild late autumn syrup can be fed well into October. I myself gave up the hassle and aggravation of feeding sugar syrup a couple of years ago. I now merely thoroughly soak the by now notorious 1 kg bags of sugar until the contents, not merely the paper, of the bag are quite damp — and just before the bag gets to the disintegration stage it gets plunked right on top of the clustering bees. Six bags - is a good start. The colonies are checked again at the end of October and perhaps four more bags are donated. I would not now return to the chore of syrup feeding in autumn because the new method of feeding works a treat. Don't believe me? See if I care!!

THE MONTHLY ROUND - September

The feeding of colonies should be well underway by the middle of this month. For the conventional beekeeper sugar syrup in the proportions 1 lb. sugar to make up a pint of liquid will give the bees not too much work to do to evaporate the surplus fluid prior to storing and sealing it as sugar syrup honey. Syrup having the proportions stated may be easily made up with cold water if the beekeeper is patient. Merely measure the amount of water required relative to the intended sugar quantity, i.e. 15 lbs of sugar needs approximately 10 pints of water. If the water is poured into a big enough container and the sugar poured in afterwards stirring the syrup all the time, when all the sugar has been added the contents of the container will look cloudy and grainy. Leave the syrup to settle for about 30 minutes. After this time the fluid will be clear but a layer of precipitated sugar will be seen on the base of the container. Disperse this layer and stir for a few minutes. Leave for at least 10 minutes or even two days. The precipitated sugar layer will be much thinner at this stage and one more "leathering" will produce a slurpy, oily-looking, life-giving and very acceptable supplementary feed for the bees entering winter. As I stated in the August Monthly Round feeding saturated sugar in 1 kg bags as a canopy over the wintering bees saved me a great deal of work last autumn. This year will be more of the same, however care must be taken that the bag does not collapse due to taking on too much water. A sheet of newspaper placed on the frame tops as a "carpet" for these soggy bags will safeguard the bees against a sugar avalanche. A mild September/early October can be of great benefit to the bees, not only by allowing them access to the last of the late blossoming plants like ivy or late heaths, but because the older almost senile bees will die in the field during this period, thus eliminating excessive mouths to feed by natural wastage outside the hive. This could have a critical effect on a colony's survival in late winter/early spring, especially with a very strong colony. In unfavourable late autumns these older bees, not being able to fly freely, will remain in the colony consuming precious stores before dying around mid-winter, to fall to the hive floor, and at best restrict through ventilation and at worst block up the hive entrance and cause the demise of the colony if the beekeeper does not clear the entrance on a regular basis as mentioned in the August issue.

The hives should all be checked for weather-tightness. The roof should be secure, if it has a shallow rim (less than 6"!) a couple of building bricks back and front will make it secure against virtually all winds apart from typhoons and "twisters." The hive, should be level and tilted slightly forward to allow any moisture penetration to drain out of the entrance. Mouse guards, if used, should be fitted before the end of October. A full width entrance no higher than $3/8$ inch will eliminate the need to fit fiddly mouse guards. A strip of wood of the correct thickness wedged into the entrance to leave a $3/8$ inch high gap at the floorboard may be used to modify the conventional high entrance of the commercially produced floor board entrances. The underside of many such floorboards i.e. $5/16$" - $3/8$" edge strips. If these strips are complemented by a strip of similar thickness along the back edge, the floor board may then be inverted and the beekeeper will have an "instant" ideally-sized entrance, obviating the need to fit supplementary mouse guards.

Equipment should be cleaned and repaired in the ensuing months and stock taken of needs for ordering at the start of the spring next year. The "heather" beekeeper will still hopefully be frantically cutting comb and pressing his honey and swiping throngs of wasps, which in good summers invade even the tightest of "bee tight" honey houses. The presence of wasps, although a pest during extraction and honey handling, will be a sure sign that nature has recovered and that 1988 was indeed the year of consolidation after the past trio of dreadful to mediocre seasons since the idyll which was 1984.

THE MONTHLY ROUND - October

By the middle of the month sugar syrup feeding should be virtually over —for the well organised beekeeper— ordinary mortal type beekeepers will still be guddling along or perhaps not even have started feeding —shame!

The progressive lazy beekeeper or the involved "extra busy" beekeeper will have checked the hives by the end of this month — to find that the bees have demolished the canopy of wetted sugar donated at the beginning of September and converted it into "sugar honey." The afore mentioned species of beekeeper will then merely pop on another couple of wetted bags per colony and leave the hives until late December when the "canopy" will be re-established by the placing of 4-6 further dampened bags of sugar over the cluster. This sugar will last the bees up to the end of April. By feeding a couple of jars fills of light sugar syrup each week after the first major pollen intake has been noted, the colony will develop quite steadily using the water in the syrup to metabolise pollen for brood rearing and also to augment the moisture needed to liquify the "rock" sugar. A colony so managed will be in no danger ever of dying of "isolation" or any other kind of starvation. Bees don't die of *pollen starvation*. They can die of pollen deficiency though!

The coming winter months offer the beekeeper the opportunity to increase the pleasure of keeping bees by reading some of the "standard" works like "The Hive and the Honey Bee"; "Honey Farming"; "Beekeeping at Buckfast Abbey"; "The World of the Honeybee"; yes, and even Maeterlinck's "The Life of the Bee" — these books and many more besides are available for the asking in the Moir Library.

The social season is upon us also, the monthly/two weekly meetings of the local associations are a marvellous opportunity to renew lapsed friendships and swop bee talk. The ready listener can learn much from the casual conversation and passing remark, to say nothing of the wealth of extremely Important information which speakers, especially the natural teachers among us, can impart during talks and lectures—support your local association and of course the National Association. You really do need those organisations — more so now that beekeepers are virtually on their own for problem-solving and economical disease diagnosis. It goes almost without saying that every local association should have its own microscope for disease diagnosis. A dissecting microscope is necessary to check for acarine — the technique of removing the head and front pair of legs to expose the big thoracic trachea is astonishingly simple to perform. Using this type of microscope, slides of bee parts can be examined, making the beekeeper more aware of just how highly specialised the honeybee really is.

Braula coeca and the dreaded *Varroa* mite can be visualised in all their "glory" as individual specimens with a dissecting microscope. The microscopic diagnosis of *Nosema* and the brood diseases as well as pollen identification unfortunately require another type of microscope — a compound microscope, preferably with an oil immersion objective to give a magnification of at least x1300 with which it will be just possible to see the rods *of Bacillus larvae* (AFB); the lanceolate *Melissococcus pluton* (EFB) and other secondary bugs like "*bacterium euridice*" which is usually present in an EFB infestation.

Pollen can be easily seen in detail at x400 — I use "seen" rather than "identified" deliberately. Pollen identification is an extremely skilled procedure. The obvious pollens like the, to coin a phrase "microscopic Forget-me-not" or the enormous (relatively speaking) grain of the Evening Primrose are easy, however, to "crack" some of the *Prunus* species, apple, plum, cherry etc., without prior knowledge is not so simple. The first-time viewer of pollen "sub-microscope" will be dazzled by the profusion of shape and form of the pollens — the nearest substitute is the kaleidoscope.

Nosema spores can be easily identified if present (hopefully not!) at x600 and look like fat rounded rice grains. Diagnosis is one thing, treatment is another — Bailey's "Infectious Diseases of the Honey-bee" and Henrek Hansen's "Brood Diseases" should be an integral part of each association book list AFB and EFB are notifiable by law and any colonies suffering from AFB must be destroyed. However, EFB can be treated using sodium sulphathiazole, terramycin and streptomycin *under supervision.*** The standard treatment for *Nosema* is to complete a Bailey Comb Exchange and fumigate or destroy the old frames. Acarine is no longer a problem where miticides are being used to treat for Varroa. Fumigation of spare comb using 80% Acetic Acid will go a long way to keeping disease at bay. This procedure will eliminate EFB, *Nosema,* Chalk brood, and Wax moth larvae and eggs from combs. If an association wishes to be autonomous the two microscopes mentioned are a must, as is someone among the membership who can use, or is prepared to learn how to use microscopes, and also to teach their use, and to make up specimen slides. As well as microscopes I would recommend that each association ultimately be in possession of an anti-varroa heating cabinet and the necessary number of wire mesh cassettes for holding the bees. Using the heating cabinet procedure *Varroa* can be held in check on a two year cycle, implementing drone brood sacrifice and false floor procedures to complement the cabinet treatment. By combining these tried and tested methods, which are of course biological or biotechnical, the beekeeper can honestly state that habitual chemotherapy against *Varroa* is not used — resulting in uncontaminated honey which is what the honey eating public expects.

By constructing heating cabinets in advance of Varroasis being found in the UK beekeepers can experiment at leisure on the use of these cabinets, finding put the problems which might arise in their use and eliminating these problems before the disease is found here. (Since varroa has been found in the UK this has not become a preferred method of treatment.)

"Treatment may only be carried out by Department of Agriculture and Fisheries Scotland (DAFS) staff if they decide destruction of colony is not justified, in 1988 - now carried out by Bee Inspectors who are part of the Rural Payments and Inspectorate Directorate (RPID) - Editor's note.

THE MONTHLY ROUND November

What can one say about beekeeping in November in Britain? Not much yet! The honeybees have their heads down, and all they want to do is slumber — but imagine yourself trying to sleep peacefully with between one and six hedgehog-sized bloodsucking parasites scurrying about your body while your hands were tied together. The analogy is not quite true, but only in respect of the tied hands. The bees cannot use their legs to remove Varroa mites from their bodies — and proportionally speaking that is the order of size of the Varroa/host bee body size relationship. (This is the prospect for wintering our bees when we get Varroa in Britain.)

The dormant period as such will be a thing of the past for the progressive beekeeper who wishes to keep his bees Varroa-free without resorting to drugs in Britain. Since as far back in time as 1977 beekeepers in the Soviet Union have been treating their bees against Varroa by shaking them from the combs, queen and all, into mesh cassettes and heating them to 48°C for around 12 minutes. The Varroa females are unable to maintain their hold on the bee at this temperature and fall like rain from the bees. Once the procedure has been perfected by the operator, a colony going into the winter can be 100% freed from the mites. Late autumn or very early spring has been found to the optimum time for anti-Varroa procedures; chemotherapeutic as well as non-chemotherapeutic.

At the present, "state of the art" treatment against Varroa has to be complemented by brood destruction. The mites in the sealed cells still cannot be reached and killed — so the brood carrying them has to be sacrificed to kill off the next generation of mites before they emerge with the new bees. Despite the bees being extremely roughly handled, very few queen bees are either killed or damaged by the heating procedures. A great advantage of the method, is that any colonies found to have old or dud queens can be re-queened on the instant. Stores can be arranged optimally in the brood chamber and the population of bees equalised — the Russians budget for about 2 kg of bees/colony and calculate hive needs at 2.5 kg of stores per 300gr of bees. This works out at approximately 21,000 bees covering around 6/7 frames with a total stores complement of 36.6 lbs. Check your colonies around the start of October and you will be pleasantly surprised to find that the colonies which were at honey gathering strength in July/August will now be clustering on around 6/7 combs, also the traditionally acceptable weight of stores. ie. 30-40 lbs. which beekeepers have been using as a rule of thumb, has now been demonstrated to have a sound scientifically researched basis. By using the Russian method, the loss of valuable sugar feeding, due to feeding a queenless hive, can be avoided, and the beekeeper can be certain that what he feeds is good stock rather than feeding blind in traditional winter feeding procedures. Who will change their ideas? Those who don't change when we get Varroa will finish up collecting stamps or selling/attempting to sell, contaminated honey to an increasingly suspicious honey eating public. The beekeeper who can categorically state that no chemotherapy is utilised will rate high in credibility and sell more honey A weekly check of out-apiaries is all that is required after the bees have been settled in their horizontally orientated hives, which have been tilted slightly forward for drainage of any water ingress, and whose roofs have been secured against 'lift off'. The overwintering hive should ideally be exposed to the south and sheltered from the prevailing wind by a hedge or wall/fence. Exposure to the maximum amount of winter sunshine is of paramount importance, as is not siting the hive in a damp hollow or frost trap. The first hard winter where these parameters have been ignored will drive home to the 'depleted' beekeeper just how damaging poor attention to overwintering can be. At time of reading this, beekeepers on the fringes of Siberia are still shaking bees from combs into cassettes in ambient temperatures of 3-7°C, to heat-treat them against Varroa — quite a thought isn't it!

All the honey-handling equipment should now be washed and cleaned and preserved, ready for use in the active season next year.

Queen excluders can now be easily freed from brace comb and residual propolis. The careful beekeeper can do himself a lot of good here by saving the brace comb in a moth-tight container and selectively removing propolis and saving every scrap — it is a very valuable commodity, and habitual saving of scraps can soon build a fair amount. The extractor should be washed down with cold water, and tinned or moving parts lightly coated with vaseline, or some such substance. Benches and woodwork should be washed down in a strong solution of bicarbonate of soda and water to inhibit fungal growth.

The books should be flying thick and fast between the Moir Library and the beekeeper at the present time — we can never know too much about our bees and learning from the experiences of others is a sensible way to make progress.

THE MONTHLY ROUND - December

The bees in a well-stocked, healthy colony will be clustered well down almost out of sight, even in a single brood box at this time of year, except on the odd bright warm day which can occur even in December when the colony will be tempted to transient expansion before reforming tightly as the chill of evening makes itself felt again. These 'passing' periods of cluster expansion complement the feeding of candy on the 'canopy' of sugar bags above the cluster, since instead of consuming precious natural stores during this period of winter activity the bees work on the candy or crystal sugar, even elaborating it into liquid stores to be placed in the cells immediately under the candy or sugar crystal.

Colonies fed either large thick slabs of candy covering four or five frames or given the 'canopy' of between four to six pre-dampened sugar bags around mid-December and having an initial complement of adequate stores (30-50 lbs honey and pollen), with a good population of healthy bees in a well located, weather-proof hive, will, all things being equal, never again be in danger from isolation starvation. The coverage of the supplementary feed is extremely important because this gives the bees the facility to construct tunnels within the feed. These pathways allow the bees to move between combs on the warmer days during winter without being exposed to the still relatively hostile environment above the frame tops of the combs on which they are clustering.

This year was the year of growth and consolidation for the progressive beekeeper. New queens should have been bred in abundance and new colonies established to make good the losses of 1985 and 1986. 1987 was not such a disaster for beekeepers as the previous two, however the weather patterns of that year were not particularly conducive to easy queen rearing. Although from mid-July onward, 1988 failed to come up to its early promise, it was just right for the beekeepers who commenced queen rearing in May. June was glorious, at least in West Central Scotland, and the success rate for Queen mating was so good that very few 'misses' occurred. My own queen rearing procedures were virtually over by mid-June except for the token complement of queens reared during the later weeks into mid-July — these queens were however a bonus and showed a greatly reduced mating success rate due to 'critical dates' being missed because of prolonged poor and unsettled weather in early July. I have advocated an early start to queen rearing for many years in this column, and history has proved the effectiveness of the management procedure for myself anyway. Strong queenless nucleus stocks made up from mid-April onward will allow the beekeeper to produce new queens at the earliest possible time and also buy time if there are failures. Queen rearing started in June, if failures occur, leaves the beekeeper little time to re-start. I feel that the attitude to queen rearing which dictates mid to late summer as the optimum time in traditional beekeeping is the major contributing factor in the high failure rate in queen rearing attempts by the hobbyist beekeeper. This is the conclusion I have reached after conversations with many beekeepers over a number of years. To many beekeepers queen rearing is an amalgam of apprehension coupled to a pre-judgement that the attempts will be failures anyway. By deciding to rear more queens than are needed the beekeeper can afford to be, if not prodigal, at least relaxed in his approach to rearing — lost a few — who cares! There are plenty more in the pipeline! If the bare minimum are attempted to be reared anxious beekeepers can turn into beekeepers with paranoia as failures occur and time gets short. I know of an excellent book which describes virtually 'failsafe' procedures for queen rearing, but I'm sayin' nothin'! (Another chance for Eric to promote his book - Editor.) How did I get into talking about queen rearing in a December issue anyway?!

The Varroa floor insert should now be a matter of routine and the folded paper insert should be checked for Varroa. In the future, beekeepers will be well advised to have the floors of their hives modified to accommodate the recommended mesh covered insert which is used on the continent — this device with the base under the mesh lightly oiled (vegetable oil not engine!) to snare any live Varroa which might fall from the bees and prevent it climbing back on board again, is virtually standard hive equipment in most European countries where the 'beast' is rampant.

The cleaning of excluders should be finished by this time of year, as should most of the other equipment maintenance. Out apiaries should be checked on a routine weekly basis for disturbance to hives. After storms or high winds, a visit the next day is well worth the effort considering the cost of colony replacement.

All we need now is a mild, stress-free winter and an early blossoming of the snowdrop in mid-January with the crocus following on in late February and then the beekeeping active period begins again — almost before we realise it — hopefully!

BEES AND THINGS POST VARROA - December

The time is fast approaching when your 'make or break' late winter oxalic acid treatment(s) should be performed, which will hopefully be complementary to the formic acid treatment applied earlier in the season, preferably during the June Gap or in mid-April. The key words "Integrated Pest Management System" (IPMS) translate to - "Hit the mite with all in your treatment armoury, regularly and according to time of year."

However, IPMS alone will not save our bees from the impending potential doomsday scenario of spring 2009, if it becomes a repeat of spring 2008. Colonies covering around 3 brood frames (not 3 frames of brood!!, which survived the 2008 winter, even into early April, just dwindled away in the spring due to unfavourable weather conditions which denied the bees access to the critically important crocus and willow pollen. Even worse, combined with the poor flying weather in many locations, especially in the West Central area the willow catkins bloom was sparse. As a matter of fact, this shrub has 'missed' in the Glasgow area for the past 4 years. Global warming spin-off? Perhaps the relatively mild winters of recent years are affecting our native plants adversely, maybe they need the sharp spells of frost of yesteryear to stimulate that "vital spark."

Any answers out there?

The miserable 'bee weather' throughout 2008 and especially that in late summer and in August/September will also unfortunately work against the bees in the coming spring. The pollen normally harvested from the lime, heather and balsam stored over winter is a marvellous 'power house' of protein to promote a steady population build up for the early summer nectar flows from OSR, sycamore, hawthorn, chestnut and other early secondary sources - this year in most areas in Scotland the blossom of these important plants came and went with hardly a bee visit. This will of course result in a 'short-fall' of much needed brood food for the late winter/ early spring build up. All is not lost however!

Many beekeepers in the South of England, Clive de Bruyn among them, advocate the insertion in early spring of pollen or pollen substitute supplements in patty form. Under normal circumstances we in Scotland never needed to resort to pollen substitutes - but times are a changing and if we don't change to meet new challenges by moving with the times - we eventually won't need to – as beekeepers anyway!!

There are many recipes out there for patties - why not at least make the effort to anticipate that your bees might need a helping hand - and don't forget to provide adequate winter stores. If you were late feeding syrup - a couple of thoroughly dampened sugar bags or a slab of candy might just make that life or death difference. Many beekeepers this summer had dismal queen rearing successes and this is causing some to over winter queens which are coming into their 3rd winter. If the queens in question have proved to be exceptional performers there is indeed the temptation to 'go for broke'. However, the older well proven queen is a gamble - because by demonstrating high fecundity and an excellent egg laying rate she is also in danger of having 'shot her bolt' relative to her residual stored sperm and could just become a liability which might well result in the loss of the colony. I heard, in the pre Varroa era, on quite a number of occasions, beekeepers relating stories of their 'advanced yeared' queens based on two important parameters;

1 They had never SEEN a swarm emerge from the particular hive.

2 They didn't mark their queens. Thus, unless the queen has a 'date in her mouth' these tales cannot be confirmed or denied. Only by marking a queen and noting her individual 'signature' and continuous presence by taking dated photographs annually could these anecdotal stories be verified. Since Varroa incursion into these islands such stories have tailed off! Personally, I am of the opinion that queen rearing should be persevered with even to the extent of sacrificing a honey harvest to ensure that every colony going into the winter is headed by a current year queen.

There are abundant tales of beekeepers noting Varroa mites riding 'jockey' on queens - these mites are not there just for the thrill of the chase - they are sucking the life blood from the unfortunate host and simultaneously debilitating her. Such a circumstance occurs in the hive mostly unseen by the beekeeper frequently, more frequently perhaps than beekeepers want to admit. High losses of colonies overwintering with queens entering their second winter are being noted by observant beekeepers worldwide as one result of Varroa predation. There has been a subtle change in attitude to overwintering 'madame' rather than 'mademoiselle' queens since Varroa arrived in the different countries -

almost like an imperceptible 'wind of change'. In America for example – the beekeepers' Bible, (dare I use the word?) "The Hive and the Honey Bee" has in edition after edition over the years since it first appeared in 1853 advocated overwintering colonies with queens of mature years - even up to the 1975 edition and I quote, " . . . but it seems foolish to replace queens (each year) on a calendar basis." Management for Honey Production", page 373, 7th line from top, 1975 edition. The 1992 revised edition however has done a 'head stand' and quotes, " . . . that is why re-queening colonies at least once a year is very important." "Activities of Bees", page 349, 11th line from foot of page, 1992 edition.

Varroa incursion into the U.S occurred in the early 1980s, probably around 1982-3. Varroa was officially recorded in October 1987 in virtually every State. Incidentally beekeepers in the UK, especially beekeepers in the South of England had been importing U.S queens right up to the summer of 1987. There is absolutely no doubt in my mind that the mite was a 'gift' from America! The mite, when nobody is looking for it, takes about 3 - 4 years to become clinically observable. The Americans did not expect to have to deal with Varroa until 1990 - the then American experts had to later admit that they were caught with their "proverbial pants down." Over the years since 1987 the philosophy on queen overwintering changed subtly. I like to think that it was the book, "The Swarm Trigger Discovered", (*Eric promoting his book again – Editor*) which was widely read in the U.S. and Canada in the mid 1980s, which assisted that 'Earth Shift'. Karl Pfeferle, the German author of "Beekeeping using Single Walled Hives, and Varroa", advocated that only by working with current year queens, which he reared in nucleus hives could successful beekeeping be carried out in the long term in the face of Varroa. If you are unconvinced note the ages of your overwintering queens from now on and surprise yourself.

POLLEN — SIMPLY FOOD FOR THE BEES?

Beekeepers are well aware of the importance of pollen. As a food for bees and other insects it is indispensable.

Honeybees are particularly dependent on pollen since it is their only source of protein, which of course all animals need for tissue renewal and growth — and of course reproduction. Without pollen bee colonies are unable to sustain brood rearing and eventually the colony population dwindles and the colony dies. Feeding sugar will sustain colonies in periods of dearth for a particular limited time — but without pollen the colony is doomed. Pollen has other attributes apart from its critical importance to bees. Without pollen the plant kingdom would be totally different — pollen is the male component of the fertilisation process in plants. Pollination, be it by insects or by wind, is the decisive factor which causes a flower to become a "fruit"; without pollination a flower is merely a frustrated masterpiece of nature.

Pollen is also a curse for many of our fellow men (women!). During the summer months world-wide when the flowering plants are dehiscing, people who are allergic to pollen suffer chronic hay-fever and extreme misery. In America alone, there are 1.5 million people who dread dry summers, due to acute hay-fever.

Much research has been carried out on the medicinal benefits of pollen —and so rages a controversy. There are medics who swear by pollen as a virtual panacea, and other medics who swear equally vehemently that it is nothing of the kind.

A few of the studies into pollen are given below as is a "potted history" of its use in ages past. Since time immemorial primitive peoples have made use of pollen in their diet. The Spanish Jesuits believed that the Aztec, Maya and Inca legendary "Fountain of Youth" was based on the eating of pollen by these tribes. Even today the surviving members of the North American Indian tribes use pollen. To the Navajo Indians pollen is the single most sacred item in the universe — a symbol of life, fertility, peace and plenty.

In 1945 the first scientific studies were carried out on pollen in medicine and since then over 1030 scientists have published more than 1500 scientific papers on the benefits of pollen as a food supplement. In Russia Dr. Nikolai Tsitsin of the Russian Institute of Longevity ascribed the long life and excellent health of study subjects to the habitual consuming of pollen.

In 1949 Professor Alain Callais of the French Agriculture Academy reported that 35g of pollen satisfied the daily needs of humans and 25g was sufficient to maintain life in emergency. In 1950 an American scientist, Dr. S. Janis, described the antibiotic properties of pollen. In 1959, in the Cook County Hospital in Chicago, Doctors I. M. Bush and A. Zamm successfully treated groups of patients suffering from inflammation of the prostate gland caused by zinc deficiency. In the same year Dr. Erik Ask-Upmark at the University of Uppsala in Sweden cured bacterial prostasis, which had resisted all traditional treatments, using pollen extract.

In 1966 Dr. L J. Denis in Belgium achieved an astonishing libids stabilisation effect on patients threatening impotence with whom he had been carrying out long-term pollen therapy. This was the first indication of the high levels of sex hormones present in pollen. In 1970 scientists at the Zagreb University demonstrated the presence of the sex hormones oestrogen, testosterone, epitestosterone and androsterone in pollen.

An analysis of pollen carried out at the Bonny Laboratory in Geneva showed that the constituents of pollen were as follows:

— 35% protein

— 40% higher sugars

— 5% fats

— 3% minerals and trace elements

— amino acids, nicotinic acid, folic acid, biotin

— vitamins, enzymes, hormones, antibiotic substances and other as yet unidentified substances.

There is yet another side to pollen in the fields of commerce and crime: In the oil industry fossil pollens are used like a "dipstick" and locate the drill bit geologically. Such knowledge can help make the difference between a well that makes money and one that doesn't Exxon's senior research associate Lew Stovar states that if, for example, you are core drilling in Sumatra, Borneo or Malaysia, you look for a prehistoric mangrove. That means pollens from mangroves, palms and such. In Korea you would be looking for temperate species like oak, willow and others in sediment as old as 25 million years. In criminology pollen has also played a decisive role. Can a grain of pollen convict a man of murder? The place is Vienna, Austria, the year 1959; police are questioning a young man accused of murder — he claims at the time of the crime he was climbing a steep, sandy mountain. His boots are given to a palynologist, Dr. Wilhelm Klaus, together with a geographical survey. The boots are carefully scraped, they have already been well cleaned by the suspect — less than 1g of dirt remains. It contains 1200 pollen grains. They include spruce, willow, a plant called filipendula (meadow sweet), and a 20-million-year old hickory grain. These grains speak not of high, dry land, but of a river basin, and because of the ancient hickory grain a specific swampy outcrop 20km from Vienna. The suspect was confronted, he admitted he had lied and led the police to a shallow grave — in damp ground.

Hazel and birch pollen on a newly greased gun barrel refuted a murderer's claim that the gun had remained unused since the previous March. A grain of Atlantic cedar pollen embedded in the ink of a signed and dated document proved it was a forgery, written in October and not in June as claimed. Even the Turin shroud contained some 56 pollens, some of which came from plants unique to Anatolia and Palestine — further deepening the controversy surrounding the origin of the shroud.

So, the "simple" grain of pollen is really not such a passive object after all.

The origins and constituents of nectar and honey can also write an interesting story. Beekeeping is indeed a many faceted interest

Sources

1. National Geographic

2. Deutches Bienen Journal

WHO WOULD BE A BEEKEEPER?

This beekeeping year has been a year of 'dramatics'. In the West the early summer flows failed to realise their potential. Apart from the odd day when the wind was light enough for bee flight, most of the period was bedevilled by strong wind above 15 mph (24 kph) at which effective honeybee foraging activity ceases. In Scotland anyway! The June Gap struck with a vengeance this year, even spring-sown rape had a struggle to break the dearth, again due to the high winds.

Despite the dearth, or perhaps more likely due to it, many colonies swarmed, however, these swarms were, in many cases, not reproductive swarms. They were 'hunger swarms'. The writer was called out to a number of swarms around the west of Glasgow and was struck by the lethargy of the bees after hiving. They didn't seem interested in leaving the hive. After experience with the first couple of swarms and subsequently feeding 1:1 sugar syrup and noting the dramatic positive change in behaviour all the other swarms were hived on combs filled with 1:1 sugar syrup — the hum of contentment could be heard yards away and the speed at which the swarm entered the hive had to be seen to be believed. It was reminiscent of the classic Hollywood cattle stampede to water, after the herd's long haul over the parched prairie.

During late June early July, despite being fed and avidly feeding on water saturated sugar in one kilo bags, brood rearing came to a grinding halt in the majority of hives in particular apiaries. My city apiaries kept ticking over. The apiaries in the rural areas, devoid of ornamental trees and flowers virtually stopped their brood rearing dead. These colonies had sugar aplenty — but no pollen. The same type of situation occurred in Scotland in the summer season of 1985. The weather was so poor then that the bees could not get out for pollen. Sugar kept the adults alive at that time but breeding activity was negligible. That year the colonies hardly reached the expected colony size for overwintering and some beekeepers united hives in a desperate attempt to increase population size for the winter.

A good number of my own colonies went into that winter with the size of population normally associated with mid-March, after overwintering. That was a trying winter for the bees and their keepers. Although the spring and summer of 1986 were not much to write home about, the winter period December, 1985 — late March, 1986, was relatively mild. So, the beekeepers who 'bit the bullet' and took the calculated risk of leaving the 'less than optimum sized' colonies to take on the winter rather than unite with the loss of perhaps valuable queens won out.

The hiatus in brood-rearing occurring this summer would be felt dramatically in the colonies later in the season at the tail end of the summer nectar flow as the older bees began to waste. For a number of years now the writer has maintained around 30-50, 4-frame nucleus stocks during the active period for queen rearing purposes — within the city precincts where there is at least some access to pollen on the 'better' days. These nucs are a great stress-reliever for the beekeeper and a marvellous device for the restoration to honey-gathering strength of colonies which for whatever reason did not build to the desired proportions for the imminent nectar flow.

Even in seasons where things mostly go well, like 1989, 1990 and 1991, nucs with spare queens are 'money in the bank'. The hobbyist of course does not have need of 30-50 nucs. but, scaled down to suit the particular need will give the beekeeper a "cushion" against the ever present unexpected event. At time of writing the late summer nectar-bearing plants are beginning to show, viz lime, privet, willowherb, bell heather; there is also clover for the lucky few, as well as other not so major nectar producers. What they will bring in is in the lap of the great weather maker (or breaker). I wait to be surprised or horrified.

TO DRIFT OR NOT TO DRIFT

A controversy has been either simmering away or even festering away in beekeeping circles in Britain for many years. My first encounter with the dispute was made while reading Brother Adam's book "Beekeeping at Buckfast Abbey." A dramatic photograph of a tower block hive at the end of a straight row of hives was used to demonstrate the effects of the controversial phenomenon — drifting.

According to Br. Adam's wisdom straight rows of beehives were not to be encouraged because bees got lost on a continuous and significant scale, entering the wrong hive or favouring a hive located at a particular position in the row.

Over the years I voiced my doubts about the substance of the statement because I had kept my colonies in straight rows since my earliest days in beekeeping — indeed it is among the most cost-effective "labour-wise" method of working bees — especially in migratory beekeeping — which I have practised since year one. I had never noticed any significant preference for a particular hive or population depletion of specific hives.

When I was younger, more energetic and had more of the fanatical interest in honeybee husbandry than I possess today (the fanatical aspect has diminished but the interest is still keen) I checked each and every colony for rate of spring development from the beginning of March — a quick lift of roof and crownboard to note the number of frames occupied by the developing colony, every week. I almost knew every bee by its first name! I watched colony development, noted population growth, checked sealed brood quantity and quality (for pepper-pot or poor laying pattern). I checked hive activity, observing and fostering colonies whose bees foraged in low temperatures or other borderline conditions, like wind or drizzle or both. I counted returning bees to check proportions of foraging population carrying nectar/water relative to those carrying pollen — on an hourly basis. I watched bees returning to hives set in rows in windy conditions and noted of course that the wind did not give the bees an easy ride but the classic arrow-like flight to the hive here was unmistakable despite the fact that the bee might get buffeted by the wind at the hive front.

It is less easy to check brood chamber development by mere roof and crownboard removal after the first super goes on, however, by practising the classic "nine-day method" from mid-May onward the development of colony populations continued to be monitored. The observations proved than colonies which showed superior quantities always did better than colonies which were either slower in developing or tardier in foraging in less that favourable conditions. I made a habit of migrating my bees perhaps three or four times each season, deliberately placing colonies which were obviously the least strong of the apiary complement at the ends of the usual (for me) straight rows of hives at each new site. I never ever noticed that these hives suddenly appreciated in size relative to the populations in the sister colonies in the apiary. Many years ago, when I wrote the "Rambler" column in the magazine, I had letters from a few beekeepers who had extremely strong views on the reality of drifting and I was most decidedly not their "flavour of the month" for postulating such heretical views, contrary to the accepted gospel.

I have read the beekeeping literature of many different countries — and had hands-on experience of beekeeping in Germany and Denmark. The striking common factor of most of the apiaries of the larger foreign beekeeper is that they place their colonies in straight lines with the entrances all facing the same way.

The European "bee house" is a classic example of straight line beekeeping. I have visited apiaries where the beekeeper had a double tier of 50-60 hives in a straight line in the home apiary, housed in a lean-to type of bee house system. One has to ask oneself, If, as the critics will have it, drifting is a significant problem in beekeeping, why does the system of apiary management which is most conducive to drifting continue to be used by beekeepers who rely on their bees for their livelihood?"

A recent visit to Denmark threw a new light on my views and observations relative to drifting. I put It to Jan Olsen that many beekeepers considered drifting of bees to be a problem — and since he also maintained his colonies in straight lines, what were his views on the subject. His unhesitating reply was if colonies of bees are habitually, generation after generation, kept in hives set in straight rows then by the laws of natural selection "these bees and their 'successful' progeny will be systematically selected for homing qualities." I loved it! Thus, beekeepers with wandering bees are obviously long-term garden beekeepers whose bees have never had to think too much about the way home and thus, tend to get lost when presented with a multi-home situation. "Who threw that?"

THE DRONE TRIGGER

The initiation of the rearing of drones in quantity in a honeybee colony is regulated by a simple evolutionary device in nature or naturally established colonies not subject to husbandry. It is an accepted fact in beekeeping that in normal queenright colonies at the end of the active season the drones are ejected from the colonies. These colonies go into winter with only the worker and queen caste present. As spring develops into summer and the honeybee hive population increases with the increase in food intake and prosperity, the colony as it grows begins to occupy an increasing number of combs or an increasingly larger area of the already occupied comb. This gradual population expansion in numbers and occupied surface area of comb is of course significant for the timing of drone rearing. But how and why? Beekeepers are aware that most overwintered bee colonies will increase in size in favourable seasons and then make swarm preparations. The swarm act is the honeybee mechanism for reproduction; each swarm is a birth, as is even the original stock left with a final selected (by the bees!) queen after the last cast has departed. The colony begins to produce drones in quantity at some instant in time between the initiation of spring population build up and the achievement of population peak. But where and when?

Consider a honeybee cast established on undrawn worker foundation, prospering. The bees will draw cells on the comb area occupied, into which they will place pollen, honey, and the queen bee her eggs. The size of the brood nest is governed by the area of comb which the colony can maintain at the correct temperature for rearing brood. Initially the brood nest is quite small, however as the emerging brood numbers begin to exceed the numbers of foragers dying in the field the brood nest grows. The observant beekeeper will have noted development patterns in small colonies or moderately-sized nucleus stocks. The bees as they grow in population tend to migrate horizontally across the combs of the hive, that is the colony does not expand downwards to cover each occupied comb before moving across to occupy the adjacent combs. Only after a particular number of combs (approx. six to seven) have been occupied do the bees begin to colonise the extreme edges at the side and bottom of the comb. The observant beekeeper will again have noticed that in the development of such colonies as those discussed, the bulk of the drone comb is usually constructed at the edges and bottom of the comb. Considering that the bees do not utilise these extreme edges of the comb until the population is of a particular size and, by virtue of this, enjoying a fair degree of prosperity, I submit therefore that there is a direct relationship between the population increase to the point where the bees are able to cover the extremes of the comb surface comfortably in early summer and the initiation of the rearing of drones in quantity.

The next section contain articles written between 2000 and 2009 on primarily how to deal/treat varroa. The recommended treatments and treatments available have changed since that time. Api-bioxal is the UK licenced Oxalic Acid treatment against varroa. MACQS a gel containing Formic Acid is now available.

Oxalic Acid, Formic Acid and Acetic Acid are dangerous chemicals and need to be handle correctly and with extreme care.

If you are going to use them, please be careful.

MULTI PURPOSE NUCLEUS FORMATION

One of the most effective biological anti-Varroa procedures being widely used in Continental Europe at present is the formation of nuclei using combination brood combs; the top half of the comb, in a standard brood comb is worker comb, while the lower half of the comb is drone comb. It is relatively simple to produce such combs in the late spring early summer. Merely insert a worker brood comb, with the lower half of the comb cut away leaving an empty space equal to half of the comb area of a standard brood comb, into a strong colony (bees covering at least eight frames!)

A frame so treated should be placed between the last brood frame containing brood and the 'pollen' comb at each side of the brood nest. The 'pollen' comb is easily identified, it is the comb on the fringes of the brood nest. If the bees are prospering the lower half of both frames will be filled with drone comb within a couple of days. Around 17 days later (20 days at the latest!) check these combs - if the bees have done their homework both combs should be-fully laid up and well sealed. These combs should now be removed to a nucleus hive with the adhering bees BUT WITHOUT the queen. The procedure should be done with all hives in an apiary at the same time, so that a number of 4 or 5 frame nucs can be made up - depending on the size of the apiary!

The removed combs should be replaced by more combs with the lower half cut away, as the second last frame on each side of the hive. These nucs should be removed to another apiary at least a mile from the parent apiary. Check the nucs at two day intervals. When all the drone brood has emerged treat the nuclei with a 3% sugar/oxalic acid solution dosing each occupied frame space with 3 - 4 ml of solution. The following day give the nucs a young caged queen, or a ripe queen cell, or a frame of eggs and open brood. Ensure the bees have adequate stores at all times.

A 'belt and braces' method of making sure that the queen is not taken when the combi-frames are removed, Is:

1. Remove the two sealed comb-brood frames,
2. Shake all adhering bees back into the parent hive.
3. Place the combi-frames in an empty spare brood box.
4. Rebuild the hive after replacing the removed combs with 'new' comb.
5. Place a queen excluder above the topmost super containing bees.
6. Place the spare brood box with the 'beeless' combi-frames on top of this super.
7. Replace crownboard and roof and leave for a minimum of a couple of hours (even a day!) for the bees to rise to the untended brood.
8. Remove the comb-frame brood box,
9. Place it on a spare floor, with crownboard and roof.
10. Remove to another site.

Set the parent hive back to normal. The combi-frame method of drone removal is only effective up to around the end of June. If done properly the removal of around 6 combi-combs up to that time will have the effect of removing a massive amount of Varroa mites from the parent colony in the sealed worker and drone cells. Research has shown that around 85% of the mites in a bee colony are in the brood cells at any given time. It is now common knowledge that the Varroa mite has a preference for drone brood. The use of this method gives the beekeeper:

1. New young colonies for the following year or nucs to reinforce honey producers at the start of the late summer/autumn nectar flows.
2. An effective method of reducing swarming.
3. An effective form of Varroa population reduction.

All the colonies can be treated in the late autumn/early winter with whatever method suits the particular beekeeper.

MORE ON OXALIC ACID SUBLIMATION

Oxalic acid is now not just some obscure treatment substance against *Varroa destructor*. The mainland European beekeepers are now well committed to its use. As editor of *"The Scottish Beekeeper"* I am privileged to receive many of the beekeeping world's leading journals.

The German language bee press has been promulgating new and varied devices for applying oxalic acid, month on month during recent years. The treatment has progressed from the highly effective but extremely labour intensive 'spray' method, to the less labour intensive but extremely effective 'trickle' method.

The November 2003 issue of the *Deutsches Bienen Journal* illustrates a device which, using a closed circuit air stream, circulates dry pulverised oxalic acid crystals through the bee colony.

The device costs around £55, however the present 'state of the art' method of treatment using oxalic acid is 'sublimation', even this method has thrown up a number of different devices, which have already been featured in the German language journals.

The more sophisticated devices for this mode of treatment require the use of a 12V car battery to power the 'sublimator'. These devices also are not cheap. Beekeepers the world over are born innovators and given the germ of an idea, ultimately common sense prevails and the ideal design for every individual need appears.

My own personal need is; cost effective, DIY friendly, effective in use and non-damaging to the bees and noncontaminating of hive produce. "*Bienemutterchen*", the house magazine of the 'Federation of Sklenar Beebreeders' carried an article which caught my attention in its February 2003 issue illustrating a simple design, to sublimate oxalic acid crystals inside the hive consisting of a 0.7 metre length of 16mm diameter copper water piping inserted into the entrance.

My instincts told me that this could be the way forward, unfortunately I could not 'raise' the editor of the magazine to get permission to translate the piece at that time so the idea remained my secret as one of the few, if any other, English speakers reading the magazine. However, I did some experimenting with the device by making one of my own. The simplicity of the design is remarkable in that it is merely a length of copper water piping closed at one end by being hammered flat with the end turned over twice (twice is important!). I loaded the device up as directed with 3g of oxalic acid crystals and sublimated them by heating the 'blind' end of the tube with a blow lamp for 3 minutes as directed. This was done outside the hive to prove that sublimation not only occurred but was seen to occur I noticed that some of the gas condensed at the end of the tube, which was still relatively cold. This condensation would obviously affect the efficiency of the device since the full charge of the 3g treatment was obviously not going into the hive. So, like other beekeeping innovators before me I 'Innovated' with different lengths and finally plumped for the device shown, which was kindly drawn for this article in a most professional manner by Ian Craig, our Education Convener. This modified design is inserted at the top of the hive, through a hole bored in the hive roof. Factory made hive roofs have an air space above the internal rebate - the tube locates into this air space. There is no need to bore holes willy, nilly in all your fine hive roofs, one or two roofs will do. It takes no time to make a couple of the sublimator devices either. When treating - I treat 5 colonies at a time - by treating a minimum of two colonies together the first device is cool enough to handle by the time the second hive has been treated. The device is heated at the "blind" end for three minutes by a blow lamp. A good guide to the correct temperature is the discolouration of the heated end of the tube.

Even if an 'Open Mesh Floor' is fitted to the hive the treatment is still effective since the gas falls through the bees and comb as it fills the brood chamber To do the job methodically merely replace the 'in situ hive' roof with the modified roof, have an empty super (only one!) above the brood box containing the bees. Load the device with the 3g of oxalic acid, tap the device to ensure the acid goes down to the "blind" end of the tube, push the device into the hole in the 'treatment' roof, heat the 'blind' end of the device for the prerequisite time and that's it! A metal heat shield should be used to safeguard against charring the wood of the modified roof. The best and most elegant component is yet to come!

Take a small block of wood, say $1/2$ inches square and about $1/4$ inches thick and drill a 16mm hole, 21mm deep into it. This hole when filled to the brim with oxalic acid crystals carries, by weight exactly 3g Loading the device is a piece of cake! The treatment is around 98% effective and costs about £0.03p/ hive.

The bad news is that it is necessary to wear a mask to safeguard against inhaling any gas escaping from the hive body when using any form of sublimation. The good news is that the recommended mask (FFP 3-S/LU 0200/ EN 149) is relatively inexpensive and easy to obtain. Gloves and safety spectacles are also a must. The sublimation treatment can be carried out at any time when the ambient temperature is above 3°C during the late autumn through to the late winter and can be repeated two or three times at 3-4 week intervals without harm to the bees. It is most effective when the colony brood level is lowest. This method may also be used on swarms (shook or natural!) prior to the queen laying in summer.

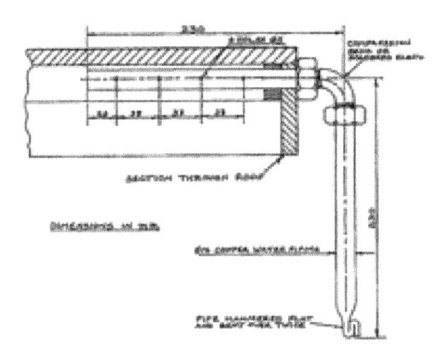

Diagram 1

OXALIC ACID SUBLIMATOR MARK II & III

The oxalic acid sublimator design which appeared in the February 2004 issue of *"The Scottish Beekeeper"* based on the German "copper pipe" sublimator has moved on. Experience gained in its use over the past winter and its use on swarms and nuclei during the summer this year has thrown up a number of what I think are improvements.

There is absolutely no doubt that the original German recommendation of 3g oxalic acid deals with the mites and does not damage the bees. The bees' tolerance of the substance is quite remarkable, even using multiple applications. However, I revisited the German device recently and did a number of new trials with it. After each use I checked the amount of crystal which had condensed in the tube and was quite astonished to find that on average approximately 1.5g of the substance did not go into the colony to be treated.

The design which appeared in the February magazine was made as a result of observations that condensation occurred in the German design, but at that time the amount of condensation was not checked. Some condensation does occur in the modified design illustrated in the February 2004 magazine, but nothing like the loss in the German design. This seems to indicate that 1.5g of oxalic acid is a sufficient dose.

I intend to reduce the oxalic acid dose in my treatments and monitor closely - monitoring is the key to successful application of any treatment. However, using the premise that less material to heat and a shorter gas path should increase the effective sublimation and reduce condensation to a minimum, further modifications to the sublimator have been made, as can be seen on the following illustration. There are now two devices, one of which is suitable for use at the hive entrance, while the other is for 'top' use as for the 'February' design. Both devices are similar except that the pipe of the device for use at the hive entrance is somewhat flattened. This is to accommodate my hive entrances which are all 7.88 mm high, to keep mice and shrews out during the dormant period. To load the devices merely remove the cap from the free end of the bend, insert a short length of 16 mm diameter pipe into the bend and load the oxalic acid into it as described in the original February article - but only use 1.5g acid crystals. Replace the cap with the blank fitted and tighten with a spanner. An extra refinement is the blanking off of the short pipe outlet with tape to stop the oxalic acid charge from falling out! If a number of these devices are made up (they cost about 60p each, I now have 30!) they can be loaded at home prior to use, thus saving time and aggravation if the weather turns inclement on site. It is also quite important to slacken back the blank end of the compression bend to avoid the coupling 'seizing' after use, A metal 'heat shield' is required to prevent the hive front/roof from scorching in the heat of the gas blow torch.

The time taken to sublimate the crystals using the new devices is around 90 seconds. By setting up a number of hives for each 'cycle' (I do ten at a time!) even the busy commercial beekeeper, using the hive entrance device could treat thirty colonies in just over an hour at a cost of 1.5 pence per colony - even doing a couple of treatments the cost Is still only 3 pence. The treatment if done during the late autumn, early winter or late January when little or no brood Is present will kill around 95 - 98% of the phoretic mites.

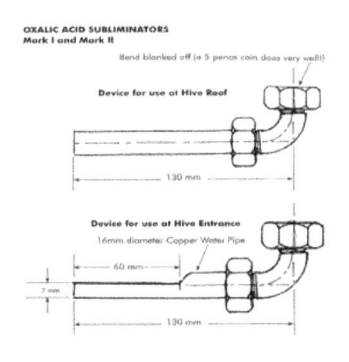

Diagram 2

ALTERNATIVE ANTI-VARROA HIVE CLEANSING RE-VISITED

The Varroa mite has hopefully made us all better beekeepers, because if we have not become better beekeepers many of us will become non-beekeepers very soon or will have already become non-beekeepers. The mite is a hard task mistress and will mercilessly root out the 'let alone' beekeepers and also the 'in denial' folk. Beekeepers can fool themselves in the short to medium term about the mite being 'somebody else's problem' but they can't fool the mite - it will always have the last laugh! Being 'in denial' does not necessarily only mean that the beekeeper is of the opinion that the mite has passed him/her by, it implies also that the beekeeper being aware of the mite's presence in the colonies will not deviate from the accepted wisdom of the potentially less than effective treatments handed down by the 'powers that be', who for whatever reason are themselves 'in denial' and have steadfastly for many years refused to discuss, let alone authorise well tried mainland European alternative methods.

The drip, drip of the continuous reports of the unusually high colony losses worldwide is beginning however to have effect and the new 'hive cleansing' substances like oxalic and formic acid are now being viewed with a less jaundiced eye.

There is an encouraging number of beekeepers now expressing an interest in treating their bees with oxalic acid, which when applied at the correct time of the year has an astonishing 'cleansing' effect on the colonies, considering the negative reaction of many, in the mid 90s, when *The Scottish Beekeeper* magazine was 'slated' for its coverage of the positive reports and translations about the substance from the German language 'bee press'. It is now time however, in my opinion to move on. The 'cleansing effect' of oxalic acid is unfortunately only felt by the phoretic mites. There is an imperative need for a much deeper 'cleansing effect' if the colonies are to remain healthy with their immune systems intact and clear of the debilitating viruses which appear to act as catalysts for colony demise. Formic acid is such a 'deep cleanser', which penetrates the membrane of the sealed brood cell and kills the mites before they can debilitate the developing bees. The sealed brood of colonies not treated with formic acid are completely at the mercy of the mites, which have 'free access' to the hapless developing bees as they suck the larval/nymph haemolymph causing the lesions into which the lethal viruses penetrate. By applying formic acid at the correct time and at the correct dosage the potential devastation caused by a heavy viral infestation can be largely avoided. The ideal time to apply formic acid is when there are large numbers of newly sealed brood in the hive, in Scotland, this will be initially from around the 10th April as the colonies accelerate the rate of spring build up in anticipation of the first major nectar flow, which occurs throughout May from; sycamore, gean, OSR, hawthorn, chestnut, dandelion, whitebeam and other sources. Another critical time where formic acid application will work wonders is June. The dreaded

'June Gap' in Scottish beekeeping lore can become a 'grace' instead of a 'sin', if the beekeeper removes the early summer honey and the bees are treated from the end of the first week in June. The colonies can also be 'swarm prevented' at the same time by splitting the strong colonies and making up queenless nuclei with unsealed brood, which may be treated by the oxalic acid Trickle Method, (which is able to be applied before the brood is sealed,) with impunity in the active season. The bees, queen and the remaining sealed brood can be treated with formic acid as carried out in early April, resulting in colonies which will work on the late summer flows unhindered by heavy mite infestations, which will otherwise be developing in colonies which have had untreated/ inadequately treated mite infestations over a 2 -3 year period.

On mainland Europe a most effective method against the mite is the requeening of all colonies with vigorous current year queens during the late summer/ early autumn and then carrying out the chosen anti Varroa treatment using the organic acids to 'cleanse' the hives.

OXALIC ACID TRICKLE METHOD

Oxalic acid treatments may be applied at temperatures as low as 5 C. Oxalic acid as a Spray Treatment was pioneered in Russia and perfected by Radetzki et al and reported in the May '94 issue of the *Schweizerische Bienen Zeitung* having proved to be extremely effective, however the method is quite labour intensive since the procedure entails withdrawing all the bee covered combs and spraying both sides of the comb with the acid solution. A less labour intensive procedure, the Trickle Treatment was pioneered by Italian scientists using a 10% aqueous solution in 1995. This solution strength was found however to be too strong for use in Northern Europe. Swiss researchers carried out exhaustive trials in autumn 1999, which have been well documented, using different solution strengths (see S.B. November 2000, page 266) and ultimately recommended a 3.5% oxalic acid aqueous solution (or oxalic acid/sugar syrup solution) applied at a dosage of 4 - 5 mls. of the solution trickled slowly onto the bees clustering between the combs, for Central Europe. The conventional wisdom dictates that a maximum of 50 mls of the solution be applied in the case of a colony covering 10 frames, thus 10 X 5 ml = 50 ml a colony covering perhaps 6 frames would be dosed with a maximum of 6 x 5 ml = 30 ml and so on. I used the Trickle Treatment as a prophylactic measure from December 2000, despite not having Varroa in my colonies at that time - to prove that the recommended dosage was tolerated by the bees. This treatment applied around the end of December will kill some 95 - 98% of the mites on the adult bees and dramatically improve the over-wintering survival chances of the colonies.

Mite fall should be monitored before and after treatment. The Trickle method may only be applied ONCE as a winter treatment. When using the Trickle method it is imperative that safety spectacles and acid proof rubber gloves are worn.

OXALIC ACID FUMIGATION METHOD

This extremely effective method was first tested in aerosol form at Fischermühle, Germany in 1994. Later, in 1999 again at Fischermühle, the method was improved by applying the oxalic acid in gaseous form using sublimation. The method was thoroughly tested 'in the field' during 2000 and 2001 and proved to be highly efficient (95%+ success rate, see S.B. December 2001, pp 295 - 298). The sublimating device used however was rather labour intensive and not cheap.

Further improvements in the sublimation procedure from that time on occurred in mainland Europe and in 2003 the first DIY device appeared, published in the February 2003 issue of "*Bienenmutterchen*" (see S.B. February 2004, pp 40 - 41). This was a 0.7 metre long 16mm dia. copper pipe, which was charged with 3 g oxalic acid dihydrate crystals and the tube heated using a blow lamp. This idea appealed to my Scottish thrift and I experimented with it, however I felt that the pipe as designed was too long - most of the oxalic crystals condensed at the relatively cold top end of the tube. I modified the design to a short bend (see S.B. February 2004, pp 40 - 41) which was charged with 1.5g of oxalic acid dihydrate crystals. I used this device in one of my apiaries on October 16th 2003, fumigating into the top of the hive above the clustering bees and joined the "Varroa Members Club" on that day. This bend was subsequently improved upon, in my opinion, and another similar pipe bend was designed suitable for fumigating at the hive entrance (see S.B. December 2004 pp 323-324.). I fumigate from above when there is no fondant or sugar bag feed on the hives, otherwise fumigation is done at the entrance. Fumigation is effective and quick and since there are many ways to 'skin the proverbial cat' improved designs for sublimation will continue to be produced. I fumigate every year in mid to late October and continue to monitor mite fall. Mite fall starts the following day and will continue for perhaps a couple of weeks depending on the infestation levels. Fumigation, as opposed to the Trickle Method may be applied two or three times at three to four week intervals. If more than one mite falling per 2 days is recorded on the Varroa floor insert during late November early December another cycle of fumigation around the end of the 3rd week in December will be necessary. When using the Fumigation method it is imperative that a properly designed gas mask suitable for use with oxalic acid gas is worn, safety spectacles and acid proof rubber gloves are also prerequisites.

Formic Acid

Formic acid is a spring/summer treatment. The optimum evaporation temperature is 18 - 24°C. This temperature will be achieved above the brood nest at any time when brood is being raised. There are numerous devices which are recommended for the application of formic acid and the acid solution strength varies according to the appliance being used. This variety of application methods and solution strengths, which incidentally have all been scientifically tested and found to be safe for the bees and effective against the mite and leave no residues, demonstrates the flexibility of this substance as an anti-Varroa treatment.

A simple method of application which has been used for the past 3 years and has been reported as achieving quite dramatic and reassuring results is a modified version of an extremely simple German design. The formic acid used is a 60% aqueous solution applied by a plastic veterinary syringe using a 20 ml quantity at 3 day intervals over a period of 10 days. The acid is applied to an ordinary flat synthetic kitchen sponge measuring 5"x 5", supported on a sheet of plastic garden mesh, which is fixed to the underside of a plywood carrier, approximately 8" square and ½" thick with a 6" square hole cut out of it. The plywood carrier is placed directly onto the brood frame tops, the sponge is laid in place, the 20ml of acid gently trickled onto the sponge and another flat square of plywood placed over the sponge to ensure the vapour is driven down into the brood nest as the acid evaporates. An empty super may be used to house the device. Three days later the 20ml amount is repeated and three days after that the dose is repeated again making 3-20ml doses in all. Mite fall will continue until all the sealed brood from this period has emerged; 12+ days from initial application. If more than five mites/day drop after this period the treatment must be repeated. Thereafter the device is removed and stored for future use. The simplicity of the design belies its effectiveness, which is due to the fact that when brood rearing is underway in the colony the brood nest temperature is at or close to 35°C. The optimum temperature range for formic acid evaporation is 18-24°C. Even in mid April the heat rising from the brood nest impinging on the sponge is said to be sufficient to evaporate the formic acid within the necessary time scale. During the brood rearing period around 85% of the mites infesting a colony of bees are in the brood cells thus by the use of formic acid around mid April the developing larvae and nymphs are saved from serious damage by the parasitisation and the colony is 'swept' virtually clean of mites. Combining oxalic acid and formic acid treatments within the correct time scale as an IPMS (Integrated Pest Management System) appears to be a winner. Makes sense to me! Formic acid kills the mites in the brood cells and also the phoretic mites on the adult bees and at this present time is the only substance which performs this double function. The substance used correctly and on the right time scale, dramatically reduces the incidence of bees exhibiting viral conditions like 'cloudy wing' and 'deformed wing' viruses and will certainly eliminate any 'jockey' mites seen on the bees – a condition which should never be allowed to happen, but which unfortunately seems to be quite common according to reports. Bees should be treated against the mite long before they are able to be seen on the adult bees - this level of infestation will usually result in colony death despite being treated.

My own colonies indicated negligible mite drop in late December 2007 and still do, as do the colonies being fostered for the 'Save the Scottish Honey bee Gene Pool' project in the Clyde Area Beekeepers' Association apiary. But that is another story!

This formic acid procedure is extremely easy to apply, however it is recommended that the user practice with water initially until the procedure is understood and perfected. When using the formic acid treatment it is necessary to wear industrial quality acid proof gloves, safety spectacles, a suitable mask to protect against the acid fumes and have a bucket of clean fresh water immediately available in case of accidents.

An important additional 'bonus effect' of the formic acid treatment is the eradication of acarine disease, which is caused by the microscopic spider, *Acarapis woodi* Rennie.

A NEW SLANT ON CHALK BROOD DISEASE?

Chalk Brood, (*Ascosphaerosis*), the fungal disease caused by *Ascosphaera apis* (Maasen ex Claussen) Olive and Spiltoir, is not normally a condition which affects the viability of a honey bee colony.

However, there are situations where it can be so severe that the survivability of the colony is threatened. Professor Len Heath investigated the disease thoroughly in an excellent article in *Bee World*, Vol 66, No. 1, 1985. Many theories for the incidence and spread of the disease are mooted in his excellent article and the condition has been exhaustively documented on a worldwide scale over the years. The disease seems to disappear for a period and then suddenly arise again.

In America the disease was widely prevalent in Georgia, Iowa and Florida in the 1920s. At that time beekeepers reported that they re-queened colonies on a continuous basis and the condition then disappeared. However, outbreaks of chalk brood reappeared in widely separate locations in North America within a few years in the late 1960s.

It has been suggested that the Italian breeding stock of the 1920s had a greater resistance to the disease and that this could explain the apparent disappearance of the disease in the 1920s. Another theory postulates that the genes for increased susceptibility to chalk brood were introduced into breeding stocks and transmitted with queens and package bees all over America.

It has also been suggested that its reappearance in the late 1960s was due to mutation in the fungal stock. However, parallel studies of the British and American strains of *A. apis* have shown no major and consistent morphological or physiological differences to confirm this theory.

Another possibility which was proposed was that changes in weather patterns could be responsible. This idea was deemed improbable because such changes appear to fall within the range of weather conditions in those parts of the world where chalk brood is found. Chalkbrood was found in New Zealand in 1957 by D.W.A. Seal, who was of the opinion that it had been present for some time. However, the Principle Research Officer at Wallaceville described Seal's samples as 'diseased larvae with the appearance of chalk brood'. The fungal mycelium which permeated the larvae was only tentatively identified by him as *A.apis*, He went on to state that chalk brood was seldom found In NZ, even in areas having high rainfalls, implying that it was occasionally found in such regions. Reports on most of the 200,000 honey bee colonies in NZ examined annually dating from 1957 showed no incidence of chalk brood until 1984, when it was found in Kerikeri in North Island. A suggestion regarding the NZ situation is that *A.apis* is a latent endemic condition which only manifests itself when an environmental event occurs which activates the disease to clinically observable levels.

Observations made by myself in late spring/early summer 2007 and again in mid-summer 2008 caused me to re-visit Prof. Heath's *Bee World* article to find out if he had mentioned anything about a phenomenon that came to my attention relating to chalk brood on both of these occasions.

The Clyde Area Beekeepers' Association decided in November 2006 to try to establish a project to foster the honey bee gene pool in the West of Scotland using the apiary of the Glasgow and District Beekeepers' Association. I was appointed Project Leader! The project was ambitious to say the least because at the time, early winter, there were only two surviving colonies in the apiary, however there was a sufficient amount of equipment of mixed design with which to begin.

The original two colonies were given VIP treatment and although having queens of indeterminate age came out of the winter relatively strong. As Project Leader, I decided that I would sacrifice 10 of my own colonies to get the project going. The original two colonies were fed to encourage build up - these would supply the drones for the new queens. All went according to plan - April 2007 was a beekeeper's dream and by early May I was able to make up 10 x 4 frame queenless nuclei and transport them to the CABA apiary for mating of the virgins produced. Despite May turning out to be as bad as April had been good the virgins were mated and laying by early June. All seemed to be on course for a good start to the project - then disaster struck - 6 of the 10 nucs were affected by the heaviest infestation of chalk brood I have ever seen and seemed to be 'write offs'. Desperate situations require desperate measures and inspirational thinking. The infested combs were shaken free of the adhering bees, empty drawn deep frames previously fumigated with acetic acid were used to replace the discarded (and destroyed by fire!) frames. The bees were then fed with 2.5 litres of 1:1 sugar syrup and a sponge soaked with 20 mls, 60% formic acid inserted right on the frame tops for good measure (inspirational!!). The bees were left for a couple of weeks before being examined again. That examination was a revelation and will remain with me for the rest of my life - each and every one of the previously, seemingly doomed nuclei had a full comb of healthy sealed brood and to cap it all the chalk brood condition never appeared again in the apiary. We had an Open Day on the 21st July 2007 and sold seven - 5 frame nucs bursting with bees to beginners attending that day. Every colony was opened for examination, Ian Craig the SBA President, who had been informed of the chalk brood problem at the time it occurred was also at that Open Day and can confirm not only that no chalk brood disease was present but also that the bees handled perfectly with no-one being stung or even threatened. A minor miracle!!

The CABA apiary bees exhibited negligible chalk brood this year 2008 and again the Open Day on 20th July saw another six beginners receive their 5 frame nucs again bursting with bees. Ian Craig was again present at the Open Day!

The weather in Scotland over the summer of 2008 must be a near record for continuous unfavourable weather and all colonies in the CABA apiary and my own outfit were fed more or less continuously. There was never a time when either syrup, fondant or sugar bags was not available to the bees - and it paid off by keeping the colonies in good heart.

However, one colony in one of my out apiaries slipped through the net - I came back from holiday in late June, just at the end of the June Gap, and on examining the colonies found to my horror that chalk brood was back with a vengeance in one strong nuc, which had been missed and not fed before I went on holiday. The condition was nowhere near the severity of the previous year in the CABA apiary but uncomfortably high. I did not remove any frames but I fed 3 litres of 1:1 sugar syrup to the colony - unfortunately I did not have my formic acid with me - so the bees only got the syrup. I went back the following week – with more feed and the requisite formic acid. I opened the colony expecting to find the disease condition, at best the same as the previous week, or at worst, epidemic in the colony. To my astonishment there was no trace of chalk brood and the cells that had contained the mummies had either very young larvae or were full of uncapped honey - the feeder was empty!

To cut a long story short I am convinced, despite all the science and scrutiny that has gone before that the explanation for out breaks of chalk brood has nothing to do with queen quality, genetic resistance or good housekeeping on the part of the colony. In my opinion the disease is a condition of inadequate food and more specifically the dearth of income of nectar or the lack of liquid feeding. The application of the formic acid in the 2007 case I am certain was not of any great significance. But that is not to say that used hive furniture should not be adequately sterilised before use.

At present I am unable to continue with my Chalk Brood postulation – none of my colonies now has any evidence of the disease. Perhaps interested parties having an *A.apis* presence might be moved to confirm or deny this submission.

I strongly feel that at the larval stage hunger or competition for food is a critical factor in the development of the *A.apis* in the honey bee colony.

As an aside, looking at another brood disease, European Foul Brood, which, it is accepted has associations with larval hunger and competition for food - it has been observed that the incidence of EFB is extremely low or not clinically observable in colonies which are husbanded in areas with excellent forage conditions, in fact Bailey in his book "Infectious Diseases of the Honey Bee" states on page 139, paragraph (c), the effect of beekeeping on EFB - "In localities with uninterrupted nectar flows, where colonies can grow unhindered each year, (EFB) infection may remain slight and the disease unapparent." I put forward the postulation that the same conditions hold for Chalk Brood and that by feeding colonies heavily in times of dearth will go a long way to maintaining healthy bees.

In fact, the use of antibiotics to treat EFB might just be a waste of time, effort and money.

7. The Shepherd Method of Swarm Control
Taylor Hood

In December 1936, Mr William Shepherd of Newton Mearns explained his system of Swarm control to GDBKA, an advance to that of Snelgrove. In this talk he spoke about his board based on one he saw at an Apiary Visit of Mr Steven of Kilmaurs in 1928. It was not until March 1951 he gave the details in *The Scottish Beekeeper* p44-45, of his board and tube and how it prevented swarming.

"Apparatus – A Shepherd board with queen excluder zinc fixed over 1½ by 4 inches centre hole and a piece of perforated zinc to slide and close off as much of the excluder as desired.

A shepherd tube to drain bees from the hole in the front of the board to a point ½ an inch above the hive entrance".

Shepherd found that if the queen was in the bottom brood chamber that any virgin queens produced in the top box went down the tube got mated and returned to the bottom brood chamber and the old queen was superseded.

In 1970 in the August *Scottish Beekeeper*, James Burns the Secretary of GDBKA wrote about his method of swarm control adapted from that of Snelgrove and Shepherd. He found that his method produced like Shepherd very good honey yields.

Burns method was simply find the queen and put her in a brood chamber with frames from the original Brood Chamber that did not have brood or eggs and then complete the brood chamber with drawn foundation. This brood chamber was then put with a floorboard on the original site along with a queen excluder, super/supers and crown board above this. The crown board had 3 holes (porter bee escape size. Each hole was covered with perforated zinc. One of the holes had a ¾ inch clear space except for a piece of queen excluder so that the bees could sort themselves out over a 3 day period then the space was replaced with perforated zinc. The original brood chamber containing the rearranged brood and eggs was placed above the crown board. A hole ⅝" in the front centre of this brood chamber, above the crown board previously drilled, was opened and a telescopic tube was put in place. The tube went down the front of the hive stopping half an inch off the flight hole/entrance. Two Vs were cut in the bottom of the tube to stop the loss of bees if the tube was to extend/drop down with the possible blockage of the tube. Bees in the top brood chamber build queen cells, so the top brood chamber was fed syrup during this time. Burns stated that no more needed to be done until the brood including the queen cells had hatched – he also emphasised the importance that the only way out for bees in the top box from the 3rd day on must only be via the tube. If this was the case the virgin queen went down the tube, got mated, returned and entered the bottom brood chamber superseding the old queen.

I am not sure why this method of swarm control died out, however I do believe there is merit in it and that some of Shepherd's thinking and concepts have been incorporated into some of today's modern day swarm boards. For a few pounds a Shepherd tube can be made from plastic waste/water system pipe, a 90 degree bend and pipe clips. So, I am going to give it a go and hopefully the story will continue in a future *Scottish Beekeeper* magazine. As Shepherd himself once said, "there is nothing new in beekeeping".